Emerging Issues in Fish Larvae Research

Gilthead seabream larva (42 days after hatching). Photo: Bernd Ueberschär

Manuel Yúfera
Editor

Emerging Issues in Fish Larvae Research

Springer

Editor
Manuel Yúfera
Instituto de Ciencias Marinas de Andalucía,
(ICMAN-CSIC), Campus Universitario
Río San Pedro s/n
Puerto Real, Cádiz
Spain

ISBN 978-3-319-73243-5 ISBN 978-3-319-73244-2 (eBook)
https://doi.org/10.1007/978-3-319-73244-2

Library of Congress Control Number: 2017962104

Printed on acid-free paper

This Springer imprint is published by Springer Nature
The registered company is Springer International Publishing AG
The registered company address is: Gewerbestrasse 11, 6330 Cham, Switzerland

Preface

Fish larvae are amazing organisms. They are among the smallest free-living forms of vertebrates and exhibit fascinating growth potential. In fish aquaculture, all of the phases of the life cycle from the fertilized egg to the adults are essential, and if one of them fails, the whole production cycle fails. The larval phase, however, has traditionally been considered the most sensitive period both in the wild and for cultured fish. The vulnerability at this stage has been a powerful driving force for fundamental and applied research. In this research line, the horizon is to know what is necessary to allow the fish larvae to grow and develop properly and to become healthy. The development of basic larviculture procedures based on feeding with rotifers and Artemia in the 1970s and 1980s allowed a constant supply of fry to support the increasing fish farming production during the last decades. Research on fish larvae biology has continued since then and has been aimed at increasing our basic knowledge and improving rearing methodologies. However, there are still large knowledge gaps and the rearing process is far from optimal. For years, scientific research has focused on relevant limiting factors seriously affecting survival and growth, in a sequential manner. New species have been introduced by applying similar protocols, and only some aspects related to feeding and nutrition have received scientific attention. This simplistic empirical approach is insufficient for understanding mechanisms for development and interrelations with the surrounding rearing water. During the last decade, new analytical tools have opened up new avenues for research in both physiology and the influence of environmental conditions. In the past few years, books and specialized journals have published reviews on the different disciplines related to fish larvae biology and aquaculture, providing wide descriptions of main topics like development, pathologies, nutrition, feeding, and physiology. These publications have increased our knowledge from a textbook perspective. It may thus be considered that there is overall, good knowledge on larval fish biology and on the current reality of larval rearing. Most of this information, however, is based on a small number of species, and knowledge in larval biology is advancing fast. Nowadays, there are new research ideas that have implications not only for fish larvae but also for other vertebrates, and there are new approaches to old problems, using omics methodologies, for example.

The research effort on larvae is now moving beyond the strict interest for growing fish; it is also focusing on using them as model organisms for fundamental research. Moreover, the numerous species being studied within the context of aquaculture will help to uncover both the variability and the similarities within this diverse group of vertebrates.

Instead of providing broad overviews covering general topics in larval fish biology, this book presents examples of novel research in fish larvae with very specific objectives. It refers to current advances derived from recent projects and a doctoral thesis targeted at filling specific gaps in our knowledge. By representing alternative points of view, the different chapters show how rearing and environmental conditions affect physiology and developmental processes from a molecular basis, and how these factors influence the final characteristics of late larvae and juveniles. Overall, this book will point to recent findings on the importance of environmental cycles, some specific nutrients, and the microbial environment on developmental processes. There are more emerging topics of interest, but with these few examples we hope to illustrate the dynamism of current research within this field. These are exciting times for biologists and the discipline of biology—especially when the target of the research is fish larvae.

Puerto Real, Spain Manuel Yúfera

Contents

Chapter 1
Investigating Fish Larvae-Microbe Interactions in the 21st Century: Old Questions Studied with New Tools

Ragnhild I. Vestrum, Birgit Luef, Torunn Forberg, Ingrid Bakke and Olav Vadstein

Abstract All animals need a mutualistic interaction with their microbiota for proper development and functioning. Also for the fish-microbiota interaction considerable research has been done, and especially for reared fish larvae this interaction is crucial for their viability. However, during the 1980s and 1990s a number of findings revealed at that time current methods were not suitable for studying the total microbial community and that data on composition of microbiota was biased. Several recent methodological revolutions have boosted the possibilities for addressing questions related to fish larvae-microbiota interactions that previously lacked suitable tools for proper evaluation. These methodological achievements include the development of experimental rearing systems including gnotobiotic systems for fish, new visualization tools, and molecular "omics" tools for characterizing the response of the host on a variety of levels and for characterizing both composition and activity of fish microbiota. We present and review these tools and give examples on how they have been used to improve our understanding of fish larvae-microbiota interactions. With respect to understanding, this includes in particular how the microbiota is established and maintained, what the functionality of the microbiota is and how it affects fish health, and finally how we can apply this knowledge for management of a healthy and beneficial microbiota in aquaculture settings.

Keywords Microbiome · Germ-free model systems · Imaging
Molecular methods · Omics

R. I. Vestrum · B. Luef · T. Forberg · I. Bakke · O. Vadstein (✉)
Department of Biotechnology and Food Science, NTNU Norwegian University of Science and Technology, 7491 Trondheim, Norway
e-mail: olav.vadstein@ntnu.no

Present Address:
T. Forberg
BioMar AS, Pirsenteret, 7010 Trondheim, Norway

M. Yúfera (ed.), *Emerging Issues in Fish Larvae Research*,
https://doi.org/10.1007/978-3-319-73244-2_1

1.1 Introduction

Mariculture has the possibility of being a major contributor to feeding a growing human population, but this will require solutions to significant challenges (Duarte et al. 2009). A major biological challenge will be to overcome bottlenecks with regards to producing high quality juveniles suitable for the intensive cultivation environment. We know that e.g. egg quality, nutrition and physiochemical conditions contribute to the observed problems, but the well-known observation of poor reproducibility between fully replicated tanks with equal genetic variability (e.g. full sibling groups) cannot be explained by such factors (Vadstein et al. 1993). Even though this is a complex problem with multiple dimensions, it is now demonstrated beyond any doubt, that detrimental host-microbe interactions are a major cause of these problems (Vadstein et al. 1993, 2013).

Since Robert Kock and Louis Pasteur's major discoveries in the 1880s, the relationship between man and their microbes has had primary focus on disease. However, during the last 15 years a paradigm shift has occurred, as substantial effort has been devoted to better understanding the role of our microbiota for health and normal development. The last decennium has revealed a multitude of ways in which the microbiota affects the normal development of animals, including fish (Kanther and Rawls 2010; Sekirov et al. 2010). Most of this research has been motivated by human health and therefore most studies have been done in man and mammalian model organisms. It is now important to verify the generality of these findings with other vertebrates and invertebrates, including fish as they have a core role in the evolution of vertebrates.

Until 15 years ago quantitative and qualitative studies of microbes in aquaculture systems were performed by culture-based techniques, and until the 60s it was a general consensus that the knowledge of microbes associated with animals was fairly well understood. Culture-dependent techniques mean that samples are serially diluted and spread on so-called non-selective and selective medium on agar plates. Quantitative estimates of densities are then given as colony forming units (CFU). Since the late 1970s it has been shown that the discrepancy between total counts in epifluorescence microscope and CFU is huge—typically a factor of 1000 for natural waters. This is the first problem for culture-dependent methods. For some time it was claimed that this difference was due to dead microbes, but it is now well established that this can only account for a limited part of the discrepancy (e.g. Karner and Fuhrman 1997). For man-made systems, like aquaculture systems, the culturability can be considerably higher than in natural systems, but the difference in the estimated abundance of microbes between the two methods is normally still one order of magnitude. Studies during the last two decades have shown that the difference between CFU and total counts is due to the selectivity of non-selective agar—i.e. the majority of the microbes are still not possible to culture at laboratory conditions (Hugenholtz et al. 1998). This is known as the "great plate anomaly" (Hugenholtz 2002) and points to a second problem with culture-based methods; for qualitative analysis of the composition of the microbes, the culture-based results are highly biased, as the culturability is dependent on phylogeny (Hugenholtz et al. 1998). A third drawback of culture-based methods is that it is time consuming. Pure cultures have to be established from colonies through several steps of sub-culturing, and finally the pure cultures must be identified through a substantial number of tests. For appropriate identification of species composition more than 100 pure cultures should be characterized per sample (1% resolution = 1 isolate represents 1% of the population). In reality, few studies have characterized that many pure cultures per sample, which is a clear indication of the labor involved. For one of the authors of this chapter (O.V.) these methodological limitations felt overwhelming in the early 90s. However, since then the revolution in method development has caused another type of paradigm shift. This shift is largely due to the rapid development in molecular biology and imaging methods. Method development, possibilities, and applications are covered in detail below.

Based on the knowledge of the bias in culture-based studies one question is unpleasant but impossible to disregard: Do we have to do it all over again? Or more specific, is the cultivation-based knowledge we have on host-microbe interactions in fish so biased that it counteracts the progress in our knowledge and understanding? The answers to these questions are not straight forward. First, all types of "old" knowledge is probably not equally biased. Second, even though results from high-throughput sequencing studies of fish microbiota are accumulating at a decent speed, the amount of data is still too limited to do a proper evaluation of "old" data. The severity of the problem is exemplified by the study of Fjellheim et al. (2012) who found a negative correlation between quantification of bacteria associated with larvae by plating on agar plates versus by quantitative PCR of rDNA. Thus we are convinced that we have to do a lot of the work over again.

The aim of this chapter is to give an overview of new methods and how they can be used to speed up and improve our understanding of fish-microbiota interactions. With respect to understanding, this includes in particular how the microbiota is established and maintained, what the functionality of the microbiota is and how it affects fish health, and finally how we can apply this knowledge for management of a healthy and beneficial microbiota in aquaculture settings. We will describe the methods with focus on applications without going into technical details, and give examples of applications for studying host-microbiota interactions.

1.2 Experimental Designs for Studying Fish-Microbiota Interactions

Traditionally, experiments with fish are run in cultivation tanks. These are normally just a downscaling of aquaculture production systems, and therefore realistic systems from an aquaculture perspective. However, the degree of experimental control for fish-microbe studies is limited. Below we will describe different experimental systems, going from high relevance/low control to low relevance/high control.

We want to stress that it is important to keep this range of experimental systems, as high control systems can be used to study mechanisms and hypotheses which then can be verified in systems where the fish live a more normal life, i.e. low control systems. We will try to highlight pros and cons within the different systems.

1.2.1 Traditional Rearing Tanks

Traditional rearing tanks can have variable volumes and can be run in traditional flow-through systems (FTS) or as recirculating aquaculture systems (RAS), where the water is reused after removal of waste and toxic substances. Tank volumes depend on the size of the fish and the population size needed for experimental reasons. From an ethical point of view and due to legislation the number of animals used in an experiment should be as low as possible, but especially in applied research the experimental systems should be relevant for a practical setting. This might create a conflict between ethics and the practical relevance of new knowledge. The use of FTS versus RAS has clear implications for fish-microbe studies, as for RAS there will also be a recirculation of microbes. In some situations this can be an advantage (maintaining the effect of the fish on the microbiota), whereas in other cases a disadvantage (a high background concentration of microbes). FTS versus RAS is therefore an important part of the experimental design. Traditional rearing tanks have limited possibilities to control the microbiota in the rearing water. This includes import and export of microbes, but particularly microbial growth in the rearing tanks. For example, in studies with probiotics, it might be difficult to predict the probiotic microbes' ability to establish in the system as the competitive situation with background microbes will vary depending on the species present and system design. However, we have shown that it is possible to control and stabilize the microbiota in rearing tanks through a selection regime against some microbes, and favoring others (r- and K-selection) (Attramadal et al. 2014).

1.2.2 Systems with Rearing of Single Individuals

One of the drawbacks of using normal rearing tanks is that you have limited possibilities for controlling import and export of microbes. Another problem is that one fish may affect the other individuals in the same tank. This is important as moribund individuals may infect healthy individuals, and that there is a continuous sharing of microbes by defecation and re-ingestion (Reitan et al. 1998). A way out of this problem is to rear single individuals in small units. Size and type of units may vary from e.g. 40 ml in plastic cups (Forberg et al. 2016) to 2 ml in 24-multiwell plates (Fjellheim et al. 2010; Sandlund and Bergh 2008). Size of the system may vary dependent on species and the length of the experiment. These types of systems can be run without water exchange for yolk sac stages and early

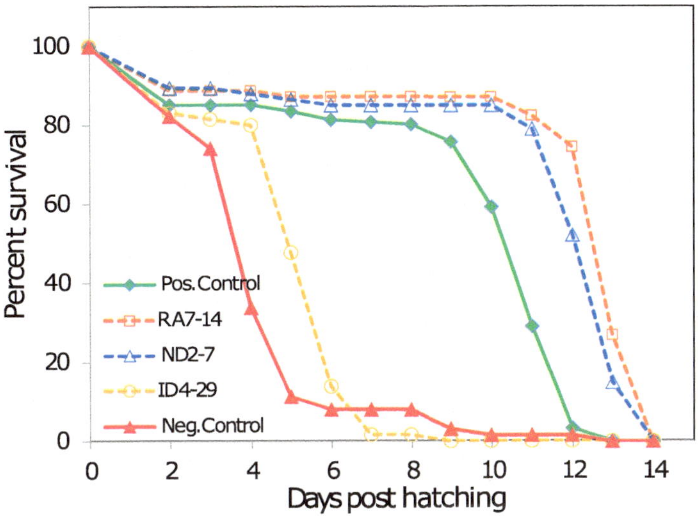

Fig. 1.1 Survival of yolk-sac larvae of Atlantic cod (*Gadus morhua*) incubated together with different bacteria, including background bacteria (positive control), a pathogen (negative control) and three probiotics candidates, versus time. Each treatment included 72 individuals. Data from Fjellheim et al. (2010)

first feeding (Fjellheim et al. 2010; Sandlund and Bergh 2008) or with an irregular water exchange for longer experiments (Forberg et al. 2016). In systems with water exchange, there are possibilities for maintaining some degree of control of the microbes entering the system by e.g. adding a defined or described microbiota. It is, however, important to keep in mind that the way by which the renewal of the water takes place may create different types of selection regimes for the microbiota in the rearing unit (e.g. continuous *versus* periodic dilution rate).

An example of the use of 24-multiwell plates with single individuals in 2 ml of water is shown in Fig. 1.1. In this experiment, unfed yolk-sac larvae were exposed to either probiotic candidates, a pathogen (negative control) or only background bacteria (positive control). As shown in the figure, you get reliable results with a limited number of individuals (in this case 72 individuals per treatment), and the reproducibility is very good (unpublished results). This simple system easily revealed that one probiotic candidate actually was pathogenic (ID 4-29). In systems with rearing of single individuals, it is possible to reduce the number of individuals used for the experiment to a minimum, and confidence intervals for the survival can be calculated from surviving individuals and the total number of individuals based on binomial statistics (Box et al. 1978). Moreover, chi-square and exact tests can be used for statistical analysis.

The pros of these systems should be evident based on the information above. The down side is mainly the simplicity of the system and for some species the difficulty of rearing the fish for sufficiently long periods. The simplicity includes lack of interactions between individuals of fish, which is an advantage for some types of experiments. This clearly emphasizes the advantage of having a possibility for doing experiments in different types of systems, and selecting the systems based on the aim of the study.

1.2.3 Germ-Free and Gnotobiotic Systems

If traditional rearing tanks is on one end of the experimental system scale, germ-free and gnotobiotic systems are on the other end. A germ-free system has no microbes at all, and in a gnotobiotic system the biota in the system is known (= gnotos) and predefined. This can only be achieved by first generating germ-free test animals or fish, which are then deliberately colonized by one or several bacterial strains known to the researcher. Gnotobiotic systems were first developed for mammals such as rats, but also for fish several gnotobiotic systems have been published (Table 1.1). In general, the procedure of creating germ-free fish involves a first step with surface disinfection of the eggs and subsequent hatching in an environment without microbes. The subsequent rearing under germ-free or gnotobiotic conditions may be in the presence or absence of antibiotics. When rearing the fish with antibiotics, you can only use antibiotic resistant bacteria and there is also a possibility of negative effects on larvae due to long term exposure to antibiotics (Moullan et al. 2015). As documented in Table 1.1, many different techniques have been used to surface disinfect the eggs, and gnotobiotic systems have been attempted for several fish species.

The main advantage of using these systems is the high degree of control of the microbial environment. This is appealing for many types of experiments with a reductionistic approach—a strategy with a long history of success in natural science. The main disadvantage of germ-free and gnotobiotic systems is the complexity in performing these experiments. A reproducible method for achieving germ-free larvae is needed, and avoiding contamination during the experiment is also crucial (i.e. maintain them germ-free or gnotobiotic). In the cases were the fish is fed live prey, the prey also needs to be germ-free. When using *Artemia* as live feed it is possible to hatch them from disinfected cysts, and this makes the process more straightforward. However, when using e.g. rotifers a germ-free culture has to be established, they have to be reared germ-free, and therefore must be fed germ-free feed (microalgae or yeast). Moreover, this production line has to be kept throughout the germ-free or gnotobiotic fish experiment. One problem not given much attention in the literature, is the maintenance of the gnotobiotic condition. When only one microbe is added it is a presence/absence problem, but when several species are added in a given ratio it is not straight forward to maintain this ratio. This has to do with the selection regimes in the rearing unit and due to renewal of water (see also Sect. 1.2).

Results generated from one fish species may not be easily transferrable to another, as germ-free zebrafish (*Danio rerio*) seem to be less developed than conventional fish (Rawls et al. 2004), while germ-free sea bass larvae have more

Table 1.1 Overview of published gnotobiotic systems for fish (reproduced from Forberg and Milligan-Myhre 2017)

Species	Sterilizing technique	Methods used to detect bacteria	Citation
Atlantic cod (*Gadus morhua*)	Glutaraldehyde, rifampicin and ampicillin	Flow cytometry, PCR and culture-based	Forberg et al. (2011)
Atlantic halibut (*Hippoglossus hippoglossus*)	Glutaraldehyde and antibiotic mixture	Culture based	Verner-Jeffreys et al. (2003)
Platyfish (*Xiphophorus maculatus*)	Laparatomy of surface disinfected pregnant females	Culture based	Baker et al. (1942)
Red drum (*Sciaenops ocellatus*)	Hydrogen peroxide disinfection of eggs	Culture based	Douillet and Holt (1994)
Atlantic salmon (*Salmo salar*)	PVP-I scrubbing of fish and aseptic collection of eggs and aseptic fertilization/ PVP-I disinfection of fertilized eggs	Culture based	Trust (1974), Lesel and Lesel (1976)
Sheepshead minnow (*Cyprinidon variegatus*)	Antibiotic mixture	Culture based	Battalora et al. (1985)
European sea bass (*Dicentrarchus labrax*)	Glutaraldehyde, rifampicin and ampicillin	PCR and culture	Dierckens et al. (2009), Rekecki et al. (2012)
Threespine stickleback (*Gasterosteus aculeatus*)	PVP-I and bleach soak	PCR, microscopy and culture-based	Milligan-Myhre et al. (2016)
Tilapia (*Tilapia macrocephala*)	Formaldehyde/Hydrogen peroxide, sodium hypochlorite and antibiotic/antifungal mixture	Culture-based	Shaw and Aronsen (1954), Situmorang et al. (2014)
Turbot (*Schophthalmus maximus*)	Antibiotic solution	Culture-based	Munro et al. (1995)
Zebrafish (*Danio rerio*)	PVP-I and bleach	Culture-based and PCR	Rawls et al. (2004), Pham et al. (2008)

developed digestive tracts than conventional larvae (Rekecki et al. 2009). This discrepancy suggests that the mechanisms underlying host responses may differ with species, highlighting the need to study effects of gnotobiology in more than one species. The progress achieved with these gnotobiotic systems is included in the text below on studies of the host.

1.3 Imaging

Light and fluorescence imaging allows studying and surveying relatively large areas of tissues and biofilms. Confocal laser scanning microscopy in combination with different florescent labeling methods has become an indispensable technique to investigate spatial and temporal distribution of microbial consortia, largely due to its improved focusing depth. However, light- and laser-based microscopy techniques are generally limited in spatial resolution to approximately 0.2 μm, due to the wavelength of light. Recent developments in super-resolution optical microscopy have found ways to get around this diffraction limit, and will likely continue to increase the level of spatially resolved detail that can be obtained under dynamic conditions (Betzig et al. 2006; Klar et al. 2000; Rust et al. 2006). Yet these techniques still leave uncertainty as to the cells' ultrastructure. One of the big advantages of modern Transmission electron microscopy (TEM) is its ability to study precise spatial relationships of macromolecular complexes while retaining the context of the surrounding cellular and tissue architecture. Correlative light and electron microscopy, often referred as CLEM, combines the best of both worlds (Müller-Reichert and Verkade 2012). By combining, correlating and thus integrating different state-of-the-art microscopic modalities at different scales and levels of resolution we can gain a better understanding of key functions carried out by microorganisms in complex microbial communities; i.e. the versatility of fluorescent markers and the high spatial resolution of the TEM, to analyse dynamic subcellular architecture at high resolution.

The field of fish larvae-microbe interactions is currently being driven to a large extent, by sequencing-based methods such as metagenomics and metatranscriptomics (see below). Complementary tools are of interest to deeper probe and understand microbe-host and microbe-microbe interactions and also the microbial activities. Here, we give a short overview of different microscopic techniques which are available to study fish larvae-microbe interactions (Fig. 1.2). Until now, mainly light and fluorescence imaging and conventional TEM are used to study host-microbe interactions. Some of the techniques we discuss here are new to the field, but applying them could give new insights into multispecies microbial communities, their respective intra- and interspecies interactions, as well as their interactions with the host tissue.

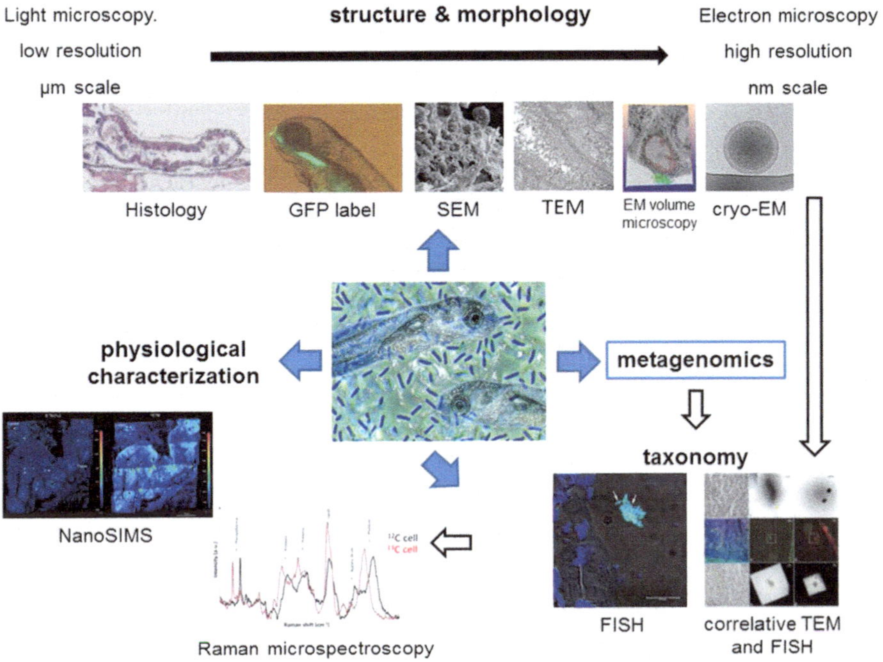

Fig. 1.2 Combining state-of-the-art imaging methods for the characterization of fish larvae-microbe interactions. Characterization of fish larvae-microbe and microbe-microbe interactions by low to high resolution imaging techniques, identification of bacteria by correlating microscopy with fluorescence in situ hybridization (FISH). 16S rRNA sequences recovered via clone library analysis and metagenomics provide information on FISH probe design. Where appropriate, FISH imaging approaches can be combined with Raman spectroscopy or NanoSIMS in order to determine identities of bacteria and their activities on a single cell level. Abbreviations: Transmission electron microscopy (TEM), electron microscopy (EM), scanning electron microscopy (SEM), green fluorescent protein (GFP), cryogenic electron microscopy (cryo-EM), nanoscale resolution secondary ion mass spectrometry (NanoSIMS). Center image from KJ Playfoot; GFP labeled *V. anguillarum* in zebrafish from O'Toole et al. (2004); cryo-EM image from Luef et al. (2015); correlative EM and FISH from Knierim et al. (2011); Raman spectrum from Stecher et al. (2013); NanoSIMS figure from Berry et al. (2013); EM volume microscopy image (FIB-SEM) was provided by DM Jorgens

1.3.1 Imaging from Low to High Resolution

In order to gain spatiotemporal insight into fish larvae microbial community organization, genetically-encoded labeling of microbes such as probiotics or pathogens, or fluorescence in situ hybridization (FISH) of bacteria can be applied. Fluorescence imaging allows studying and surveying a relatively large part of the larvae, e.g. the digestive tract. The digestive tract is often the main focus for studying fish larvae-microbe interactions, because it is a crucial interface between the host and its digestive environment and therefore represents an important putative portal of entry for microorganisms. For example, green fluorescent protein (GFP) has become widely used as a biomarker for monitoring gene expression, protein localization and protein-protein interactions in living cells due to its non-invasive detection and ability to generate light independently of exogenous substrates. In a study by O'Toole et al. (2004), zebrafish were exposed to GFP-labeled *Vibrio anguillarum*, a well-known fish pathogenic bacterium, and the infection was visualized at the whole fish and single bacteria cell levels using light microscopy, allowing new insight into the infection pathways of this pathogen. The ability to visually locate single bacteria during infection of a live host, presents an interesting tool for studying host-microbe interactions. For morphological structural studies, sections of biological tissues are often analyzed by using special staining techniques combined with light and electron microscopy. Rekecki et al. (2009) performed a histological study by light microscopy on cultured germ-free sea bass larvae fed germ-free live feed. Morphometrical analysis clearly revealed that germ-free, static larvae were significantly larger than conventional and germ-free, rotating larvae. Stereological methods also showed that germ-free, static larvae had a more developed gastrointestinal tract compared to the two other groups. However, histological analysis showed minor variations in regional morphology of epithelial cell types observed in the gut in individual larvae between and within treatment groups. This was the first histological study on cultured germ-free marine fish larvae fed germ-free live feed. Fluorescently labeled antibodies were successfully applied on histological sections of zebrafish tissue to visualize that microbiota stimulate intestinal epithelial proliferation (Rawls et al. 2004). Bates et al. (2006) characterized the acquisition of microbiota during zebrafish development by using FISH with oligonucleotide probes targeting conserved sequences of the eubacterial 16S rRNA gene. Gunasekara et al. (2012) combined a gnotobiotic *Artemia* test system with histological and TEM monitoring tools to study the digestive tract morphology in response to feeding isogenic yeast strain (wild type an mnn9 yeast cell wall mutant yeast strain), after experimental infection with *Vibrio campbellii*. Colonization of the gut lumen by *V. campbellii* could be observed by TEM for the group of *Artemia franciscana* fed the wild type yeast. Additionally, it was also observed that *V. campbellii* caused damage to the gut epithelium. The gut epithelium remained intact in challenged *Artemia* fed mnn9 yeast. In general, conventional TEM examination offers an important tool for investigating the microbial ecology of e.g. the gastrointestinal tract, gills and skin, and determining the presence of autochthonous microbiota, but also of harmful bacteria in fish (Declercq et al. 2015; Hansen et al. 1992; Rekecki et al. 2013; Ringø et al. 2003, 2006).

In order to determine the 3D community organization, modes of interactions between microbes (i.e. macromolecular strategies), as well as microbe-host interactions, volume electron microscopy (volume EM) such as focused ion beam scanning electron microscopy (FIB-SEM) or Teneo volume scope (serial block face imaging) can be employed. The strength of volume EM is to study spatial relationships of macromolecular complexes while retaining the context of the surrounding cellular and tissue architecture within large 3D volumes (tens to hundreds of microns in each dimension). Back-scatter electron detection, the mode of imaging for volume EM, requires a higher amount of heavy metal staining than typical for conventional TEM in biological samples. Due to this heavy staining and the nature of back-scatter electron detection, the resolution obtainable by volume EM is limited to 3–4 nm per pixel. To the authors' knowledge, so far these techniques have not been applied to the field of fish larvae-microbe interactions. But Armer et al. (2009) imaged transient blood vessel fusion events in relation to the surrounding tissue microenvironment in zebrafish by correlative volume EM.

In order to elucidate the ultrastructure of microbes at a much finer level (2–3 nm resolution) and thus provide detailed insight into the inner workings of each cell, cryogenic Transmission Electron Microscopy (cryo-TEM) and cryogenic Electron Tomography (cryo-ET) can be utilized (e.g. Milne and Subramaniam 2009). Since this cryogenic approach results in unstained frozen-hydrated samples and because low dose data acquisition schemes minimize radiation damage, ultrastructural details are preserved with high fidelity. Therefore, macromolecular complexes and cells can be visualized in a "close-to-native" state; however, the area that can be studied is typically limited to individual cells. Cryo-TEM provides 2D images, with a comparably high electron dose, thus at a high signal-to-noise ratio, whereas cryo-ET provides 3D ultrastructural information on cellular and subcellular organization, thus at a low signal-to-noise ratio. Cryogenic sample preservation, coupled to 3D image reconstructions, will enable high-resolution morphological comparison of microbes, e.g. sub-cellular features, organelles, cell-cell associations and cell surfaces, which are not easily detectable by applying other approaches.

Correlating cryo-florescence microscopy can help bridge the gap in resolution between light and electron microscopy (e.g., Plitzko et al. 2009; Sartori et al. 2007). Under cryogenic conditions, features of interest are localized by their fluorescent signal by light microscopy before zooming in on the structures of these features by cryo-TEM.

1.3.2 Linking Cellular Morphology and Phylogenetic Identity

One of the key challenges in the high-resolution spatial study of multi-species consortia is how to link cellular morphology and ultrastructure of their community members to their respective phylogenetic identity. Linking metagenomics, metaproteomics, and metabolomics information with near intact morphology, ultrastructure, and high quality 3D images of spatial interactions and networks is enabling us to start tackling this challenge.

Traditionally studying identity and the spatial organization of microorganisms has been accomplished by FISH imaging, but optical techniques are diffraction limited and hence, ultrastructural features and/or macromolecular strategies cannot be observed by FISH imaging alone. Knierim et al. (2011) reported a strategy to combine FISH labeling and imaging with whole-mount cryogenic or resin-embedding section EM.

1.3.3 Linking Microbial Populations to Specific Metabolic Processes

Measuring activity patterns of microbes in their natural environment is essential for understanding ecosystem functioning and the complex interactions of microorganisms with eukaryotes. Approaches that allow measuring substrate uptake by specific populations—even at the single cell level—and visualize the spatial organization of the communities enhance our understanding of metabolic processes. For example, by combining FISH using fluorescent and/or halogenated probes with (1) Raman microspectroscopy, which is a non-destructive spectroscopic method that produces a chemical "fingerprint" of the abundant molecular bonds in individual microbial cells, or (2) nanoscale resolution secondary ion mass spectrometry (NanoSIMS; see review Wagner (2009) and references therein), specific microorganisms that have metabolized the substrate of interest can be identified and visualized with high resolution imaging and isotope mapping. Stecher et al. (2013) developed a single cell isotope labeling tool by applying these techniques that allow detailed insights into substrate partitioning and niche competition among individual intestinal microbiota members, including incoming pathogens.

The rise of single-cell approaches opens unprecedented opportunities in the field of environmental microbiology. It is now possible to detect metabolically active single cells in environmental samples, quantify their activity and identify them phylogenetically (Fig. 1.3; Berry et al. 2013, 2015). Moreover, microbial interactions within mixed populations can be identified and allow the analysis of numerically rare community members that despite low abundance could have important ecological and metabolic functions within an ecosystem (Berry et al. 2015). This recent study shows that incubation of samples containing complex microbial communities with D_2O subsequently allows microbiologists to identify individual metabolically active bacterial and archaeal cells by Raman micro-spectroscopy without any further sample pre-treatment. Heavy water could also be used to label microbial cells living within or attached to plants, animals and possibly even humans. Although consumption of high amounts of heavy water is toxic or even lethal to plants, animals and humans, the D_2O concentrations required for activity labeling of microbial communities are below the toxicity threshold for body water deuteration of several important model animals (Berry et al. 2015). Consequently, it should become possible to perform in vivo activity labeling experiments with microbial gut communities in aquaculture species by providing heavy water for a short time period.

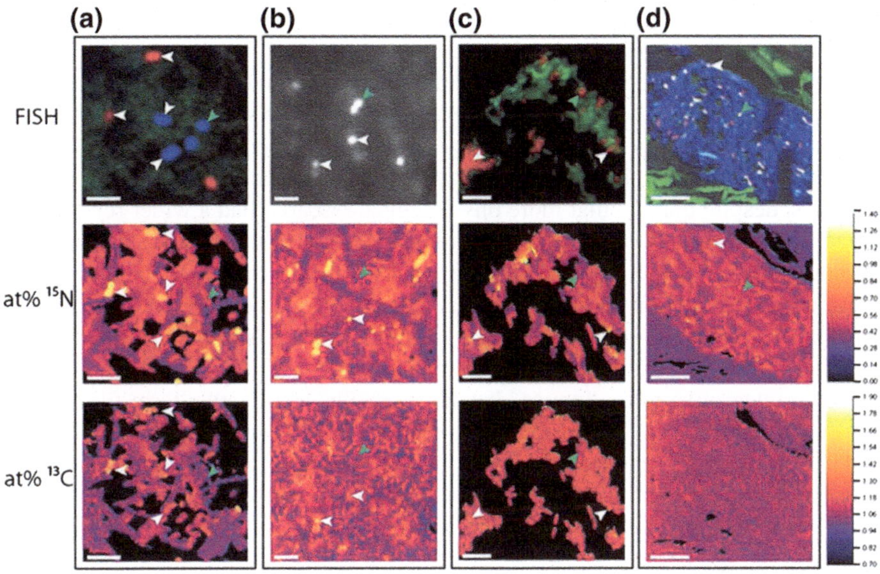

Fig. 1.3 Stable isotope-labeled threonine was given in i.v. to mice and combined fluorescence in situ hybridization (FISH) with high-resolution secondary ion mass spectrometry imaging (NanoSIMS) to characterize utilization of host proteins by individual bacterial cells. Per-cell C and N isotope composition was quantified. Representative NanoSIMS (at% ^{13}C and ^{15}N) and FISH images 8 h after i.v. injection of ^{13}C, ^{15}N threonine. **a** *B. acidifaciens* (blue) and *Ruminococcaceae* OTU_5807 (red) cells (bar: 5 μm; prepared without embedding; other bacteria shown in green), **b** *Akkermansia* spp. (white) cells (bar: 2 μm; prepared without embedding), and **c** *Lactobacillaceae/Enterococcaceae* spp. (red) cells (bar: 5 μm; prepared without embedding; other bacteria shown in green). **d** Semithin sections of lumen contents with *Lachnospiraceae* OTU_11021 (yellow/white, overlap of green and red signals), all other bacteria (blue), and autofluorescent dietary fibers (green) (bar: 10 μm). Exemplary cells with or without significant enrichment (white or green arrows) are indicated. No enrichment in carbon is visible in **d** because of dilution by unlabeled carbon in the resin (from Berry et al. 2013)

1.4 Molecular Methods to Study the Host

During rearing of fish larvae, numerous interactions between the microbiota in the system and the fish will take place. The microbes are delivered to the fish both through the water, the feed and the microalgae in the system (De Schryver and Vadstein 2014). The microbes are normally found on the skin, on the gills and in the gut of the fish. The relationship between a fish (the host) and its microbiota, can be commensal (neither harmful nor beneficial effects for host), mutualistic (host and microbe receive some sort of benefit) or pathogenic (harmful effects for the host). Interactions with pathogens or some opportunistic bacteria might cause infections or even death, but a majority of the bacteria surrounding the fish are harmless or beneficial for the larvae (Olafsen 2001; Ringø and Birkbeck 1999). These commensal or mutualistic bacteria might help in protecting the fish from infection from other bacteria in the rearing water, and may also contribute directly by providing the fish with proteolytic enzymes and vitamins and by breaking down xenobiotics in the intestine of the fish (Hansen and Olafsen 1999; Xu and Gordon 2003).

The molecular revolution has had implications both for studies on microbes, and for the characterization of host responses. Until 10 years ago most studies concerning fish-microbe interactions used mainly growth and survival as host response variables, despite the fact that more physiological research had a wider repertoire of methods. Growth and survival are powerful measurements as they are integrative and thus capture a wide variety of primary and secondary responses. However, only considering these two response variables does not allow a deeper understanding of mechanisms involved in host-microbe interactions. By using molecular biology techniques, this limitation has been overcome.

As an integrative term, "omics" has been used as a collective term for methods that include genomics, transcriptomics, metabolomics and proteomics. The genomes of several different species of fish, including species relevant for aquaculture, are now sequenced, and even though the function of many genes is not yet known there is a huge amount of information available. For example, the sequencing of the Atlantic cod (*Gadus morhua*) genome (Star et al. 2011) revealed that the cod has a unique immune system compared to other teleost fishes, a discovery that will improve the analysis of cod research in the years to come. An overview of the different "omics" is given in Fig. 1.4. Genomics reveal the total gene inventory of a species, and thus sets the limits for what a species possibly can do. In terms of

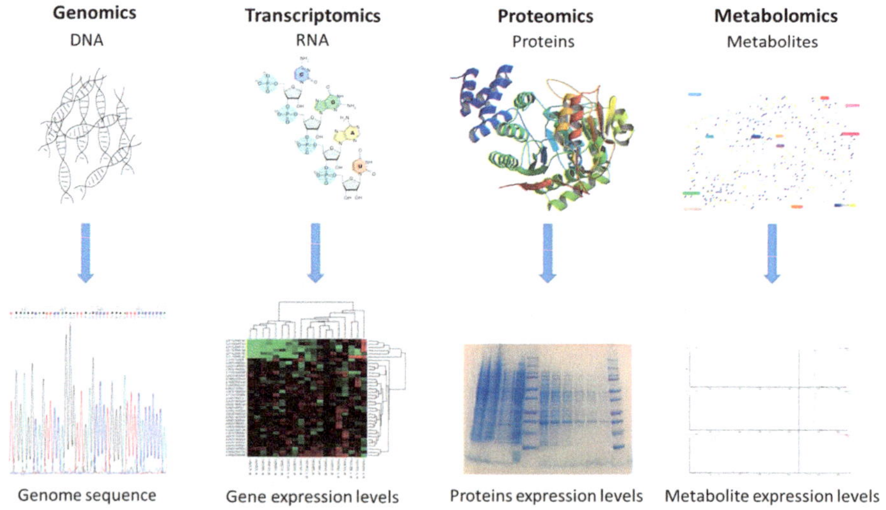

Fig. 1.4 Use of "omics"-based approaches for the investigation of host responses in host-microbe interactions. Host responses can be investigated based on DNA, RNA, proteins or metabolites. The choice of starting material decides the "omics" approach used, and hence what knowledge is gained. For more details, see text

studying host responses genomics is thus not that relevant, but it serves as the basis for all the other functional "omics" tools as all transcripts, proteins and metabolites are coded for in the genome.

1.4.1 Transcriptomics

Transcriptomics is the "omics" tool so far used the most in studies of host responses in fish. The transcriptome refers to the complete set of RNA transcripts that are produced in a specific cell, tissue or organism, at a specific developmental stage or physical condition, and transcriptomics is the study of these transcripts. The genome of an organism is relatively stable, but the transcriptome is dynamic and changing with e.g. the developmental stage of the organism, the physiological condition and also the environmental conditions. This makes transcriptomics an excellent tool for understanding the relationship between the genotype and phenotype of an organism (Qian et al. 2014). In the early days, quantitative PCR (qPCR), suppression subtractive hybridization PCR (SSH-PCR) and microarrays were the tools of choice. Microarrays and qPCR are still powerful tools, but have the drawbacks that they rely on fluorescent dyes for labeling DNA and it is necessary to know the sequence of the genome, or at least the genes of interest, and you are thus limited to a defined set of chosen genes.

The work of Rawls et al. (2004) is an example of how the techniques mentioned above have been used on a germ-free zebrafish model. By conducting a broad, functional genomics-based analysis using microarrays they identified 212 genes that were differently expressed in fish depending on the presence or absence of bacteria. Fifteen of the genes that were upregulated by bacteria were involved in DNA replication and cell proliferation. These findings indicate that epithelial proliferation of the host is associated with the microbiota in the gut, which corroborates the results from the morphological observations. In the same study they also used qPCR to reveal that some of the host responses showed microbial specificity. Similarly Forberg et al. (2011, 2012) used qPCR and SSH-PCR to investigate the host responses of Atlantic cod larvae reared in different microbial conditions using a gnotobiotc system. They identified several genes that were influenced by the microbial content of the rearing water. They also found that dead bacteria can affect the gene expression of certain genes, and observed a positive effect of dead, probiotic bacteria on the survival of the cod larvae.

The techniques mentioned above, have been used extensively for more than a decade, however these tools have some limitations (as mentioned above; review by Qian et al. 2014). With the advance of high throughput sequencing (HTS), a new method for mapping and quantifying transcriptomes has evolved, namely RNA-sequencing (RNA-seq) (Wang et al. 2009). RNA-seq is also known as whole transcriptome shotgun sequencing (WTSS). This technique uses HTS to sequence cDNA directly from the RNA sample of interest. The differences between probe-dependent methods and probe-independent methods, were illustrated by

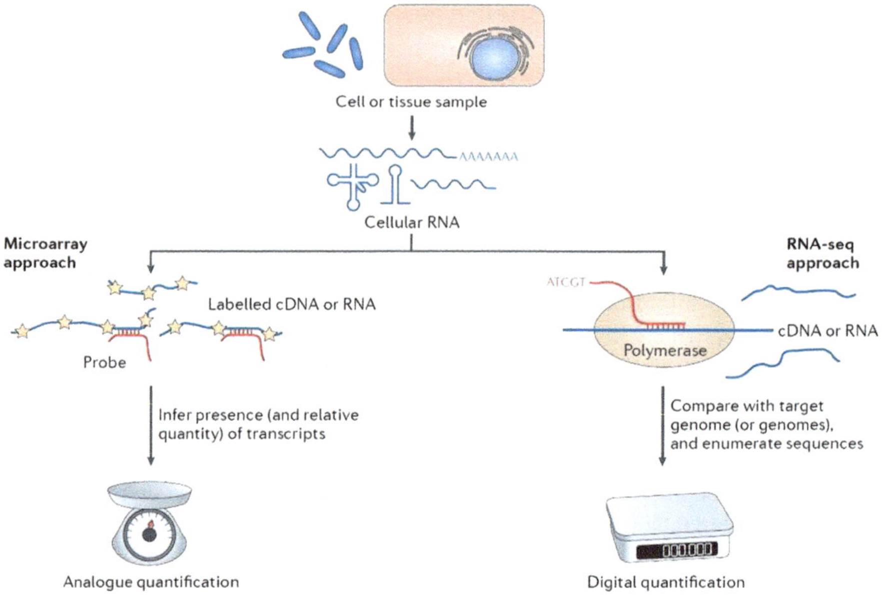

Fig. 1.5 Differences between probe-dependent and probe-independent methods for gene expression analysis (from Westermann et al. 2012). Notice the difference in how the changes in the transcriptome are detected

Westermann et al. (2012) and are shown in Fig. 1.5. In probe-dependent methods like microarray analysis and qPCR the DNA or RNA is labeled using fluorescent dyes, while probe-independent methods like RNA-seq do not involve this kind of labeling step.

For aquaculture production, the main use of RNA-seq has so far been for studying transcriptome changes after pathogen infection or after vaccination. For example, in a study by Sarropoulou et al. (2012) on European sea bass (*Dicentrarchus labrax*) reared in aquaria with natural seawater, transcripts were investigated by RNA-seq after oral vaccination against *V. anguillarum*. *V. anguillarum* is a well-known fish pathogen, but the cellular response of the host against this pathogen, is largely unknown. They successfully sequenced the transcriptome of head kidney and hindgut of vaccinated and unvaccinated fish, unraveling differentially expressed genes in the two groups of fish.

Another very important aspect in aquaculture is the effects various feed sources have on the fish, and RNA-seq can be used to unravel these. In a study by Penglase et al. (2015) one group of cod larvae were fed a diet consisting of copepods and another was fed the normal rotifer/*Artemia* diet, which is common in aquaculture. It is well known that cod feeding on natural zooplankton, like copepods, have less deformities and better growth than cod fed rotifers and *Artemia* (review by Hamre 2006), and several studies concerning the nutritional value of the rotifer/*Artemia* diet have been conducted. By using RNA-seq Penglase et al. (2015) also showed that the different diets have different effects on the redox system in cod larvae. They found that cod larvae fed a rotifer/*Artemia* diet had different expression of approximately half of the genes involved in the redox system of the cod, compared to the copepod-fed cod larvae. They suggested that the rotifer/*Artemia* diet causes an oxidative stress response in the larvae and that this might be one reason for the observed dietary-induced differences in larval growth.

One important limitation in using RNA-seq for studying host-pathogen interactions is that it is normally used for studying one species at a time. Westermann et al. (2012) hypothesized that by using a dual approach to the RNA-seq method, one could potentially manage to monitor the expression of genes at different time points throughout an experiment from both the host and the pathogen at the same time without having to physically separate them. Indeed, in a study conducted by Westermann et al. (2016) they infected human (HeLa) cells with *Salmonella enterica*, and sequenced the transcriptomes in the cells at 4 and 24 hours post-invasion. At those time points both human and bacterial transcriptomes were present in the cells, and the dual RNA-seq captured both types. The distinction between human and bacterial transcriptomes was done by in silico analysis. To our knowledge, this method has not yet been applied in aquaculture research. However, Zhang et al. (2016) used zebrafish as a model system when studying immunological responses to human cancer cells, which indicates that it should be possible to study host-pathogen interactions in aquaculture species by this method, as well.

1.4.2 Proteomics

Another "omics" technology that has arisen with the molecular revolution is proteomics. The proteome is the set of all proteins that are expressed in a cell, a tissue or an organism, and proteomics is the study of these proteins. Compared to genomics and transcriptomics, this approach also captures information on changes in the organism's protein activity. This is measured as post-translational modifications, which are important for the function of the cell and give information on the physiological state of the organism—an aspect that is not covered by transcriptomics. According to a review by Rodrigues et al. (2012), many studies actually show a lack of correlation between the transcriptome data and the protein expression.

The common workflow when using a proteomics-based approach is first sample preparation i.e. protein extraction, then protein separation and quantification, and the last step is protein identification and characterization. A simplified outline of the most common workflows is presented in Fig. 1.6. Depending on whether using a gel-based method, like two dimensional electrophoresis (2-DE) or difference electrophoresis (DIGE) for the separation and quantification of proteins, or a method based on mass spectrometry (MS), the experiment will be classified as either "gel-based" or "gel-free". More detailed information on methods and protocols is out of the scope of this chapter.

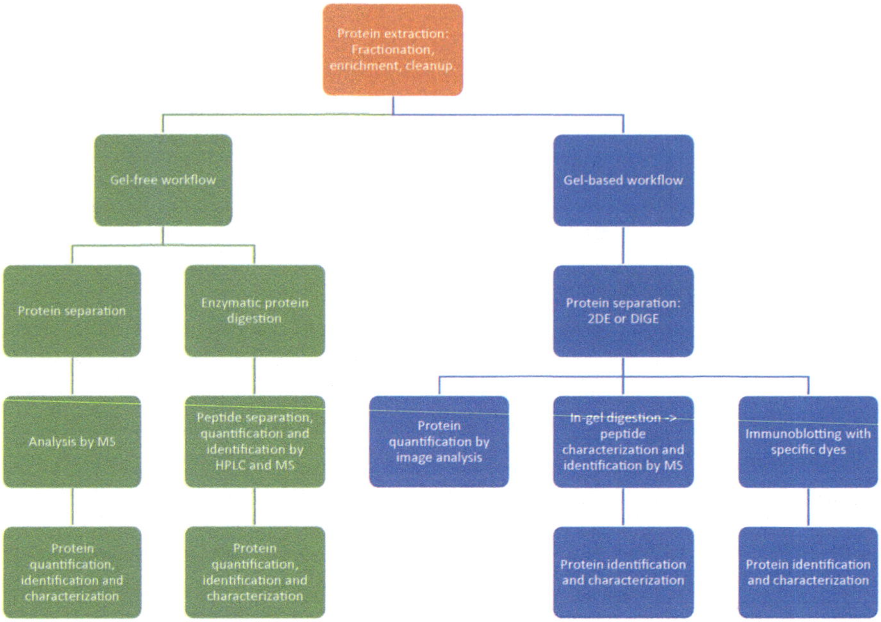

Fig. 1.6 Simplified outline of the workflow of proteomics experiments. Abbreviations: 2DE: 2 dimensional electrophoresis; DIGE: difference electrophoresis; MS: mass spectrometry; HPLC: high performance liquid chromatography

In aquaculture, one of the main challenges is to avoid various diseases, and pathogen detection and health management are popular and necessary research fields. Many of the most common diseases in marine organisms are caused by bacteria that are inherent in the environment (Toranzo et al. 2005), and the mucosal surfaces of the fish, including the gills, the skin and the gut, are very often the sites of infection. It is therefore of great importance to unravel how the mucus with all its proteins react upon contact with pathogens or opportunistic bacteria. Basic research is needed in order to gain more knowledge about this topic, and for understanding the mechanisms of infection, including host responses. Rajan et al. (2011) created a proteome reference map for skin mucosa in Atlantic cod, with the aim of establishing a benchmark for studies in this research area. The skin mucus was analyzed by using 2D gels, liquid chromatography combined with mass spectrometry (LC-MS) for identification of the spots from the 2D gel, and cloning of key immune competent genes. They created a reference map of the cod mucosal proteome, and the proteins were grouped in 8 different clusters based on their biological functions (Fig. 1.7). Later, Rajan et al. (2013) investigated and compared the skin mucus proteome of healthy cod to that of cod infected by *V. anguillarum*. They identified a group of proteins that were differently expressed in infected cod compared to healthy cod, and these proteins appeared to be closely linked to immune responses (Rajan et al. 2013).

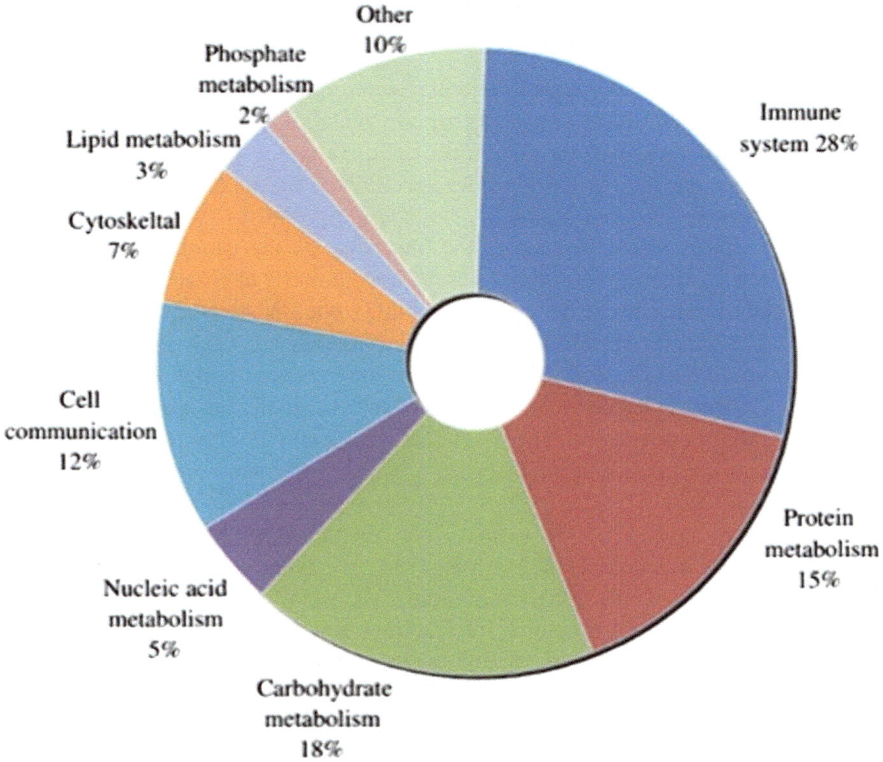

Fig. 1.7 Protein reference map divided in groups for the cod (*Gadus morhua*) mucosal proteome. Proteins were identified by using liquid chromatography combined with mass spectrometry (LC-MS), and clustered according to the biological processes they belong to (from Rajan et al. 2011)

Rajan et al. (2011, 2013) conducted their studies at a time when the genome of cod was not yet sequenced. Having information on the genome sequence available when the studies were performed would have simplified their analyses of the proteome immensely. Most other fish species used in aquaculture are not model organisms, and their genome sequence is not known. Researchers therefore face the same challenges as Rajan et al. did in 2013. However, the cost and amount of labor needed for whole genome sequencing, is rapidly decreasing, and we believe that the genome sequence of more and more aquaculture species will become available in the years to come.

1.4.3 Metabolomics

The last of the "omics" approaches we will discuss in this chapter, is metabolomics. Metabolomics is defined as the study of the metabolome, i.e. the set of small molecules (metabolites; smaller than 1500 Dalton) that are present in a cell, tissue or organism at a specific time and under specific conditions (Nielsen et al. 2006). The metabolites are the end products of metabolic processes in an organism. The metabolome can therefore be seen as the link between the genotype and the phenotype. When comparing metabolomics to the "omics" approaches already discussed, metabolomics might be the most complex one. When studying the level of expression of a specific metabolite, one has to remember that this level is based on the activity of all enzymes that are involved in the synthesis and conversion of that specific metabolite, and that the metabolic network in a cell consists of uncountable reactions all tightly connected. These connections mean that even a small change or perturbation in the proteome might cause a significant change in the metabolome (Nielsen et al. 2006).

When analyzing part of the metabolome, metabolite profiling and metabolic fingerprinting are techniques that are often used. According to Nielsen et al. (2006), metabolite profiling is the analysis of a specific set of metabolites; an analysis which provides direct physiological information, whereas metabolic fingerprinting is an unspecific analysis of a sample and the information obtained can only be used for grouping of different samples. However, the two notations that are most commonly used for classifying approaches in metabolome analysis are untargeted and targeted metabolomics. Untargeted metabolomics aims at detecting all metabolites in a sample, including unknowns, while targeted metabolomics aims at measuring defined groups of already annotated metabolites (Roberts et al. 2012). The most important methods used in metabolomics are nuclear magnetic resonance (NMR) and mass spectrometry (MS), often combined with chromatography techniques.

Compared to MS, NMR spectroscopy has a much lower sensitivity. However, it is highly reproducible between measurements, cost-effective, and can be performed without difficult and labor intensive sample preparations and separation steps. It is also possible to sample scans over a long time period, as the samples will not be destroyed during the measurements (Smedsgaard 2006). The main drawback compared to MS, is that more material is needed for the analysis (typically mg). NMR is most often used to solve the structure of molecules and by doing so, specific compounds in the sample can be identified, but the metabolites need to be present in rather high concentrations. Compounds that can be identified by NMR are e.g. known amino acids, some carbohydrates and phosphorus containing compounds e.g. ATP.

Currently, the use of metabolomics in aquaculture is mainly to study the effects of various environmental factors or different feed regimes. Chauton et al. (2015) for example, used ^{1}H NMR spectroscopy to study the effects that temperature and diet have on the early development of cod larvae. In this study cod larvae from two different experiments, one in which the larvae were exposed to various temperatures and one where the larvae were fed one of two different diets, were compared. They found that the larvae had metabolic differences throughout their life stages correlating to the temperature and diet they had been given. In the nutrition experiment, NMR data showed that the larvae fed zooplankton had higher levels of creatine, betaine/TMAO, taurine and choline than those fed rotifer/*Artemia* diet. This might be correlated both to the nutritional value of the different diets, and the difference in growth between the two groups (the larvae fed zooplankton were significantly larger).

For studies of microbe-host interactions and the host responses in fish, metabolomics is at present not a widely used method. For mammals however, germ-free models have been used in metabolomics studies in order to investigate the role of the host's microbiota in various physiological functions of vertebrates (review by Montalban-Arques et al. 2015). At least one study has used NMR-based metabolomics to investigate microbe-host interactions and host responses in fish. Asakura et al. (2014) investigated the microbial dynamics in the gut of various fish species following changes in feed type using 16S rRNA gene sequence analysis, and combined this with NMR-based metabolomics to investigate the metabolic dynamics as well (Fig. 1.8). Montalban-Arques et al. (2015) argues that it should therefore be possible to use metabolomics to study other microbe-host interactions in fish as well, such as effects of probiotics or infection in particular.

Even though NMR is suitable for metabolome studies, MS-based methods are currently mostly used. As MS is a more sensitive method than NMR, it is possible to detect metabolites with low abundances. This is a clear advantage when studying

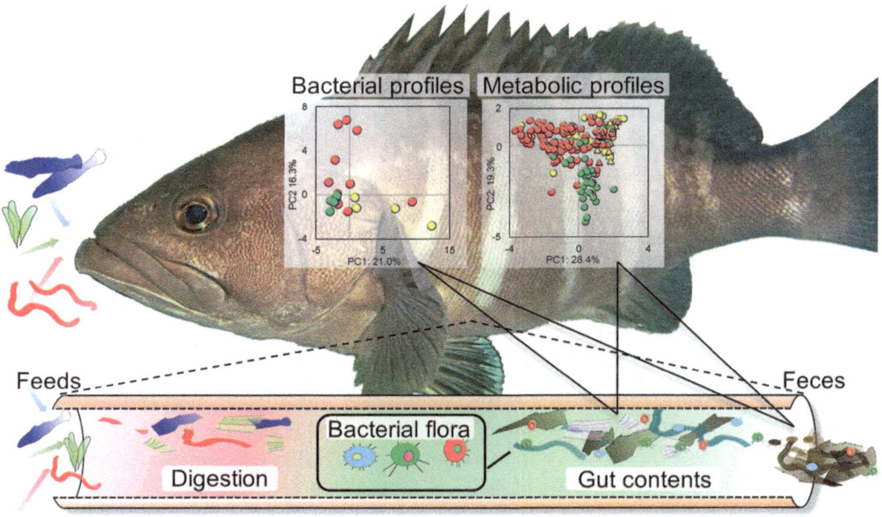

Fig. 1.8 Illustration of the evaluation strategy used by Asakura et al. (2014) to investigate metabolic and microbial dynamics in the intestine of various fishes (from Asakura et al. 2014)

fish larvae, as samples are either small or need to be obtained by pooling a large number of individuals. However, the sample preparation is much more tedious than when using NMR, and there is a high risk of destroying samples. Both methods have their advantages and disadvantages. The best suited method has to be chosen based on the experimental system and according to the information wanted in the relevant study.

To date, not many studies using MS-based methods have been performed on aquaculture relevant fish species. However, zebrafish are often used as a model organism. Elie et al. (2015) used LC-MS to study potential biological effects of polycyclic aromatic hydrocarbons (PAHs) and oxygenated PAHs (oxy-PAHs) in zebrafish. Their approach was to use untargeted metabolomics, and they aimed at uncovering the metabolites that varied between two independent groups of fish. The groups compared were zebrafish exposed to PAHs, zebrafish exposed to oxy-PAHs and a control group which was not exposed to any organic pollutants. They found that metabolic perturbations were associated with both types of substances, and that 63 metabolites were altered when the fish were exposed to one or the other substance. The alterations found were related to protein biosynthesis, oxidative stress, neural development and vascular development and cardiac development function. What makes this study particularly interesting is their comparison of these metabolomics data, with transcriptomics and genomics data from previous studies with zebrafish exposed to the same compounds (Goodale et al. 2013; Gurbani et al. 2013; Jayasundara et al. 2015). Through this they nicely showed how the different "omics" approaches can be used to complement each other.

1.5 Molecular Methods to Study the Microbiota

Most of the studies of fish-associated microbes published before year 2000 were based on approaches involving culturing of bacteria. Culture-dependent methods (Nayak 2010) led to the identification of a number of bacteria that are able to colonize fish mucosal surfaces, and increased understanding of bacterial adhesion to fish mucus (reviewed by Hansen and Olafsen 1999). Furthermore, culture-dependent approaches were used for characterizing the microbiota of fishes, and genera like *Vibrio*, *Pseudomonas*, *Achromobacter*, *Corynebacterium*, *Flavobacterium*, *Lactobacillus*, *Plesiomonas*, *Moraxella*, *Clostridium*, and *Micrococcus*, were often identified as abundant members of fish gut microbiota (Hansen and Olafsen 1999, and references therein). However, due to the bias in cultivation-based methods (see Sect. 1.1.) abundant members of the culturable microbiota are not necessarily abundant members of the original community. The need for better methods for investigating the composition, dynamics, and functionality of microbial communities was obvious. Indeed, the development of new DNA-based technologies, first the PCR-based methods in the 70s, and later the high resolution "omics" techniques based on high-throughput DNA sequencing technologies, has led to a revolution in the field of microbial ecology (Muller et al. 2013).

1.5.1 The Development of Culture-Independent Methods Based on the 16S rRNA Gene

Analysis of the 16S rRNA gene sequence has been the gold standard for studies of bacterial phylogeny, diversity, and taxonomic identification for the last decades. Carl Woese pioneered this work by using small subunit rRNA gene sequences for phylogenetic analysis, and discovered the third domain of life; the Archaea (Woese and Fox 1977). The 16S rRNA gene encodes the small subunit of ribosomal RNA in prokaryotes. The rRNA folds up in conserved secondary and tertiary structures and is an important structural and functional component of the 30S ribosomal subunit. The DNA sequence exhibits a mosaic pattern of highly conserved and variable sequence regions. It is therefore possible to design broad-coverage PCR primers targeting conserved regions, whereas variable regions reflect the diversity among the organisms under study. Furthermore, the gene is functionally conserved, universally present, and only very rarely subjected to horizontal gene transfer. These properties make the 16S rRNA gene highly suitable as a marker for diversity and phylogeny. Also several databases, reference data sets, and user-friendly online tools, e.g. for classification, are available. Examples of such are the Ribosomal Database Project (Cole et al. 2014) and SILVA (Quast et al. 2012). A drawback with the 16S rRNA as a marker gene for microbial diversity is the fact that the copy number varies from 1 to 15 among bacterial genomes, as documented in the rrnDB database (https://rrndb.umms.med.umich.edu/; Klappenbach et al. 2001). This leads to an overestimation of taxa with high copy numbers, such as Gammaproteobacteria, and to underestimation of taxa with low copy numbers. However, analysis tools have recently been developed for correcting for variable copy numbers in 16S rDNA sequencing data sets (PiCrust; Langlille et al. 2013).

The application of culture-independent approaches for studying microbial diversity accelerated when fingerprinting methods for analysis of 16S rRNA PCR products representing microbial communities were developed in the 1990s. Muyzer et al. (1993) developed a protocol for analyzing the sequence diversity of 16S rRNA PCR products by denaturing gradient gel electrophoresis (DGGE). Since then, particularly DGGE, but also other fingerprinting techniques, such as terminal restriction fragment length polymorphism (tRFLP) and Sanger sequencing of clone libraries, have been quite extensively applied for studying fish-associated microbiota (reviewed by Llewellyn et al. 2014). DGGE allows for determining the 16S rRNA sequence for specific bands by PCR reamplification and Sanger sequencing, but the success of obtaining taxonomic information varies and depends on strength and resolution of the bands in the profiles.

Publications based on such 16S rRNA-based culture-independent methods for studying fish-associated microbiota started to accumulate from the early 2000s and provided new insight regarding the diversity of fish microbiota (reviewed by Nayak 2010). Such studies demonstrated that a large proportion of the gut microbiota in Atlantic salmon (*Salmo salar*) and coho salmon (*Oncorhynchus kisutch*) is unculturable (Navarrete et al. 2009; Romero and Navarrete 2006), and identified members of the fish gut microbiota previously not recognized. For example, Holben et al. (2002) found that a *Mycoplasma* bacterium was abundant in the gut of Atlantic salmon. Kim et al. (2007) analyzed 16S rRNA amplicons by both DGGE and clone library sequencing, and found that the gut of rainbow trout (*Oncorhynchus mykiss*) harbors a larger bacterial diversity than previously recognized, and that gut mucus communities differed from those of the gut content. DGGE analysis of 16S rRNA amplicons is restricted when it comes to providing detailed taxonomic information, but is well suited for investigating dynamics of microbial communities. Also the number of samples that can be compared is limited to what is run on one gel, as gel-to-gel variability can be a problem. We used DGGE analyses of 16S rRNA amplicons to investigate factors structuring the microbiota associated with cod larvae. In an experiment where cod larvae were fed different live feed diets in replicate rearing tanks, diet was found not to be a major determinant of the cod larval microbiota (Bakke et al. 2013). We later performed an experiment where cod larvae were reared with different water qualities (unpublished). Interestingly, the water seemed to affect the larval microbiota more heavily than the diet (Fig. 1.9).

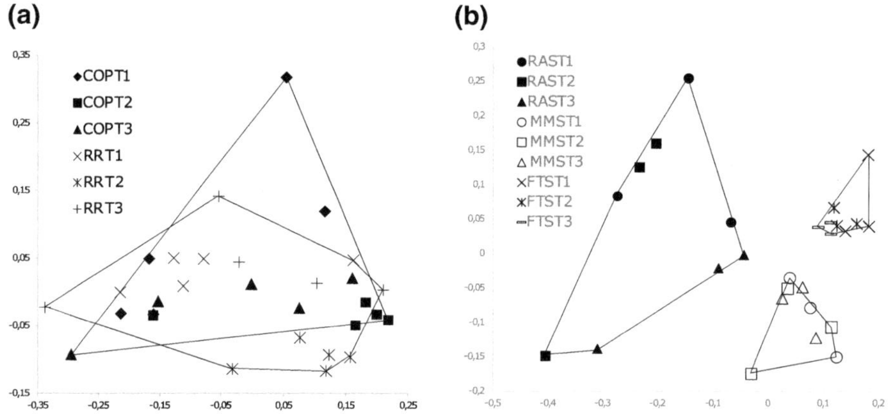

Fig. 1.9 Non-metric multidimensional scaling (NMDS) ordination based on Bray-Curtis similarities between DGGE profiles for comparison of the microbiota associated with Atlantic cod (*Gadus morhua*) larvae at the age of 17 days post hatching. **a** Cod larvae reared with distinct live feed diets (modified from Bakke et al. 2013). **b** Cod larvae reared with different water microbiota (based on unpublished data, but see Attramadal et al. 2014). The NMDS plots in two dimensions are based on ranks of the similarities in the similarity matrix for all samples. Each point represents one sample, and points close to each other are more similar than points that are far apart. Abbreviations: COP: Copepod live feed diet; RR: Rotifer live feed diet; RAS: Recirculating aquaculture system; MMS: flow-through systems including a biofilter for microbial maturation of rearing water; FTS: Flow through system; T1–T3: Replicate rearing tanks 1–3

1.5.2 High Throughput Sequencing and "Omics" Approaches

Although the analysis methods described and exemplified above have contributed substantially to a better understanding of the composition and dynamics of microbial communities colonizing fishes, they still have deficiencies, particularly when it comes to resolution and taxonomic information. The development of new high-throughput sequencing (HTS) technologies enabled assessment of the diversity of microbial communities at a new and far more detailed scale. By indexing 16S rRNA PCR products (amplicons) representing different microbial communities, it is possible to pool amplicons, and obtain large number of sequencing reads for many samples simultaneously (Andersson et al. 2008). The first HTS technology employing this approach was the 454 pyrosequencing technology. One run with the latest version of this Roche 454 technology typically produced around one million sequence reads with an average length of around 700 base pairs (reviewed by Goodwin et al. 2016). The 454 pyrosequencing technology was eventually outcompeted by the Illumina sequencing platforms, which provided considerable higher output of sequencing data per run to lower costs (Goodwin et al. 2016). This improved the sequencing depth and the resolution in microbial community analysis even further (Caporaso et al. 2012). A number of pipelines have been developed for processing and analysing 16S rRNA sequence data sets. Examples are USEARCH (Edgar 2013), QIIME (Caporaso et al. 2010), and Mothur (Schloss et al. 2009). Furthermore, databases for 16S rRNA gene sequences, such as SILVA (Quast et al. 2012) and the Ribosomal Database Project (Cole et al. 2014), make it possible to assign taxonomy to the sequencing data obtained.

Roeselers et al. (2011) were the first to examine the diversity of fish microbiota by HTS. They analyzed 16S rRNA amplicons representing zebrafish gut microbiota both by the pyrosequencing technology and by Sanger sequencing of clone libraries. Pyrosequencing revealed the presence of new bacterial phylotypes that had not been detected by Sanger sequencing of clone libraries. This was mainly due to differences in resolution of the methods. Moreover, the gut microbiota was surprisingly similar between wild and domesticated fish, and the authors concluded that a core set of bacterial taxa was characteristic to the zebrafish gut microbiota. We investigated the ontogeny of the microbiota associated with developing cod larvae (Bakke et al. 2015) by pyrosequencing of 16S rRNA amplicons, and found that the early larval microbiota was highly dissimilar from the microbial communities associated with water and live feed, and that the larval microbiota underwent a discontinuous succession in developing larvae (Fig. 1.10). Whereas the old cultivation-based methods revealed some tens of strains associated to a fish, the HTS methods have revealed hundreds of bacterial taxa (e.g. Bakke et al. 2015; Hansen and Olafsen 1999).

More recently, the application of Illumina sequencing technology has led to better coverage and a more detailed description of fish gut microbiota. Ghanbari et al. (2015) reviewed studies where HTS of 16S rRNA amplicon libraries were employed to characterize fish gut microbiota, and remarked that diversity of the gut

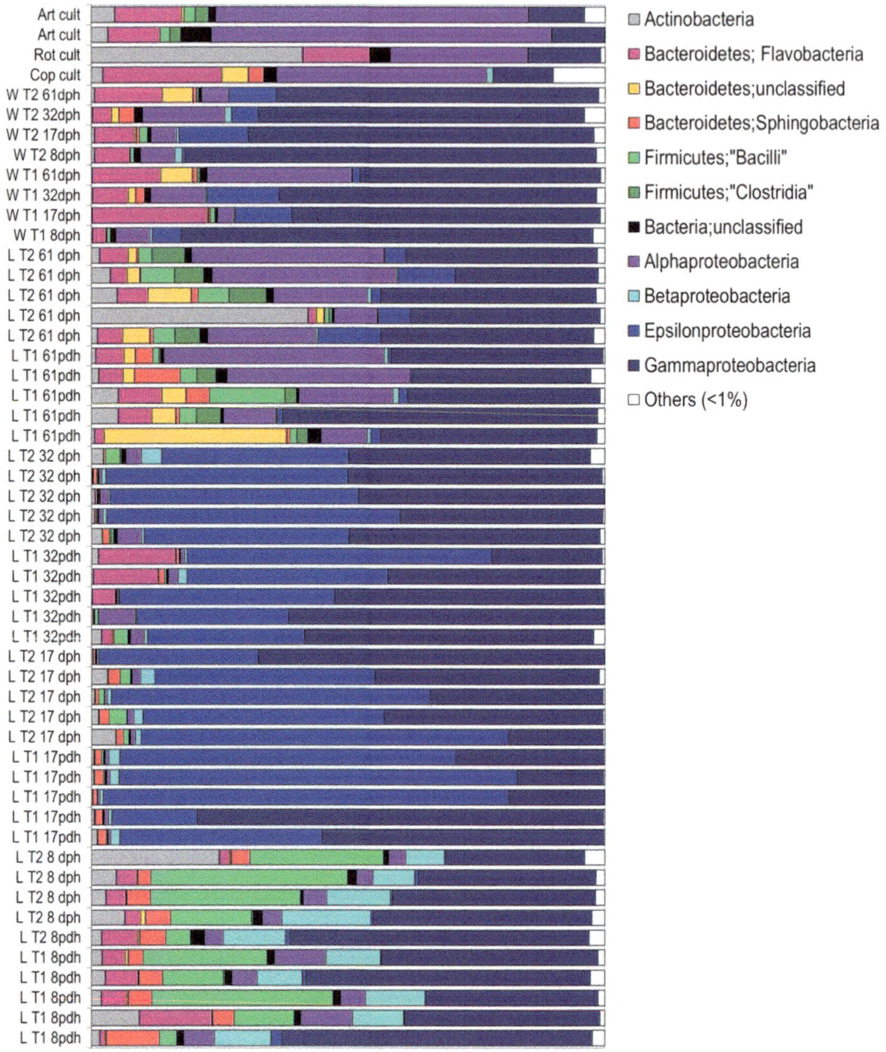

Fig. 1.10 Relative abundance of bacterial phyla and families represented in the v4 16S rDNA amplicons obtained from individual Atlantic cod (*Gadus morhua*) larvae, water and live feed samples at the phylum level. Bars labeled D8, D17, D32 and D61 represent cod larva individuals at the ages 8, 17, 32 and 61 days post hatching, respectively. Bars labeled L as first letter are larvae, whereas bars labeled W and F represent water and live feed samples, respectively. T indicates replicate tanks, and dph refers to days post hatching (from Bakke et al. 2015)

microbiota varies with trophic level: planktivorous fish, followed by omnivorous and then herbivorous had the highest diversity, while carnivorous fishes had the lowest diversity of the gut microbiota. Bolnick et al. (2014) examined the effects of diet and sex on gut microbiota of threespine sticklebacks (*Gasterosteus aculeatus*) and Eurasian perch (*Perca fluviatilis*). They Illumina sequenced 16S rRNA amplicons, and found that sex, as an example of a genotype marker, modulated the effects that the diet had on the gut microbiota. Llewellyn et al. (2016) performed Illumina sequencing of 16S rRNA amplicons to investigate the diversity of the gut microbiota of wild and reared Atlantic salmon at different life stages (parr, smolt, adult) from widespread geographical localities (Eastern Canada, Western Ireland, and West Greenland). They found that individuals at the same life stage had surprisingly similar microbiota composition, even if they were separated by large geographical distances. Thus, life stage appeared to be a more important determinant than geography for structuring the gut microbiota.

The examples described above illustrate how the development of DNA sequence-based methods has allowed for increasingly better characterization of fish-associated microbial communities. This in turn has led to improved understanding, not only of the composition, but also of the factors influencing the formation, the structure and dynamics of fish microbiota. Still, few studies have addressed the functional potential of the fish microbiota, despite the fact that HTS technologies can be used to characterize the functional genes encoded (metagenomics) and expressed (metatranscriptomics) by a microbial community (Aguiar-Pulido et al. 2016, see also text below). Furthermore, analysis tools have been developed to predict the functional potential of microbial communities from 16S rRNA sequence data: PiCrust (Langille et al. 2013) and Tax4Fun (Aßhauer et al. 2015).

In metagenomics, total DNA from the community under study is extracted and sequenced by a random shotgun strategy. For metatranscriptomics, a more complex and expensive sample preparation protocol is required. In bacterial cells rRNA is abundant, and mRNA constitutes less than 5% of the total RNA. Moreover, the bacterial mRNA lacks a polyA-tail and is highly unstable. Therefore, laborious and expensive protocols are needed to enrich and stabilize the mRNA extracted from microbial communities, before reverse transcription and DNA sequencing is performed (Bashiardes et al. 2016). This reduces the number of samples that are included in metatranscriptomic analysis, and it has only rarely been applied for studying host-associated microbial communities (Bashiardes et al. 2016). A few studies have applied a metagenomics strategy to investigate the functional potential of fishes (Ghanbari et al. 2015). By doing a metagenomics profiling of the microbiota of turbot (*Scophthalmus maximus*), Xing et al. (2013) revealed that quorum sensing and biofilm formation were important gut microbiome functions. The functionality of microbial communities can also be examined in terms of the proteins and the metabolites they synthesize by proteomics (metaproteomics for communities) and metabolomics, respectively. These technologies are described in more detail above (Sects. 1.4.2 and 1.4.3). Metaproteomics has not been extensively applied for studying host-associated microbiomes. There are a number of technical challenges, among them the interference of host-encoded proteins (Xiong et al. 2015). Similarly, metabolomics analysis of host-associated microbiota is not straight forward, and a number of challenges remain to be solved (Smirnov et al. 2016).

1.6 Perspectives and Conclusions

Above, we have tried to illustrate the huge potential the newly developed imaging and "omics" methods have for investigating fish-microbiota interactions. The methodological revolution covers both the development of methods per se, and the fact that it is now possible to implement many of these new methods in unspecialized labs or purchase them from commercial companies. Thus, these methods are to a large extent available for all labs. Commercially available methods are e.g. high-throughput sequencing for characterizing the composition of the microbiota, and transcriptome studies by either high-throughput sequencing or microarrays.

Despite the fact that many methods are now easily available for the research community, our ambitions should be a mechanistic understanding and not only an observation of phenomena. Therefore, the 21st century calls for multi- and interdisciplinarity. The price for this is a need for communicating across disciplines and funding of larger projects, and the prize is a faster build-up of knowledge on a more mechanistic level (and more fun). A possible consequence of this ambition is that various species of fish may be considered as adequate model systems and that people working with fish-microbe interactions (often with an aquaculture perspective) may contribute with fundamental knowledge on host-microbe interactions. I.e. there will be a build-down of the existing barriers between applied and fundamental research. It is important that we intensify studies on other animals than mammals, with the aim to generate the fundamental and general knowledge on host-microbe interactions.

From an applied perspective, the fundamental knowledge regarding fish-microbe interactions can be used to establish knowledge-based microbial management methods for aquaculture. Today the aquaculture industry faces huge problems associated with infections and disease. This does not only have a negative effect on the economic sustainability, but also the societal and environmental sustainability and thus the credibility of the aquaculture industry. Today, there are e.g. huge differences in the use of antibiotics between countries, exemplified by Vietnam using 700 g per metric ton fish produced and Norway using 1 g per metric ton (Defoirdt et al. 2011). These differences are complex, but even for the Norwegian salmon industry the antibiotic use has gone from 600 g per metric ton in the mid-1980s with huge disease problems, to 1 g and less from late 1990s and onwards (NORM/NORM-VET 2011). This change is due to a multitude of countermeasures including the development of efficient vaccination programs, optimization of nutrition and implementation of hygiene rules. Another example of the potential of microbial management is our success at NTNU/Sintef in various experiments with turbot. By taking microbial management seriously, we were able to run a variety of experiments with average survival of 38 ± 9% (range 25–55%, N = 10) until metamorphosis (Reitan et al. 1997). The experiments were done during a time period where survival in the aquaculture industry was typically less than 10%. A substantial part of the reason for this success was that we developed a method for selecting against opportunistic bacteria in the rearing water, the so-called "microbially matured water" concept (Salvesen et al. 1999; Skjermo et al. 1997; Vadstein et al. 1993). These two examples, the very low use of antibiotics in the Atlantic salmon production in Norway and the "microbially matured water" concept, demonstrate that there is a huge potential for reducing both infection problems and the consequences of overuse of pharmaceuticals like antibiotics, in the aquaculture industry. This will, however, require a knowledge-based development of microbial management tools, which are based on a proper understanding of the fish-microbiota interaction—especially in healthy fish.

In conclusion, we are convinced that we will have to do a lot of the work on fish-microbiota interactions over again, as the new molecular tools have clearly demonstrated the bias in cultivation-based methods. However, with respect to methods we are in a broad sense in a much better situation now than three to four decades ago. The range of experimental systems available makes it possible to optimize the experiment to the question/hypothesis addressed, and in a way which reduces the use of experimental animals. We think that the use of different experimental systems should strive for increased evenness as they all have advantages and limitations and at the moment traditional rearing tanks are dominating in aquaculture research. Furthermore, the potential of combining new and old imaging tools has not been explored much in studies of fish-microbe interactions. The visualization possible with these tools and the possibility of determining spatial organization of biofilms on mucosal surfaces will be of high value for understanding multispecies microbial communities, their respective intra- and interspecies interactions, as well as their interactions with host tissues. The use of "omics" in aquaculture research is growing rapidly, and the results we have seen so far are just the beginning. By whole genome sequencing and more (species) specific protocols, the methods will become more available, and by combining the different "omics" approaches the information gained will increase immensely in the years to come. These methods have substantially improved our understanding of fish larvae-associated microbes. To date, this is particularly true for descriptions of species present. However, in future research the description of responses and dynamics of fish microbiota should not be restricted to the composition and the taxonomic information, but also include the functionality of the microbes. We believe that at this stage good hypothesis, clever design of experiments and efficient interdisciplinary teams are the limiting factors for further progress—not methods.

Acknowledgements This work was a part of the ERA-Net COFASP project MicStaTech "Water treatment technology for microbial stabilization in landbased aquaculture systems" funded by the Research Council of Norway (Contract 247558), the Research Council of Norway project MicroMucus: "Microbial contributions to the Atlantic salmon (*Salmo salar*) skin mucosal barrier" (NRC project no. 262929/F20), a People Marie Curie Actions International Incoming Fellowships FP7-PEOPLE-2013-IIF (project number 625655) to B.L., and a Ph.D. scholarship to R. I. V. from the Faculty of Science, NTNU.

References

Aguiar-Pulido V, Huang W, Suarez-Ulloa V et al (2016) Metagenomics, metatranscriptomics, and metabolomics approaches for microbiome analysis. Evol Bioinform Online 12(Suppl 1):5. https://doi.org/10.4137/EBO.S36436

Andersson AF, Lindberg M, Jakobsson H et al (2008) Comparative analysis of human gut microbiota by barcoded pyrosequencing. PLoS ONE 3(7):e2836. https://doi.org/10.1371/journal.pone.0002836

Armer HEJ, Maiggi G, Png KMY et al (2009) Imaging transient blood vessel fusion events in zebrafish by correlative volume electron microscopy. PLoS ONE 4:e7716. https://doi.org/10.1371/journal.pone.0007716

Asakura T, Sakata K, Yoshida S et al (2014) Noninvasive analysis of metabolic changes following nutrient input into diverse fish species, as investigated by metabolic and microbial profiling approaches. PeerJ 2:e550. https://doi.org/10.7717/peerj.550

Aßhauer KP, Wemheuer B, Daniel R et al (2015) Tax4Fun: predicting functional profiles from metagenomic 16S rRNA data. Bioinformatics 31(17):2882–2884

Attramadal KJ, Truong TMH, Bakke I et al (2014) RAS and microbial maturation as tools for K-selection of microbial communities improve survival in cod larvae. Aquaculture 432:483–490

Baker JA, Ferguson M, Tenbroeck C (1942) Growth of platyfish (*Platypoecilus maculatus*) free from bacteria and other micro'organisms. Exp Biol Med 51:116–119. https://doi.org/10.3181/00379727-51-13854

Bakke I, Coward E, Andersen T et al (2015) Selection in the host structures the microbiota associated with developing cod larvae (*Gadus morhua*). Environ Microbiol 17(10):3914–3924

Bakke I, Skjermo J, Vo TA et al (2013) Live feed is not a major determinant of the microbiota associated with cod larvae (*Gadus morhua*). Environ Microbiol Rep 5(4):537–548

Bashiardes S, Zilberman-Schapira G, Elinav E (2016) Use of metatranscriptomics in microbiome research. Bioinform Biol Insights 10:19. https://doi.org/10.4137/BBI.S34610

Bates J, Mittge E, Kuhlman J et al (2006) Distinct signals from the microbiota promote different aspects of zebrafish gut differentiation. Dev Biol 297:374–386

Battalora MSJ, Ellender RD, Martin BJ (1985) Gnotobiotic maintenance of Sheepshead minnow larvae. Prog Fish Cult 47(2):122–125. http://dx.doi.org/10.1577/1548-8640(1985)47<122:GMOSML>2.0.CO;2

Berry D, Mader E, Lee TK et al (2015) Tracking heavy water (D_2O) incorporation for identifying and sorting active microbial cells. Proc Natl Acad Sci USA 112:e194–e203. https://doi.org/10.1073/pnas.1420406112

Berry D, Stecher B, Schintlmeister A et al (2013) Host-compound foraging by intestinal microbiota revealed by single-cell stable isotope probing. Proc Natl Acad Sci USA 110:4720–4725

Betzig E, Patterson GH, Sougrat R et al (2006) Imaging intracellular fluorescent proteins at nanometer resolution. Science 313:1642–1645

Bolnick DI, Snowberg LK, Hirsch PE et al (2014) Individual diet has sex-dependent effects on vertebrate gut microbiota. Nat Commun 5. https://doi.org/10.1038/ncomms5500

Box GE, Hunter WG, Hunter JS (1978) Statistics for experimenters: an introduction to design, data analysis, and model building. Wiley, New York

Caporaso JG, Kuczynski J, Stombaugh J et al (2010) QIIME allows analysis of high-throughput community sequencing data. Nat Methods 7(5):335–336

Caporaso JG, Lauber CL, Walters WA et al (2012) Ultra-high-throughput microbial community analysis on the Illumina HiSeq and MiSeq platforms. ISME J 6(8):1621–1624

Chauton MS, Galloway TF, Kjørsvik E et al (2015) 1H NMR metabolic profiling of cod (*Gadus morhua*) larvae: potential effects of temperature and diet composition during early developmental stages. Biol Open 4(12):1671–1678. https://doi.org/10.1242/bio.014431

Cole JR, Wang Q, Fish JA et al (2014) Ribosomal database project: data and tools for high throughput rRNA analysis. Nucleic Acids Res 42(D1):D633–D642

Declercq A, Chiers K, Van den Broeck W et al (2015) Interactions of highly and low virulent *Flavobacterium columnare* isolates with gill tissue in carp and rainbow trout. Vet Res 46:25–41

Defoirdt T, Sorgeloos P, Bossier P (2011) Alternatives to antibiotics for the control of bacterial disease in aquaculture. Curr Opin Microbiol 14(3):251–258

De Schryver P, Vadstein O (2014) Ecological theory as a foundation to control pathogenic invasion in aquaculture. ISME J 8(12):2360–2368

Dierckens K, Rekecki A, Laureau A et al (2009) Development of a bacterial challenge test for gnotobiotic sea bass (*Dicentrarchus labrax*) larvae. Environ Microbiol 11(2):526–533

Douillet PA, Holt GJ (1994) Surface disinfection of red drum (*Sciaenops ocellatus* Linnaeus) eggs leading to bacteria-free larvae. J Exp Mar Biol Ecol 179(2):253–266

Duarte CM, Holmer M, Olsen Y et al (2009) Will the oceans help feed humanity? Bioscience 59 (11):967–976

Edgar RC (2013) UPARSE: highly accurate OTU sequences from microbial amplicon reads. Nat Methods 10(10):996–998

Elie MR, Choi J, Nkrumah-Elie YM et al (2015) Metabolomic analysis to define and compare the effects of PAHs and oxygenated PAHs in developing zebrafish. Environ Res 140:502–510

Fjellheim AJ, Klinkenberg G, Skjermo J et al (2010) Selection of candidate probionts by two different screening strategies from Atlantic cod (*Gadus morhua* L.) larvae. Vet Microbiol 144 (1):153–159

Fjellheim AJ, Playfoot KJ, Skjermo J et al (2012) Inter-individual variation in the dominant intestinal microbiota of reared Atlantic cod (*Gadus morhua* L.) larvae. Aquac Res 43 (10):1499–1508

Forberg T, Arukwe A, Vadstein O (2011) A protocol and cultivation system for gnotobiotic Atlantic cod larvae (*Gadus morhua* L.) as a tool to study host microbe interactions. Aquaculture 315(3–4):222–227

Forberg T, Milligan-Myhre K (2017) Gnotobiotic fish as models to study host-microbe interactions. In: Schoeb T, Eaton K (eds) Gnotobiotcs, 1st edn. Academic Press, United States, chapter 6

Forberg T, Sjulstad EB, Bakke I et al (2016) Correlation between microbiota and growth in Mangrove Killifish (*Kryptolebias marmoratus*) and Atlantic cod (*Gadus morhua*). Sci Rep 6. https://doi.org/10.1038/srep21192

Forberg T, Vestrum RI, Arukwe A et al (2012) Bacterial composition and activity determines host gene-expression responses in gnotobiotic Atlantic cod (*Gadus morhua*) larvae. Vet Microbiol 157(3):420–427

Ghanbari M, Kneifel W, Domig KJ (2015) A new view of the fish gut microbiome: advances from next-generation sequencing. Aquaculture 448:464–475

Goodale BC, Tilton SC, Corvi MM et al (2013) Structurally distinct polycyclic aromatic hydrocarbons induce differential transcriptional responses in developing zebrafish. Toxicol Appl Pharm 272(3):656–670

Goodwin S, McPherson JD, McCombie WR (2016) Coming of age: ten years of next-generation sequencing technologies. Nat Rev Genet 17(6):333–351

Gunasekara RAYSA, Defoirdt T, Rekecki A et al (2012) Light and transmission electron microscopy of *Vibrio campbellii* infection in gnotobiotic *Artemia franciscana* and protection offered by a yeast mutant with elevated cell wall glucan. Vet Microbiol 158:337–343

Gurbani D, Bharti SK, Kumar A et al (2013) Polycyclic aromatic hydrocarbons and their quinones modulate the metabolic profile and induce DNA damage in human alveolar and bronchiolar cells. Int J Hyg Envir Heal 216(5):553–565

Hamre K (2006) Nutrition in cod (*Gadus morhua*) larvae and juveniles. ICES J Mar Sci: J Conseil 63(2):267–274

Hansen G, Olafsen J (1999) Bacterial interactions in early life stages of marine cold water fish. Microb Ecol 38(1):1–26

Hansen G, Strøm E, Olafsen J (1992) Effect of different holding regimens on the intestinal microflora of herring (*Clupea harengus*) larvae. Appl Environ Microbiol 58:461–470

Holben WE, Williams P, Saarinen M et al (2002) Phylogenetic analysis of intestinal microflora indicates a novel Mycoplasma phylotype in farmed and wild salmon. Microb Ecol 44(2): 175–185

Hugenholtz P (2002) Exploring prokaryotic diversity in the genomic era. Genome Biol 3(2): reviews0003–0001 https://doi.org/10.1186/gb-2002-3-2-reviews0003

Hugenholtz P, Goebel BM, Pace NR (1998) Impact of culture-independent studies on the emerging phylogenetic view of bacterial diversity. J Bacteriol 180(18):4765–4774

Jayasundara N, Van Tiem Garner L, Meyer JN et al (2015) AHR2-mediated transcriptomic responses underlying the synergistic cardiac developmental toxicity of PAHs. Toxicol Sci 143(2):469–481

Kanther M, Rawls JF (2010) Host–microbe interactions in the developing zebrafish. Curr Opin Immunol 22(1):10–19

Karner M, Fuhrman JA (1997) Determination of active marine bacterioplankton: a comparison of universal 16S rRNA probes, autoradiography, and nucleoid staining. Appl Environ Microb 63 (4):1208–1213

Kim DH, Brunt J, Austin B (2007) Microbial diversity of intestinal contents and mucus in rainbow trout (*Oncorhynchus mykiss*). J Appl Microbiol 102(6):1654–1664

Klappenbach JA, Saxman PR, Cole JR et al (2001) rrnDB: the ribosomal RNA operon copy number database. Nucleic Acids Res 29(1):181–184

Klar T, Jakobs S, Dyba M et al (2000) Fluorescence microscopy with diffraction resolution barrier broken by stimulated emission. Proc Natl Acad Sci USA 97:8206–8210

Knierim B, Luef B, Wilmes P et al (2011) Correlative microscopy for phylogenetic and ultrastructural characterization of microbial communities. Environ Microbiol Rep 4:36–41

Langille MG, Zaneveld J, Caporaso JG et al (2013) Predictive functional profiling of microbial communities using 16S rRNA marker gene sequences. Nat Biotechnol 31(9):814–821

Lesel R, Lesel M (1976) Obtention d'alevins non vésiculés axéniques de salmonides. Ann Hydrobiol 7(1):21–25

Llewellyn MS, Boutin S, Hoseinifar SH et al (2014) Teleost microbiomes: the state of the art in their characterization, manipulation and importance in aquaculture and fisheries. In: Sime-Ngando T, Lafferty KD and Biron DG (ed) Roles and mechanisms of parasitism in aquatic microbial communities, Frontiers Media, Lausanne, 109 https://doi.org/10.3389/978-2-88919-588-6

Llewellyn MS, McGinnity P, Dionne M et al (2016) The biogeography of the Atlantic salmon (*Salmo salar*) gut microbiome. ISME J 10(5):1280–1284

Luef B, Frischkorn KR, Wrighton KC et al (2015) Diverse uncultivated ultra-small bacterial cells in groundwater. Nat Commun 6:6372. https://doi.org/10.1038/ncomms7372

Moullan N, Mouchiroud L, Wang X et al (2015) Tetracyclines disturb mitochondrial function across eukaryotic models: a call for caution in biomedical research. Cell Rep 10(10): 1681–1691

Milligan-Myhre K, Small CM, Mittge EK et al (2016) Innate immune responses to gut microbiota differ between threespine stickleback populations. Dis Model Mech 9:187–198

Milne JLS, Subramaniam S (2009) Cryo-electron tomography of bacteria: progress, challenges and future prospects. Nat Rev Microbiol 7:666–675

Montalban-Arques A, De Schryver P, Bossier P et al (2015) Selective manipulation of the gut microbiota improves immune status in vertebrates. Front Immunol 6:512. https://doi.org/10.3389/fimmu.2015.00512

Muller EE, Glaab E, May P et al (2013) Condensing the omics fog of microbial communities. Trends Microbiol 21(7):325–333

Munro PD, Barbour A, Birkbeck TH (1995) Comparison of the growth and survival of larval turbot in the absence of culturable bacteria with those in the presence of *Vibrio anguillarum*, *Vibrio alginolyticus*, or a marine *Aeromonas* sp. Appl Env Microbiol 61(12):4425–4428

Muyzer G, De Waal EC, Uitterlinden AG (1993) Profiling of complex microbial populations by denaturing gradient gel electrophoresis analysis of polymerase chain reaction-amplified genes coding for 16S rRNA. Appl Environ Microb 59(3):695–700

Müller-Reichert T, Verkade P (eds) (2012) Methods in cell biology, vol 111, Correlative light and electron microscopy. Academic Press, United States

Navarrete P, Espejo R, Romero J (2009) Molecular analysis of microbiota along the digestive tract of juvenile Atlantic salmon (*Salmo salar* L.). Microb Ecol 57(3):550–561

Nayak SK (2010) Role of gastrointestinal microbiota in fish. Aquac Res 41(11):1553–1573

Nielsen J (2006) Metabolomics in functional genomics and systems biology. In: Villas-Bôas SG, Roessner U, Hansen MAE et al (eds) Metabolome analysis: an introduction. Wiley, Hoboken, New Jersey, pp 1–14

NORM/NORM-VET (2011) Usage of antimicrobial agents and occurrence of antimicrobial resistance in Norway. Tromsø/Oslo 2012. ISSN:1502-2307 (print), 1890-9965 (electronic)

Olafsen JA (2001) Interactions between fish larvae and bacteria in marine aquaculture. Aquaculture 200(1):223–247

O'Toole R, von Hofsten J, Rosqvist R et al (2004) Visualisation of zebrafish infection by GFP-labelled *Vibrio anguillarum*. Microb Pathog 37:41–46

Penglase S, Edvardsen RB, Furmanek T et al (2015) Diet affects the redox system in developing Atlantic cod (*Gadus morhua*) larvae. Redox Biol 5:308–318

Pham LN, Kanther M, Semova I et al (2008) Methods for generating and colonizing gnotobiotic zebrafish. Nat Protoc 3(12):1862–1875

Plitzko J, Rigort A, Leis A (2009) Correlative cryo-light microscopy and cryo-electron tomography: from cellular territories to molecular landscapes. Curr Opin Biotechnol 20:83–89

Qian X, Ba Y, Zhuang Q et al (2014) RNA-Seq technology and its application in fish transcriptomics. OMICS 18(2):98–110

Quast C, Pruesse E, Yilmaz P et al (2012) The SILVA ribosomal RNA gene database project: improved data processing and web-based tools. Nucleic Acids Res 41(D1):D590–D596

Rajan B, Fernandes JMO, Caipang CMA et al (2011) Proteome reference map of the skin mucus of Atlantic cod (*Gadus morhua*) revealing immune competent molecules. Fish Shellfish Immun 31(2):224–231

Rajan B, Lokesh J, Kiron V et al (2013) Differentially expressed proteins in the skin mucus of Atlantic cod (*Gadus morhua*) upon natural infection with *Vibrio anguillarum*. BMC Vet Res 9 (1):103. https://doi.org/10.1186/1746-6148-9-103

Rawls JF, Samuel BS, Gordon JI (2004) Gnotobiotic zebrafish reveal evolutionarily conserved responses to the gut microbiota. Proc Natl Acad Sci USA 101(13):4596–4601

Reitan K, Natvik C, Vadstein O (1998) Drinking rate, uptake of bacteria and microalgae in turbot larvae. J Fish Biol 53(6):1145–1154

Reitan KI, Rainuzzo JR, Øie G et al (1997) A review of the nutritional effects of algae in marine fish larvae. Aquaculture 155(1):207–221

Rekecki A, Dierckens K, Laureau S et al (2009) Effect of germ-free rearing environment on gut development of larval sea bass (*Dicentrarchus labrax* L.). Aquaculture 293:8–15

Rekecki A, Gunasekara RAYSA, Dierckens K et al (2012) Bacterial host interaction of GFP-labelled *Vibrio anguillarum* HI-610 with gnotobiotic sea bass, *Dicentrarchus labrax* (L.), larvae. J Fish Dis 35(4):265–273

Rekecki A, Ringø E, Olsen R et al (2013) Luminal uptake of *Vibrio (Listonella) anguillarum* by shed enterocytes—a novel early defence strategy in larval fish. J Fish Dis 36:419–426

Ringø E, Birkbeck T (1999) Intestinal microflora of fish larvae and fry. Aquac Res 30(2):73–93

Ringø E, Olsen RE, Mayhew TM et al (2003) Electron microscopy of the intestinal microflora of fish. Aquaculture 227:395–415

Ringø E, Sperstad S, Myklebust R et al (2006) Characterisation of the microbiota associated with intestine of Atlantic cod (*Gadus morhua* L.): the effect of fish meal, standard soybean meal and a bioprocessed soybean meal. Aquaculture 261:829–841

Roberts LD, Souza SL, Gerszten RE et al. (2012) Targeted metabolomics. Curr Protoc Mol Biol 98:30.2:30.2.1–30.2.24. https://doi.org/10.1002/0471142727.mb3002s98

Rodrigues PM, Silva TS, Dias J et al (2012) Proteomics in aquaculture: applications and trends. J Proteomics 75(14):4325–4345

Roeselers G, Mittge EK, Stephens WZ et al (2011) Evidence for a core gut microbiota in the zebrafish. ISME J 5(10):1595–1608

Romero J, Navarrete P (2006) 16S rDNA-based analysis of dominant bacterial populations associated with early life stages of coho salmon (*Oncorhynchus kisutch*). Microb Ecol 51 (4):422–430

Rust M, Bates M, Zhuang X (2006) Sub-diffraction-limit imaging by stochastic optical reconstruction microscopy (STORM). Nat Meth 3:793–796

Salvesen I, Skjermo J, Vadstein O (1999) Growth of turbot (*Scophthalmus maximus* L.) during first feeding in relation to the proportion of r/K-strategists in the bacterial community of the rearing water. Aquaculture 175(3):337–350

Sandlund N, Bergh Ø (2008) Screening and characterisation of potentially pathogenic bacteria associated with Atlantic cod *Gadus morhua* larvae: bath challenge trials using a multidish system. Dis Aquat Organ 1(3):203–217

Sarropoulou E, Galindo-Villegas J, García-Alcázar A et al (2012) Characterization of European Sea Bass transcripts by RNA SEQ after oral vaccine against *V. anguillarum*. Mar. Biotechnol 14(5):634–642

Sartori A, Gatz R, Beck F et al (2007) Correlative microscopy: bridging the gap between fluorescence light microscopy and cryo-electron tomography. J Struct Biol 160:135–145

Schloss PD, Westcott SL, Ryabin R et al (2009) Introducing mothur: open-source, platform-independent, community-supported software for describing and comparing microbial communities. Appl Env Microbiol 75(23):7537–7541

Sekirov I, Russell SL, Antunes LCM et al (2010) Gut microbiota in health and disease. Physiol Rev 90(3):859–904

Shaw ES, Aronson LR (1954) Oral incubation in Tilapia macrocephala. B Am Mus Nat Hist 103 (5):375–416

Situmorang ML, Dierckens K, Mlingi FT et al (2014) Development of a bacterial challenge test for gnotobiotic Nile tilapia *Oreochromis niloticus* larvae. Dis Aquat Organ 109(1):23–33

Skjermo J, Salvesen I, Øie G et al (1997) Microbially matured water: a technique for selection of a non-opportunistic bacterial flora in water that may improve performance of marine larvae. Aquacult Int 5(1):13–28

Smedsgaard J (2006) Analytical tools. In: Villas-Bôas SG, Roessner U, Hansen MAE et al (eds) Metabolome analysis: an introduction. Wiley, Hoboken, New Jersey, pp 83–145

Smirnov KS, Maier TV, Walker A et al (2016) Challenges of metabolomics in human gut microbiota research. Int J Med Microbiol 306(5):266–279

Star B, Nederbragt AJ, Jentoft S et al (2011) The genome sequence of Atlantic cod reveals a unique immune system. Nature 477(7363):207–210

Stecher B, Berry D, Loy A (2013) Colonization resistance and microbial ecophysiology: using gnotobiotic mouse models and single-cell technology to explore the intestinal jungle. FEMS Microbiol Rev 37:793–829

Toranzo AE, Magariños B, Romalde JL (2005) A review of the main bacterial fish diseases in mariculture systems. Aquaculture 246(1–4):37–61

Trust TJ (1974) Sterility of Salmonid Roe and practicality of hatching gnotobiotic Salmonid fish. Appl Environ Microb 28(3):340–341

Vadstein O, Bergh Ø, Gatesoupe FJ et al (2013) Microbiology and immunology of fish larvae. Rev Aquacult 5(Suppl. 1):S1–S25

Vadstein O, Øie G, Olsen Y et al (1993) A strategy to obtain microbial control during larval development of marine fish. In: Reinertsen H, Dahle LA, Jørgensen L et al (eds) Fish farming technology. Balkema, Rotterdam, pp 69–75

Verner-Jeffreys DW, Shields RJ, Birkbeck TH (2003) Bacterial influences on Atlantic halibut *Hippoglossus hippoglossus* yolk-sac larval survival and start-feed response. Dis Aquat Organ 56(2):105–113

Wagner M (2009) Single-cell ecophysiology of microbes as revealed by Raman microspectroscopy or secondary ion mass spectrometry imaging. Annu Rev Microbiol 63:411–429

Wang Z, Gerstein M, Snyder M (2009) RNA-Seq: a revolutionary tool for transcriptomics. Nat Rev Genet 10(1):57–63

Westermann AJ, Förstner KU, Amman F et al (2016) Dual RNA-seq unveils noncoding RNA functions in host–pathogen interactions. Nature 529(7587):496–501

Westermann AJ, Gorski SA, Vogel J (2012) Dual RNA-seq of pathogen and host. Nat Rev Microbiol 10(9):618–630

Woese CR, Fox GE (1977) Phylogenetic structure of the prokaryotic domain: the primary kingdoms. Proc Natl Acad Sci USA 74(11):5088–5090

Xing M, Hou Z, Yuan J et al (2013) Taxonomic and functional metagenomic profiling of gastrointestinal tract microbiome of the farmed adult turbot (*Scophthalmus maximus*). FEMS Microbiol Ecol 86(3):432–443

Xiong W, Abraham PE, Li Z et al (2015) Microbial metaproteomics for characterizing the range of metabolic functions and activities of human gut microbiota. Proteomics 15(20):3424–3438

Xu J, Gordon JI (2003) Honor thy symbionts. Proc Natl Acad Sci USA 100(18):10452–10459

Zhang B, Shimada Y, Hirota T et al (2016) Novel immunologic tolerance of human cancer cell xenotransplants in zebrafish. Transl Res 170(89–98):e83. https://doi.org/10.1016/j.trsl.2015.12.007

Chapter 2
Environmental Cycles and Biological Rhythms During Early Development

Francisco Javier Sánchez-Vázquez and José Fernando López-Olmeda

Abstract The environment in which aquatic organisms inhabit is not constant and many factors such as light and temperature display cyclic and predictable variations. Animals present biological rhythms in many of their physiological variables to adapt to these cyclic changes, timing their functions to occur when the possibility of success is greater. Besides, the underwater photo-environment is complex since light characteristics (i.e. intensity and spectrum) are variable and depend on the absorbance properties of the water column. Fish development is hence highly influenced by all of these factors, affecting processes such as survival, growth, hatching, sex determination/differentiation and occurrence of malformations. The aim of this chapter is to review how fish development and performance is shaped by environmental factors, paying special attention to the effects of environmental cycles of light and temperature, the characteristics of illumination and the role of the developing biological clock.

Keywords Biological rhythms · Environmental cycles · Photoperiod
Light spectrum · Temperature · Circadian clock

2.1 Introduction

Aquatic organisms inhabit a highly dynamic environment which is subjected to cyclic events with different periodicities: tides (every 12.4 h), day/night alternations (24 h), lunar phases (29.5 h) and seasons (365 days). Changes in water level, photoperiod/light and temperature are driven by geophysical cycles such as the Earth's rotation on its axis and around the sun, and the gravitational force of the moon. Animals have adapted to these dependable changes by evolving a biological clock to keep track of time and anticipate to these periodic events. In fish, the development of an endogenous clock happens very early during larval development

F. J. Sánchez-Vázquez (✉) · J. F. López-Olmeda
Department of Physiology, Faculty of Biology, University of Murcia, Murcia, Spain
e-mail: javisan@um.es

M. Yúfera (ed.), *Emerging Issues in Fish Larvae Research*,
https://doi.org/10.1007/978-3-319-73244-2_2

and requires the input of environmental time cues such as light/dark cycles. Actually, light influences strongly fish development from embryo to larval stages. In this chapter we will review how the environment shapes the development and performance of fish larvae, discussing current research issues aimed at understanding the role of environmental cycles (i.e. light and temperature) driving the developing biological clock of fish larvae. Particular attention will be paid to the impact of daily thermocycles and light with different wavelengths on both fish larvae and their prey, as well as new findings on embryo development and hatching rhythms. Finally, the epigenetic changes provoked by environmental cycles during early development and in broodstock during reproduction will be introduced and discussed.

2.2 Light and Temperature Cycles

Light intensity/photoperiod and temperature change seasonally as the Earth rotates around the Sun. Besides, the axial rotation of the Earth causes linked cycles of light and temperature, generating a daily photo-/thermo-cycle: during the day (light phase) the temperature raises (thermophase), while during the night (scotophase) temperature drops (cryophase) (Fig. 2.1). Contrasting with the terrestrial environment, the underwater photo-/thermo-environment is rather peculiar as we shall see.

In one hand, the light from the sun is absorbed and scattered by the water molecules and particles in suspension, changing the light intensity, polarization and spectrum. The water acts in fact as a potent chromatic filter so that light

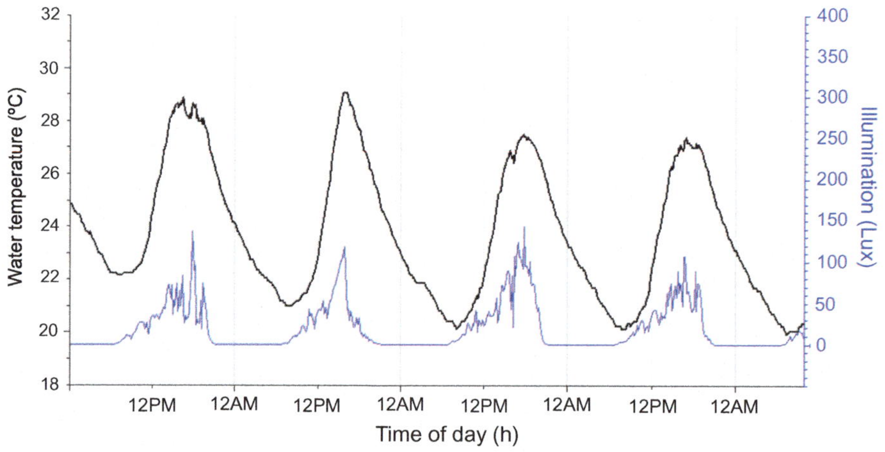

Fig. 2.1 Daily cycles of water temperature and illumination in the marine park of Santa Pola salt flats (Alicante, SE of Spain: 38° 11′ 16″N, 0° 36′ 52″W). Data was recorded in May 2013 every 10 min for 4 consecutive days using a HOBO Pendant (Onset Computer Corporation, Massachusetts, EE.UU) data logger

wavelengths below violet (λ < 390 nm) and beyond red (λ > 600 nm) are quickly absorbed, while blue wavelengths (λ ~ 450 nm) penetrates deeper in the water column. The filtering effect depends greatly on the water characteristics: oceanic water are deep blue, whereas coastal or continental waters are greenish or even yellow-brownish (Kusmic and Gualtieri 2000).

On the other hand, in the wild water temperature peaks by the end of the day and it reaches minimum values in the early hours. This daily thermocycle can serve as a powerful time cue to synchronize biological clocks, which are temperature-compensated to avoid running faster at higher temperatures and slower at lower temperatures, which would make them unreliable to keep time (Pittendrigh 1954). In fish, circadian locomotor activity rhythms entrain to daily thermocycles, although when conflicting *zeitgebers* are applied (light T = 25 h, and temperature T = 23 h), relative coordination appears, as locomotor activity increases when light and thermophase coincide, suggesting the existence of weakly coupled light- and temperature-entrainable oscillators (López-Olmeda and Sánchez-Vázquez 2009).

2.3 Clock Onset

Fish is a unique genetic model to identify the gene components of the molecular clock and investigate biological rhythms as they inhabit a wide range of environments and show great flexibility in their circadian system (Idda et al. 2012). Circadian rhythms are controlled by self-sustainable molecular clocks formed by both positive (Clock, Bmal) and negative (Per, Cry) transcriptional-translational feedback loops (Vatine et al. 2011). The positive loop is formed by Clock and Bmal, which are transcription factors that heterodimerize and activate the transcription of other genes. Among them, genes from the *per* and *cry* families form the negative loop and translocate to the nucleus to repress Clock and Bmal, closing the negative feedback loop (Vatine et al. 2011). Zebrafish *Danio rerio* is one of the most widely used fish models to investigate the entraining pathways of the circadian timing system and the importance of a light input during early development to initiate the biological clock. In this species, exposure to a single LD (Light-Dark) cycle is sufficient to start the rhythmic expression of *per1b* (Dekens and Witmore 2008) and a single LD/DL is enough to establish and set the phase of the rhythmic expression of *aanat2* (required for melatonin synthesis and phototransduction) (Vuilleumier et al. 2006).

Similar results on the impact of light in the ontogeny of the fish clock have been obtained in other fish species. In Senegalese sole *Solea senegalensis*, Martín-Robles et al. (2012) investigated the daily expression of *per1*, *per2*, *per3*, and *clock* during early development (0–4 days post-fertilization, dpf), finding an embryonic molecular clock from the first day of development and gradual entrainment of the clock machinery under LD, but not under DD (constant Dark) or LL (constant Light). In rainbow trout *Oncorhynchus mykiss* embryos, which take up to 2 months to hatch, a functional clock has been identified as early as 8 dpf, since rhythmic expression of *per1* appeared, although the effect of photic conditions was unclear (Davie et al. 2011). In gilthead sea bream *Sparus aurata* a recent paper found that clock genes (*clock*, *bmal1*, *cry1* and *per3*) expression is strongly influenced by light, and the amplitude and acrophase of the daily rhythms of expression changed in developing larvae under LD, LL or LL (0–14 days post-hatching, dph) + LD (15–30 dph) (Mata-Sotres et al. 2015).

Current research in zebrafish has further explored the effect of light wavelength in the ontogeny of the clock. The results revealed that the onset of daily rhythms of locomotor activity and clock gene (*per1b*, *per2*, *clock1*, *bmal1* and *dbp*) expression appeared earlier in larvae reared under LD-blue than in those reared under LD-white and LD-red (di Rosa et al. 2015). In European sea bass *Dicentrarchus labrax* a similar result has been found: the molecular clock (*per1*) started to oscillate in a rhythmic manner earlier (4 dph) when larvae were reared under LD-blue than in those kept under LD-white, LD-red, LL or DD (Fig. 2.2).

Taken together, these results highlight the crucial role of light during early development and suggest that a light-sensitive molecular clock is present and functional during embryogenesis, driving embryonic development and ultimately hatching rhythms, as we shall see in the following section.

2.4 Embryo Development and Hatching Rhythms

Embryogenesis in fish has been considered a continuous progression of developmental stages whose pace is basically set by temperature: a specific embryonic stage is reached faster at higher temperatures (e.g. Guerreiro et al. 2012). A key role of the biological clock, however, has been also suggested to control embryo development and hatching rhythms (Gorodilov 2010). In fact, a comparative study between zebrafish, Senegalese sole and Somalian cavefish *Phreatichthys andruzzii*, revealed that embryo development does not occur at a constant pace throughout day and night, appearing daily rhythms in developmental rate. These rhythms were synchronized to the LD cycle, influenced by temperature and with acrophases determined by the daily type of behavior of the species (Villamizar et al. 2013). Zebrafish is a diurnal species and so embryo develop faster during daytime than during nighttime, while in Senegalese sole (a nocturnal fish at juvenile and adult stages) embryos advanced their development at night and it was slowed down during the day. Thus, in this two fish species somitogenesis pace was different in embryos raised under LD regardless the temperature regime (Fig. 2.3). This temperature-independence occurs because the circadian clock must be temperature-compensated to be reliable and sustain circadian oscillations stable over time in a wide range of temperatures (López-Olmeda and Sánchez-Vázquez, 2011).

As embryo develops rhythmically, hatching do not happen at random but in a rhythmic fashion. Villamizar et al. (2013) revealed that although embryos reach a certain developmental stage ready to hatch, they wait for a particular time of day or

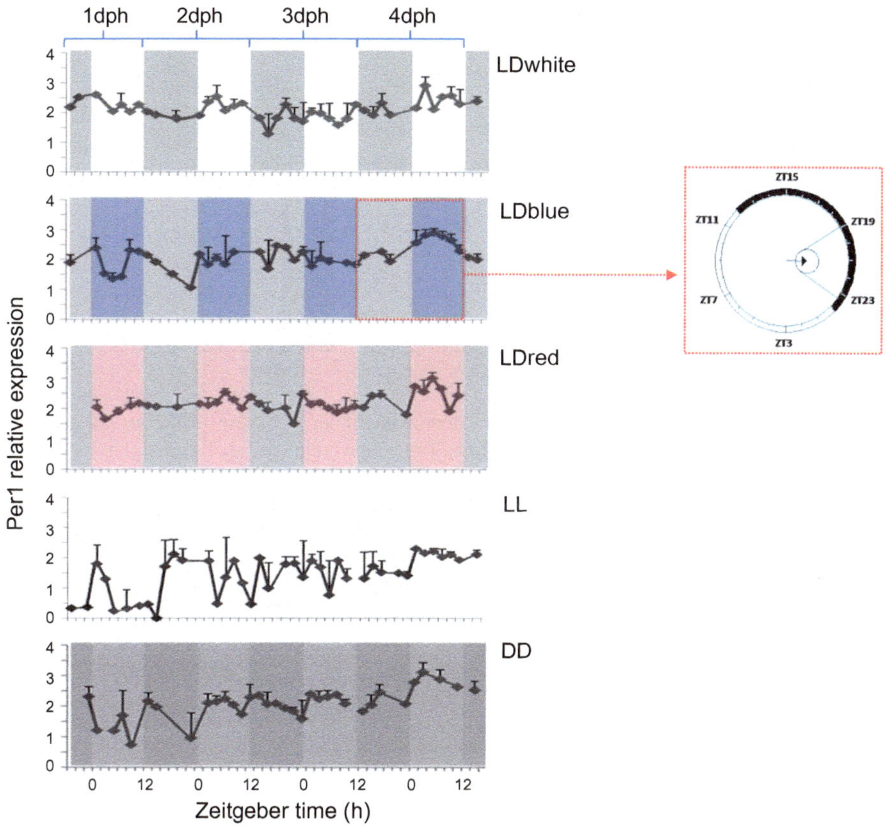

Fig. 2.2 Effect of lighting conditions in the ontogeny of the molecular clock (*per1* expression) of the developing European sea bass larvae during the first 4 days post hatching (dph). Larvae were reared in three 12 hL:12 hD cycles of different wavelengths (LDwhite, LDblue and LDred), constant light (LL) or constant darkness (DD). A significant daily rhythm was detected by Cosinor analysis at 4 dph in the LDblue group. Dark areas indicate the duration of the night. The arrow inside the circle indicates the achrophase of the rhythm (ZT21), while the inner circle and dotted lines represent the confidence intervals of the amplitude and achrophase, respectively (Sánchez-Férez et al. unpublished data)

night (so called hatching "gate" or "window"). In zebrafish the time of hatching was advanced or delayed depending of temperature (embryos hatch one-two days earlier at 32 °C than at 24 °C), but they always hatch during the day, never at night. Curiously, embryos raised at 28 °C that did not hatch the first day did not hatch at night but waited until the next day. In Senegalese sole, however, the hatching rhythm was also influenced by temperature but nocturnal: at 24 and 22 °C embryos hatch the first night, while at 18 °C they hatch the second night, but never during daytime (Fig. 2.4). This "gating" phenomenon was first observed in insects which

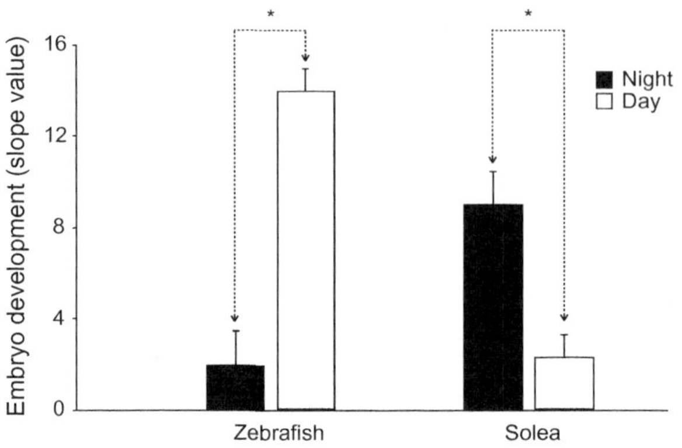

Fig. 2.3 Day/night differences in embryo development in zebrafish (diurnal species) and Senegalese sole (nocturnal) raised under LD. The slope value of the somitogenesis pace is represented. Asterisks indicate significant slope differences between day and night (ANOVA, $P < 0.05$) (modified from Villamizar et al. 2013)

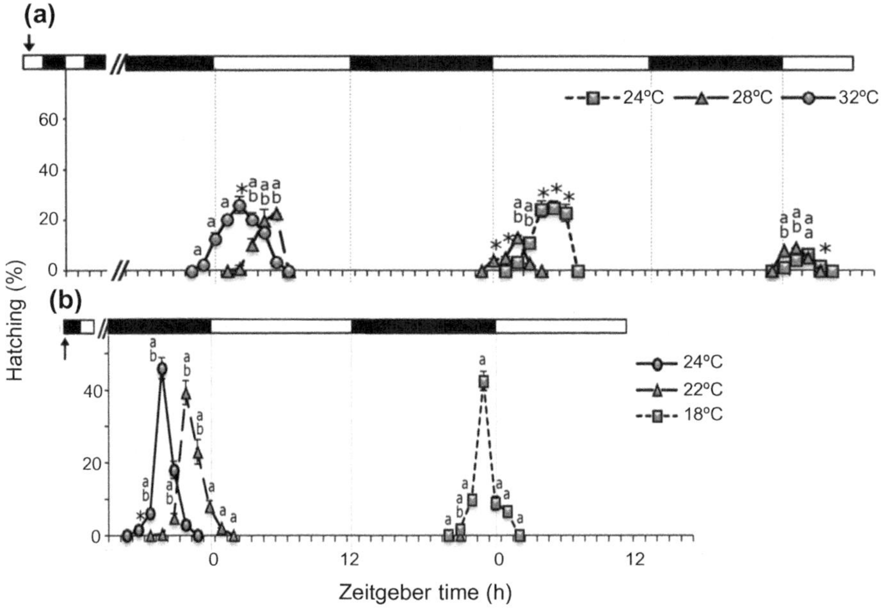

Fig. 2.4 Daily hatching rhythms in zebrafish (**a** diurnal) and Senegalese sole (**b** nocturnal) raised under LD and different temperature. White and black bars at the top of the graphs represent the light and dark phases, respectively, while arrows indicate the time of hatching (modified from Villamizar et al. 2013)

hatching occurs at a specific time of day determined by the interplay between the developmental stage and a circadian clock (Saunders 2002).

In short, hatching rhythms are apparently controlled by a clock mechanism restricting or gating hatching to a specific time of day/night (window). Hence embryos that have reached certain point of development hatch, while those that have not wait to hatch until the next available window depending on the behavior of the fish species.

2.5 Larval Performance

Newly hatched larvae are particularly fragile and extremely sensitive to environmental conditions. Both water temperature and lighting conditions (light intensity and photoperiod) have been long considered key factors determining larval survival and growth (e.g. Bouef and Le Bail 1999; Han et al. 2004). In this section, however, we will focus on the role of light cycles (LD vs. LL/DD) and light characteristics (spectrum), as well as daily water temperature cycles (TC vs. constant).

2.5.1 Light Effects

As discussed in Sect. 2.2, light radiation from the sun is modified by the water column, which acts as a chromatic filter allowing blue wavelengths reach deeper in clear oceanic waters. Therefore, it should not be surprising that when fish larvae are exposed to light of different colors (blue, green or red), their performance is affected. As reviewed by Villamizar et al. (2011), various fish species such as European sea bass, Senegalese sole and Atlantic cod *Gadus morhua* grow better when reared under blue or green light, which are conditions closer to their natural environment. However, when fish larvae are kept under constant light (LL) or constant darkness (DD), negative effects are observed with decreased growth and survival, and increased malformations. A higher rate of jaw elongation and swimming bladder hypertrophy appeared in European sea bass exposed to LL (Villamizar et al. 2009). Similarly, in Senegalese sole jaw malformations were highest in larvae exposed to the LDred, LL or DD. In another study in zebrafish larvae reared in LL, DD or LDs of different wavelengths (white, violet, blue, green, yellow and red), growth was higher under LDwhite, LDblue and LDviolet (Villamizar et al. 2014). Larvae raised under LL and LDviolet showed a higher proportion of malformations, while larvae under DD or LDred produced 100% mortality by 18 and 20 dph, respectively. Most interestingly, in the same study, the gene expression of key growth factors (*igf1a*, *igf2a*) and stress indicators (*pomca* and *chr*) were upregulated in larvae under LDviolet and LDblue, supporting the higher growth and malformations detected in larvae under LDviolet.

In flatfish, Senegalese sole larvae underwent metamorphosis earlier (9 dph) under LDblue that those reared under LDwhite (10 dph), LDred (13 dph), DD (15 dph) or LL (17 dph) (Blanco-Vives et al. 2010). Moreover, striking differences in the daily pattern of behavior arise in larvae coinciding with the onset of metamorphosis since they changed from diurnal to nocturnal in LDblue and LDwhite, but not in LDred (which remained nocturnal) or LL/DD (arrhythmic) (Blanco-Vives et al. 2012). Foraging behavior changed accordingly and so sole larvae under LDblue and LDwhite showed best prey capture success rates and highest proportions of larvae with rotifers in their guts during daytime until 9 dph. Similar results were observed video recording the modal action patterns (swimming duration, orientation, capture, miss and pass frequencies) of European sea bass larvae, which showed homogeneous distribution in the tank under LDblue, while larvae under LDwhite and LL showed highest density on the tank walls, and larvae under LDred and DD showed the lowest swimming and feeding activity (Villamizar et al. 2011). Curiously enough, in that paper live prey (*Artemia)* exhibited significantly highest hatching rate and a more even distribution under LDblue, while under LDwhite, LDred and DD they preferred to aggregate in the corners or at the bottom of the tank.

Although differences in larval growth performance are usually associated with feeding performance, some observations indicate light effects appear *before* the start of exogenous feeding, because the utilization of endogenous reserves depends on lighting conditions. In European sea bass, yolk sac absorption was faster in larvae reared under LL or LD-blue (exhausted by 9 dph), than in those reared under LDred or DD (exhausted by 11 dph) (Villamizar et al. 2009). The oil globule was also larger in 3–15 dph larvae reared under DD and LDred, indicating slower development. In Senegalese sole, similar results were observed: as early as 3–5 dph larvae reared under DD and LL showed significantly larger volume of yolk sac than that of those kept under LDblue, LDwhite or LDred (Blanco-Vives et al. 2010). In haddock *Melanogrammus aeglefinus*, however, the yolk sac was significantly reduced in larvae reared under LL and high light intensity (Downing and Litvak 2002). Summarizing, these findings are of particular importance for hatcheries as standard protocols often involved keeping larvae in DD, thus impairing the use of endogenous reserves and delaying the onset of exogenous feeding.

2.5.2 Water Temperature Effects

The impact of *constant* water temperature in fish eggs and larvae is well known (e.g. Blaxter 1986), but the influence of *cyclic changes* in temperature has been neglected. As introduced earlier, water temperature hardly remains constant during the day because the axial rotation of the Earth generates a daily temperature cycle as water cools during the night (cryophase) and warms during the day (thermophase). In the last few years, such a daily thermocycle has been reported to influence fish larvae performance and normal development. In Senegalese sole, Blanco-Vives

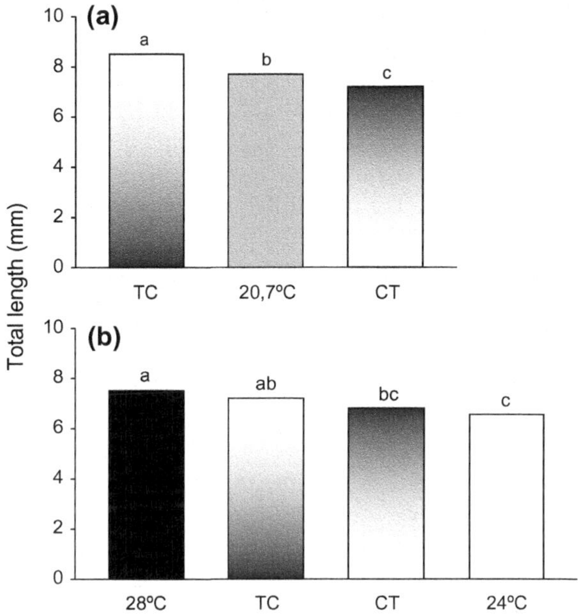

Fig. 2.5 Influence of daily thermocycles (TC/CT) versus constant water temperature in larval growth of zebrafish at 42 dpf (**a**) and Senegalese sole at 30 dpf (**b**). Fish larvae were raised under LD 12h:12h. In A, TC: 22.1 °C during the day and 19.0 °C at night, and CT: 19.2 °C day/22.0 °C night; in B, TC: 28 °C day/24 °C night, and CT: 24 °C day/28 °C night (modified from Blanco-Vives et al. 2010; Villamizar et al. 2012)

et al. (2010) investigated the effect of constant temperature (20.7 °C) versus two daily thermocycles: one in phase with the LD cycle (TC: 22.1 °C during the day and 19.0 °C at night), and an inverse thermocycle (CT: 19.2 °C day/22.0 °C night). Despite mean water temperature was similar in all three treatments (around 20.6 °C), significant differences in growth were detected and by the end of the trial at 30 dph, larvae under TC reached the largest size (8.5 mm), while those under constant temperature and CT were significantly smaller (7.7 and 7.2 mm, respectively) (Fig. 2.5). Another experiment applying daily thermocycles to developing zebrafish embryo and larvae rendered similar results (Villamizar et al. 2012), although in this case four temperature treatments were tested: two constant (24 and 28 °C) and 2 daily thermocycles: TC (28 °C day/24 °C night), and CT (24 °C day/28 °C night). Significant differences in growth appeared as early as 2 dpf and remained until the end of the trial at 42 dpf, with the greatest larval length found in the 28 °C and TC groups, and the lowest length in the CT and 24 °C. In this paper zebrafish embryogenesis was also affected by temperature, as differences in the somitogenesis pattern was observed among groups, and hatching rhythms were advanced in 28 °C and delayed in CT. Hatching occurred always during daytime, although by 3 dpf all embryos have hatched in the 28 °C and TC groups, while the 24 °C and CT group showed lowest hatching rates (45%).

In Senegalese sole, eye migration and completed metamorphosis is also influence by the temperature regime. Blanco-Vives et al. (2010) reported that larvae in the TC treatment began the process of metamorphosis earlier (eye migration at 9 dph), while those in constant temperature and CT started later (11 dph).

By 17 dph, all larvae in TC had completed metamorphosis, while larvae exposed to CT took 2-days longer to finish.

Normal larval development is also influenced by daily thermocycles as the presence of malformations differed in larvae reared under different temperature regimes. In Senegalese sole, by 9 dpf the proportion of larvae with lower jaw elongation was significantly lower in the constant temperature and CT groups, and higher in the CT group, differences remaining until the end of the trial at 30 dpf (Blanco-Vives et al. 2010). In zebrafish, different malformations (spinal curving, pericardial edema, abnormal tails and delayed yolk sac resorption) were significantly higher in larvae reared at 28 °C and in TC, while larvae in CT showed the lowest deformity rate (Villamizar et al. 2012). Such abnormalities in development probably reveal that larvae under "unnatural" environmental conditions were under stress, as their survival rate was much lower in the CT group (41%) as compared to the 28 °C and TC groups (83–77%).

European sea bass larvae exposed to daily thermocycles also exhibited different performance. By 15 dph larvae exposed to constant 21 °C or TC (21 °C day/17 °C night) showed significantly greater total length than larvae reared in constant 17 °C or CT (17 °C day/21 °C night) (Villamizar et al. unpublished data). Moreover, in this research endogenous reserve (oil drop volume) was exhausted earlier in TC, while caudal fin development (urostile inflexion) was faster in TC and 21 °C. Malformation rates were also lower in the TC and 21 °C (6.6%) than in the 17 °C (10.1%) and CT (13.3%). Foraging behavior was also enhanced as by 10 dph the proportion of larvae with *Artemia* nauplii in the gut was significantly larger in the TC and 21 °C groups (Fig. 2.6).

2.6 Sex Determination/Differentiation

Environmental sex determination in fish has been classically reported as an interaction between water temperature and genotype (Conover and Kynard 1981). Much information has been accumulated on the effects of temperature on sex determination in fish and the role of environmental, genetic and physiological factors (Devlin and Nagahama 2002). Basically, sex determination can be genotypic- (GSD) or temperature-dependent (TSD), and in many cases a rather small change in temperature (1–2 °C increase) can shift the sex proportion towards a male-biased ratio in both wild and farmed fish (Ospina and Piferrer 2008; Siegfried 2010). In addition, undifferentiated gonads are highly susceptible to environmental factors, so that temperature effects override GSD and shift sex ratio towards the opposite sex (GSD + TE) (Pandian 2014). Surprisingly, almost all trials have been performed using constant temperature regimes, despite water temperature in the wild show daily cyclic variations.

Taking this into account, Blanco-Vives et al. (2011) investigated for the first time the effects of exposing Senegalese sole larvae to daily thermocycles (TC and CT vs. constant temperature) on sex determination, gonad development and sex

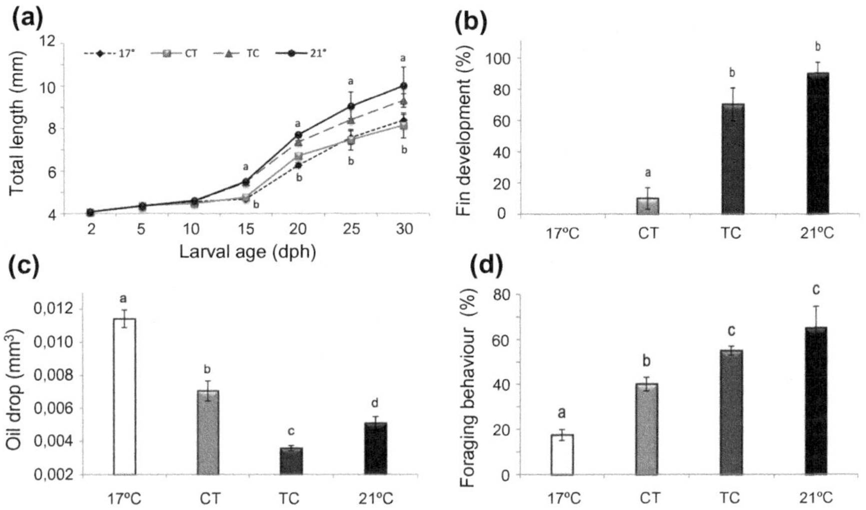

Fig. 2.6 Early development and foraging behaviour of European sea bass larvae raised under two constant temperature (17 °C or 21 °C) versus two daily thermocycles (TC: 21 °C day/17 °C night; CT:17 °C day/21 °C night). Data represent growth (larval length, **a**), caudal fin development (percentage of larvae with urostyle inflexion, **b**), endogenous reserves (oil drop volume, **c**) and exogenous feeding (percentage of larvae with *Artemia* in the gut, **d**) at 10–15 dph (Villamizar et al. unpublished)

steroid levels in juveniles. The results revealed that sex differentiation occurred earlier in larvae under TC, and at 152 dph these larvae had a higher proportion of females (71%), while those exposed to CT and constant temperature became more males (83 and 62%, respectively). These results were consistent with sex steroid concentrations, as fish in the TC group had significantly higher concentrations of estradiol, while fish in both CT and constant temperature groups showed higher concentrations of testosterone and 11-keto testosterone (Blanco-Vives et al. 2011). In zebrafish, Villamizar et al. (2012) investigated the impact in sex differentiation of two daily thermocycles versus two constant temperature regimes, finding that sex could be differentiated earlier in larvae exposed to CT and constant 28 °C by 42 dpf. In this species, however, both CT and TC produced a significantly higher proportion of females (79 and 83%, respectively), while females only accounted for 24 and 17% in the 28 and 24 °C constant temperature group, respectively (Fig. 2.7). It seems that daily thermocycles acted as female-promoters by suppressing *amh* (antimüllerian hormone) expression in testis and increasing *cyp19a* (aromatase) expression in the ovary, thus leading to androgen to estrogen conversion and feminization (Villamizar et al. 2012). A recent paper has further investigated this issue, finding daily rhythms in the expression of key genes involved in sex steroidogenesis that ultimately set sex ratio. In gonads, *cyp19a1a* and *amh* exhibited rhythms with opposite acrophases during daytime and at night, respectively, while expression of *foxl2* (forkhead box L2) in the ovary and *dmrt1*

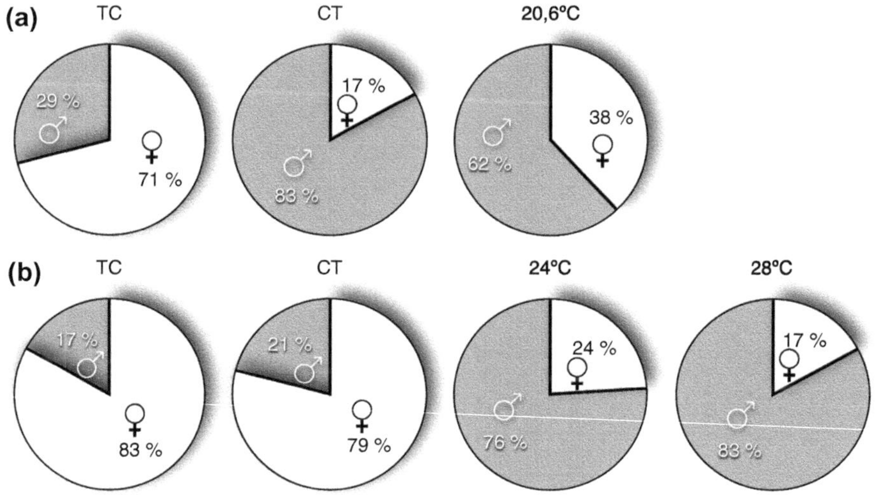

Fig. 2.7 Impact of water temperature in sex determination. Sex ratios of Senegalese sole (**a**) and zebrafish (**b**) larvae reared under natural (TC) and reversed (CT) daily thermocycles versus constant temperature regimes (modified from Blanco-Vives et al. 2010; Villamizar et al. 2012)

(doublesex and mab-3-related transcription factor 1) in testis had daytime and night time acrophases, respectively, whereas in the brain *cyp19a1b* and *cyp11b* (11beta-hydroxylase) expression peaked at night (Di Rosa et al. 2016). These findings highlight the importance of the time of day and support the hypothesis of the existence of a "window of sensitivity" at specific times of day or night. Therefore, sex determination would not only be influenced but temperature itself, but also by the time of day it rises.

In short, current findings point to the need to integrate not just the influence of environmental factors but also its *cycling* nature and the developing biological clock of fish larvae. Environmental cycles have profound effects not only in larval performance but they have irreversible long-lasting effects that ultimately shape the phenotype of adults.

References

Blanco-Vives B, Villamizar N, Ramos J et al (2010) Effect of daily thermo- and photo-cycles of different light spectrum on the development of Senegal sole (*Solea senegalensis*) larvae. Aquaculture 306:137–145

Blanco-Vives B, Vera LM, Ramos J et al (2011) Exposure of larvae to daily thermocycle affects gonad development, sex ratio and sexual steroids in *Solea senegalensis*. J Exp Zool 315: 162–169

Blanco-Vives B, Aliaga-Guerrero M, Cañavate JP et al (2012) Metamorphosis induces a light-dependent switch in Senegalese sole (*Solea senegalensis*) from diurnal to nocturnal behavior. J Biol Rhythms 27:135–144
Blaxter JHS (1986) Development of sense organs and behaviour of teleost larvae with special reference to feeding and predator avoidance. T Am Fish Soc 115:98–114
Boeuf G, Le Bail PY (1999) Does light have an influence on fish growth? Aquaculture 177: 129–152
Conover DO, Kynard BE (1981) Environmental sex determination: interaction of temperature and genotype in a fish. Science 213:577–579
Davie A, Sanchez JA, Vera LM et al (2011) Ontogeny of clock mechanisms during embryogenesis in Rainbow trout (*Onchorhynkiss mykiss*). Chronobiol Int 28:177–186
Dekens MP, Whitmore D (2008) Autonomous onset of the circadian clock in the zebrafish embryo. EMBO J 27:2757–2765
Devlin RH, Nagahama Y (2002) Sex determination and sex differentiation in fish: an overview of genetic, physiological, and environmental influences. Aquaculture 208:191–364
Di Rosa V, Frigato E, López-Olmeda JF et al (2015) The light wavelength affects the ontogeny of clock gene expression and activity rhythms in zebrafish larvae. PLoS ONE 10(7):e0132235
Di Rosa V, Lopez-Olmeda JF, Burguillo A et al (2016) Daily rhythms of the expression of key genes involved in steroidogenesis and gonadal function in zebrafish. PLoS ONE 11:e0157716
Downing G, Litvak MK (1999) The influence of light intensity on growth of larval haddock. N Am J Aquacult 61:135–140
Gorodilov YN (2010) The biological clock in vertebrate embryogenesis as a mechanism of general control over the developmental organism. Russ J Dev Biol 41:201–216
Guerreiro I, Peres H, Castro-Cunha M et al (2012) Effect of temperature and dietary protein/lipid ratio on growth performance and nutrient utilization of juvenile Senegalese sole (*Solea senegalensis*). Aquacult Nutr 18:98–106
Han D, Xie S, Lei L et al (2004) Effect of light intensity on growth, survival and skin color of juvenile Chinese longsnout cat sh (*Leiocassis longirostris* Gunther). Aquaculture 248:299–306
Idda ML, Bertolucci C, Vallone D et al (2012) Circadian clocks: lessons from fish. Prog Brain Res 199:49–57
Kusmic C, Gualtieri P (2000) Morphology and spectral sensitivities of retinal and extraretinal photoreceptors in freshwater teleosts. Micron 31:183–200
López-Olmeda JF, Sánchez-Vázquez FJ (2009) Zebrafish temperature selection and synchronization of locomotor activity circadian rhythm to ahemeral cycles of light and temperature. Chronobiol Int 26:200–218
López-Olmeda JF, Sánchez-Vázquez FJ (2011) Thermal biology of zebrafish (*Danio rerio*). J Therm Biol 36:91–104
Martín-Robles AJ, Aliaga-Guerrero M, Whitmore D et al (2012) The circadian clock machinery during early development of Senegalese sole (*Solea senegalensis*): effects of constant light and dark conditions. Chronobiol Int 29:1195–1205
Mata-Sotres JA, Martínez-Rodríguez G, Pérez-Sánchez J et al (2015) Daily rhythms of clock gene expression and feeding behaviour during the larval development in gilthead seabream, *Sparus aurata*. Chronobiol Int 32:1061–1074
Ospina-Álvarez N, Piferrer F (2008) Temperature-dependent sex determination in fish revisited: prevalence, a single sex ratio response pattern, and possible effects of climate change. PLoS ONE 3:e2837
Pandian TJ (2014) Temperature and sex ratio. Environmental sex differentiation in fish. CRC Press, Boca Raton (USA), pp 180–221
Pittendrigh CS (1954) On temperature independence in the clock system controlling emergency time in *Drosophila*. Proc Natl Acad Sci USA 40:1018–1029
Saunders DS (2002) Insect clocks. Elsevier Science, pp 576
Siegfried KR (2010) In search of determinants: gene expression during gonadal sex differentiation. J Fish Biol 76:1879–1902

Vatine G, Vallone D, Gothilf Y et al (2011) It's time to swim! Zebrafish and the circadian clock. FEBS Lett 585:1485–1494

Villamizar N, García-Alcazar A, Sánchez-Vázquez FJ (2009) Effect of light spectrum and photoperiod on the growth, development and survival of European sea bass (*Dicentrarchus labrax*) larvae. Aquaculture 292:80–86

Villamizar N, Ribas L, Piferrer F, Vera LM, Sánchez-Vázquez FJ (2012) Impact of daily thermocycles on hatching rhythms, larval performance and sex differentiation of zebrafish. PLoS ONE 7(12):e52153

Villamizar N, Blanco-Vives B, Migaud H et al (2011) Effects of light during early larval development of some aquacultured teleosts: a review. Aquaculture 315:86–94

Villamizar N, Blanco-Vives B, Oliveira C et al (2013) Circadian rhythms of embryonic development and hatching in fish: a comparative study of zebrafish (diurnal), Senegalese sole (nocturnal), and Somalian cavefish (blind). Chronobiol Int 30:889–900

Villamizar N, Vera LM, Foulkes NS et al (2014) Effect of lighting conditions on zebrafish growth and development. Zebrafish 11:173–181

Vuilleumier R, Besseau L, Boeuf G et al (2006) Starting the zebrafish pineal circadian clock with a single photic transition. Endocrinology 147:2273–2279

Chapter 3
The Digestive Function in Developing Fish Larvae and Fry. From Molecular Gene Expression to Enzymatic Activity

Manuel Yúfera, Francisco J. Moyano and Gonzalo Martínez-Rodríguez

Abstract There is a large amount of scientific literature on digestive enzymes activity during fish larval development. Nevertheless, most of this information refers to descriptive aspects of the ontogeny of hydrolyzing capacities, and only few studies looked into functional aspects and its relation with feeds and feeding activity. A comparison of published data normalized by days-degrees reveals at least three groups of species (precocious, medium and late) in relation to the time required for reaching maximum activity of digestive enzymes capacities, irrespective of the taxonomic group. Recent progress has allowed enlarging our knowledge on fish larvae digestive function. Such progress is partially based on new methodologies, but also on extended experimental designs and sampling protocols. On one side, the molecular basis of the onset of the digestive functionality has been examined in several species, cloning the mRNAs codifying the precursors of the main digestive enzymes and determining the gene expression profiles. On the other side, the daily rhythms coupling feeding and digestive enzyme production activities started to be investigated, giving a new dimension to our understanding of the digestive capacity during these early stages. Comparison between molecular expression and actual activity of the different enzymes is providing new insights on the regulation mechanisms. The connection between feeding behavior and digestive response both at genomic and biochemical levels is a required step towards the rationalization of feeding protocols with both live and inert diets, as a way to increase the nutrients utilization and the developmental performance at these early stages.

Keywords Digestive ontogeny · Proteases · Lipases · Carbohydrases
Fish larvae

M. Yúfera (✉) · G. Martínez-Rodríguez
Instituto de Ciencias Marinas de Andalucía, (ICMAN-CSIC),
Campus Universitario Río San Pedro s/n, 11519 Puerto Real, Cádiz, Spain
e-mail: manuel.yufera@icman.csic.es

F. J. Moyano
Departament of Biology and Geology, University of Almería, Almería, Spain

M. Yúfera (ed.), *Emerging Issues in Fish Larvae Research*,
https://doi.org/10.1007/978-3-319-73244-2_3

3.1 Introduction

Fish larvae demand a highly efficient enzymatic digestive machinery able to process properly the impressive amount of ingested nutrients required for supporting their high growth and development (Conceição et al. 1997; Houde 1989). Efforts to investigate the digestive capacity during the larval stage have been continuous from early years of modern aquaculture, when first techniques allowed the fish larvae to be fed and growth under semi-controlled laboratory conditions (Alliot et al. 1980; Baragi and Lovell 1986; Cahu and Zambonino-Infante 1994, 1995; Cousin et al. 1987; Hjelmeland et al. 1988; Martínez et al. 1999; Moyano et al. 1996; Pedersen and Hjelmeland 1988; Ribeiro et al. 1999; Zambonino-Infante and Cahu 1994) and particularly since sensitive quantification techniques to measure enzyme activities were adapted to fish larvae (Ueberschär et al. 1992). Advances during the last decades have been described in several reviews (Gisbert et al. 2013; Lazo et al. 2011; Rønnestad et al. 2013). Essentially, these reviews have highlighted the fundamentals of the action modes of the different enzymes, their close dependence on the development of different tissues and organs of the digestive system, as well as the progressive appearance of the different digestive enzymes following an age-dependent pattern in different fish species. In general, the description of these ontogenetic patterns pays special attention to the main events at functional level during the larval development. More specifically, they focus on the appearance of pancreatic enzymes shortly before first feeding, the enzymatic maturation of the brush border of enterocytes, and the appearance of pepsin activity in fish with stomach that indicates the transition to juvenile digestion mode. In addition, these reviews have identified some relevant topics in which more research is required for a better understanding about how the digestive system, as a cornerstone between feeding and growth, is functioning as a whole, being modulated by the interactions of the different involved processes. Several of these aspects have not yet been studied, or are starting to be investigated (Rønnestad et al. 2013), but in others, the recent progress are allowing to enlarge and change our previous concepts on the digestion in fish larvae. The digestive function is the consequence of complex interrelations among different events and processes. From food detection to digestion, absorption and catabolism of absorbed nutrients, there is a series of factors and signals regulating how feeding behavior and digestion must take place in a time framework to optimize the hydrolysis and assimilation. One primary fact to take into account is that the digestive function is closely related to feeding activity; this is obvious but sometimes ignored when analyzing digestive parameters. Feeding status and gut contents have not always been considered, in part due to the methodological difficulties associated with the experimentation with fish larvae. In this sense, there are some questions requiring deeper research effort to be properly answered. The first one is whether or not the fish larvae are eating continuously. Being this a main assumption, most feeding protocols used in larviculture traditionally tended to prolong the daytime and to feed the larvae several times a day, in order to maintain as long as possible a high concentration of live prey/

particles in the water column in order to increase the opportunities for larva-prey encounter. A second question related to a certain extent to the previous one is how much do we know about the regulation mechanisms of the digestive process in larvae. External stimuli and internal signaling are determining the feeding and digestion patterns that change on a daily basis. Thirdly, an important issue influencing every aspect of the digestive function is the interspecific and intraspecific variability. A priori, differences among species are not difficult to characterize because a high number of different species has been studied yielding specific results. Even so, these species have been studied by different research groups, under different rearing conditions and not always with the same analytical methodology. Furthermore, considering that each species has been reared and studied at a given temperature it is necessary to find a comparative way to remark the inter-specific developmental differences. On the other hand, the origin and importance of the differences among individuals is more complicated to identify, mainly when pools of larvae were used for the determinations. Not all larvae of the same egg batch are eating with the same intensity (Schaefer et al. 2017), but to what extent this variability is affecting the specific activities or if it is only a body size effect has not been yet properly examined. Analytical techniques using fluorescent substrates have allowed evaluating different activities using one single larva (Bolasina et al. 2006; Cara et al. 2007; Ueberschär et al. 1992), but it is still necessary to understand if the observed inter-individual variability reflects actual differences in the production of the enzymes or simply differences in the feeding status of the larvae. On the other hand, it is also necessary to explain certain differences observed when comparing results obtained with this methodology and with classical analytical methods. In fact, the variety of analytical techniques and ways to provide the results of enzymatic activities is an important issue to have in mind. Considering all the above mentioned aspects and sources of variability, it is necessary to build general patterns of activity throughout two temporal dimensions, the daily cycle and the whole larval stage, explaining what are the factors determining the digestion capacity and the main events occurring throughout the development to juvenile. Can we find rules to predict activity patterns under specific environmental, nutritional and feeding conditions? This is probably the main aim of the current research on digestive physiology of larval fish. Moreover, it takes an special interest considering that weaning and replacement of live prey with inert feeds is progressively becoming earlier in the development (Engrola et al. 2007, 2009; Faulk and Holt 2009; Hauville et al. 2014; Nguyen et al. 2011; Parma et al. 2013; Rodrigues et al. 2014; Saleh et al. 2013). For reaching this knowledge, it is necessary to increase the information at all levels involved in the food processing by the larval gut. However, there is an evident asymmetry since while there is a lot of information in some aspects, like in the ontogeny of pancreatic activities in many species, there is a lack in others, as in the daily patterns of activities or the regulatory mechanisms of enzyme secretion. Changes in expression levels of the mRNA codifying the different proenzymes during the ontogeny have also been examined in some species, though not as frequent as the enzymatic activity. In fact, quantitative expression pattern of at least three key proenzymes (a protease, a lipase and a carbohydrase) as a function of larval age has been described in very few species (Hansen et al. 2013; Kortner et al. 2011b; Mata-Sotres et al. 2016; Mazurais et al. 2015; Moguel-Hernández et al. 2016; Murashita et al. 2013, 2014; Parma et al. 2013; Sahlmann et al. 2015; Srichanun et al. 2013). This molecular analysis requires the previous cloning of species-specific mRNA transcripts of each enzyme precursor. This laborious and costly task has prevented in many cases obtaining results in parallel to the biochemically determined activity of the corresponding enzyme, although recent advances in sequencing techniques are making gradually easier this kind of analyses. Besides, the interpretation of results is sometimes complicated because it is common the appearance of several isoforms at mRNA level of each enzyme, and not always codifying proteins with the same function. On the other hand, circadian cycles modulate all biological and physiological processes, including the feeding activity (see Chap. 2). Recent studies on the consequences of feeding rhythms on the digestive function are changing the previously envisioned estimations of real capacity. This daily programming also exhibits important changes during the transition to feeding and digestion modes of juveniles, changes depending on the final feeding habits and that can be dramatic in some species. The potential correlations between the pro-enzyme mRNA transcript expression measurements and final biochemically determined activity of a given enzyme under specific conditions have not clearly been established yet. The digestive process can be regulated in every step from the synthesis of mRNA precursors to the final activation (and inhibition) of the synthesized protein, and it still constitutes a puzzle to elucidate, but new insights are starting to arise. Likewise, recent advances in endocrine control of appetite and the role of neural signaling and gastrointestinal hormones on feeding are starting to provide a preliminary figure of the action mechanisms in fish larvae showing the determinant involvement in the digestive function (see Chap. 4).

3.2 The Methodological Issue

Different analytical methodologies have been used in the determination of digestive capacity in larval fish based in the quantification of the activity of the different enzymes. It is possible to identify three different stages during the last decades:

1. *Late 80s and 90s*; Early works mainly based on the use of histochemical techniques, oriented to the identification of the presence of main enzymes in different tissues and not to quantification.
2. *2000–2010*; Most of the studies oriented to quantification were developed in this period using specific substrates being reactions measured in most cases using specific chromogenic substrates (Benzoylarginine Nitroanilide, BAPNA; Succinyl-L-Alanine P-Nitroanilide, SAPNA; Benzoyl-Arginine Ethyl Ester, BAEE; N-Benzoyl-L-Tyrosine Ethyl Ester, BTEE; Tertiary Amyl Methyl Ether, TAME; or Glutamyl-P-Nitroanilide, GPNA).

3. *From 2010 onwards*; Studies were still developed using the above detailed methodologies but also fluorescent dye-conjugated substrates, as well as genomic tools.

A review of more than 40 studies oriented to the evaluation of digestive enzyme activities during the initial development of different fish species reveals great similarities in the general statement and objectives among them, but also evidences some methodological issues that make difficult the comparison among results obtained by the different authors, this limiting the quality of the information provided. Some of these points are:

Time of Samplings Time profiles of digestive enzymes during the initial stages of fish development are obtained through samplings obtained along a given period, usually during the initial 30–45 days after hatching. Most authors refer the data to the age of larvae, but development of digestive biochemistry, as many other biological processes taking place in fish larvae, is a dynamic process greatly affected by water temperature. Therefore, comparisons among species should be possible only if ages are expressed as degree-days. As an example, the comparison between the moments of appearance of different enzymes in species like white seabass *Atractoscion nobilis* reared at 18 °C (Galaviz et al. 2011) to those in common snook *Centropomus undecimalis* reared at 30 °C (Jiménez-Martínez et al. 2012) are difficult. This way of expressing age of larvae is still followed by very few authors (Asgari et al. 2013; Blanco et al. 2016).

Moment of Sampling In close relation to the above mentioned, the moment of sampling is an important issue that may affect the quality of the results to a great extent. As indicated in the next section, digestion is a process showing important variations along a daily cycle and dramatic variations in the activity of enzymes have been reported during this period (Mata-Sotres et al. 2016; Navarro-Guillén et al. 2015; Zeytin et al. 2016). Almost all published studies are based in a similar schedule; one single sampling per day, taken always at the same hour during all the experiment (mainly early in the morning) and in some cases prior providing the first daily meal. Nevertheless, the aforementioned studies suggest that, at least in some species like Senegal sole *Solea senegalensis* or gilthead seabream *Sparus aurata*, maximum activity of the main digestive enzymes is reached at specific moments along the day. Consequently, the activities measured on larvae sampled only once a day may be very far from the maximum production, and conclusions about the presence and relative importance of a given activity may be highly biased.

The Way of Expressing Activities The more common way to express the activity of a given enzyme in larvae is as specific activity, that is, to relate measured units to the amount of soluble protein present in the extracts (Units/mg protein). This provides a reasonable estimation of the relative importance of the activity to the whole pool of other enzymes present in the extract and it could be useful for comparisons if the fraction of soluble protein remains constant in larvae during the sampling period. Nevertheless, for a fast growing fish larvae, this assumption may be not true, since the relative proportion of the different tissues and their composition changes quickly with growth. Hence, a more complete picture on changes taking place during the development could be obtained when expressing the activities also in relation to individual larvae or to weight units (i.e. mg). This is getting more common in some recent papers (Gisbert et al. 2009; Jiménez-Martínez et al. 2012; Solovyev et al. 2016; Toledo-Cuevas et al. 2011).

In this same sense, a lack of agreement in the definition of the units of activity makes comparisons among results obtained in different species quite difficult. Even when using the same substrate such differences arise. As an example, in Gisbert et al. (2009) the unit of α-amylase activity in larvae of common dentex *Dentex dentex* is defined as one mg of starch hydrolyzed during 30 min per mL of tissue homogenate at 37 °C measured at 580 nm, while in other papers, i.e. that aimed to study the activity in larvae of Persian sturgeon *Acipenser persicus* (Babaei et al. 2011) the unit was defined as μmol of maltose produced per min per mg.

Sensitivity of the Substrates As mentioned above, due to the limited sensitivity of chromogenic substrates assays are usually performed on extracts prepared using pools of larvae, this resulting in a loss of information about the individual variation within groups. Nevertheless, such information may be highly relevant when the heterogeneity of the response (enzyme secretion in this case) may be a part of the effect produced by the studied factor. This problem can be solved with the use of derivatives from 4-methylumbelliferone (MUB) or 7-amino-4-methylcoumarin (AMC). These substrates have been commonly used for rapid and sensitive determination of enzyme activity in soils and waters since late 90s (Marx et al. 2001; Sinsabaugh et al. 1997), but with the only exception of the early study performed by Ueberschär (1988), they have been only recently used to assess enzyme activities in larval fish (Blanco et al. 2016; Cara et al. 2007; Navarro-Guillén et al. 2015). Since individual measurements can be performed in a high amount of samples, it is possible to compare not only the medium values (as in pool extract) but also data dispersion and medians.

One of the main requirements of measures made by fluorescence is to maintain a low concentration of solutes in the reaction mixture. It is in fact only under those conditions that a proportional correlation between the actual enzyme concentration in the sample and the detected signal can be observed. If the solute concentration in relation to the substrate increases, the fluorescence-fluorogen concentration curve should reach a maximum and finally will show a decrease (Skoog et al. 1995). This is produced because of the non-radioactive deactivation process of the fluorescent response, which is mainly due to the absorption of radiation produced by other substances present in the medium not reaching the detector, or by the formation of "excimers". These molecules are produced by the union of two molecules of the fluorogen; one excited and the other still remaining in the basal state. Those excimers can be dissociated into two molecules that return to their basal state after producing a fluorescent signal at a higher wavelength. All those inconveniences can be reduced by a proper dilution of the samples (Skoog et al. 1995). The extracts used to measure digestive enzyme activities in larvae present a diverse and complex composition, with a high concentration of solutes other than proteases. Therefore, dilution of the samples is required in order to do not assign equal florescence values to very different concentrations of enzyme. These calibrations must be performed with each type of material to determine optimal range for measurement, on which the signal will be dependent only on the enzyme present, but not on substrate concentration.

Specificity of the Substrates It seems there is a general consensus about the use of specific substrates to determine the activities of the main digestive enzymes, being the main ones summarized in Table 3.1. Although some of these substrates are somewhat unspecific (i.e. hemoglobin), the rest can be considered specific and can provide accurate determinations of their target enzymes. Nevertheless, there is still a main problem in the measurement of lipase activity that reflects some methodological inconsistencies also detected when measuring such activity in adult fish. There are two types of enzymes involved in the hydrolysis of lipids, esterases and lipases, being both of them carboxylesterases. While the action of esterases is limited to the hydrolysis of solutions of short chain esters of particular monoesters as methyl butyrate or p-nitrophenyl acetate, true lipases are much more active on insoluble and aggregated substrates formed by emulsions of water insoluble long chain triacylglycerols as triolein than on soluble esters. This property, designated as "interfacial activation", can only be achieved in the presence of bile salts. However, the distinction between esterases and lipases based on chain length preference is not satisfactory since lipases also hydrolyze emulsions of partially water soluble short chain triacylglycerols as tripropionin and tributyrin. Taking this into account, in most studies performed in fish larvae, the type of substrates and assay conditions cannot differentiate between the presence of non-specific esterases or lipases.

Temperature and pH of the Assays All studies develop enzyme assays at 37 °C, but some of them specifically develop trypsin and chymotrypsin assays at 25 °C (Guerreiro et al. 2010; Suzer et al. 2007a). Authors do not provide a rationale for this difference in the methodology, but it is clear that measuring enzyme activities at different temperatures in the same samples presents inconsistencies. On the other hand, pepsin activity in fish larvae and juveniles, is usually measured by buffering

Table 3.1 Examples of common substrates used in the determination of activity of digestive enzymes in fish larvae

Enzyme	Spectrophotometry	Fluorometry
Pepsin	Haemoglobin	–
Trypsin	BAPNA	Boc-Gln-Ala-Arg-7-methylcoumarin
Chymotrypsin	SAPNA/BTEE	Ala-Ala-Phe-7-amido-4-methylcoumarin
Amylase	Soluble starch	Starch substrate labeled with BODIPY® FL
Lipase	nPmyristate/ naftilcaprilate	4-methylumbelliferyl derivatives of butyrate, heptanoate or oleate
Alkaline phosphatase	pNPP	4-methylumbelliferyl phosphate
LAP	leupNA	N-Succinyl-Ala-Ala-Pro-Phe-7-amido-4-methylcoumarin

the hydrolysis reaction at pH 2.0 or 3.0 following the methodology of Anson (1938), although in larvae of most fish species such low pH value is only attained in the juvenile stage (Yúfera et al. 2004) and eventually during limited hours within the postprandial period (Yúfera et al. 2012, 2014). Therefore, considering the real pH existing in their digestive tract, most of pepsin activity measured in studies with larvae is essentially pepsinogen and not active pepsin.

Taking into account all the mentioned above, it is suggested that studies aimed to determine the variations in the digestive enzyme profile during the initial stages of fish development should consider the following guidelines:

- To use individuals collected at different moments of the day for each sampling point. A previous assessment of diel variations of enzyme activity under the conditions used to maintain the larvae will allow the selection of the best moment to get a higher activity and lower dispersion of the data.
- Age of individuals should be expressed as degree-days.
- Provided the sensitivity and reliability of fluorescent substrates, these should be used preferably to obtain data on a certain number of separate individuals.
- Activities should be expressed both in relation to soluble protein and to individuals (normalized to weight or not). Only in this latter case it will be possible to appreciate the absolute variations in enzyme production in growing larvae.
- Assay temperatures should be the same for the different enzymes evaluated in a given study.

3.3 Advances in Molecular Biology

Determination of enzyme activities has been the most popular way for estimating the digestive capacity under any particular feeding and nutritional condition. However, the activity determined by biochemical analysis is only a part of the story. To know whether the enzymatic hydrolysis of the different macronutrient compounds is being programmed, activated and modulated, it is necessary to go in depth into the molecular basis of these processes. Analysis of the expression of the mRNA that is codifying the different enzyme precursors under the same conditions has been therefore the logical continuation to the studies focused to assess activities. It is obvious that we need first to have the sequences of the corresponding transcripts. At this respect, it is necessary to remark, as for the determination of the activity, that cloning methodology and procedures have changed in the last decade, thus providing a noticeable noise when comparing the results from different studies. The classical approach of cloning the genes of interest, one by one, either screening of custom made cDNA libraries in different type of vectors, or by the polymerase chain reaction (PCR) method using degenerated primers, designed on the conserved regions from the alignment of different closely related species to the one of interest, followed by 5′ and 3′ Rapid Amplification of cDNA Ends (RACE), has been outshined with the so called Next Generation Sequencing (NGS) approaches and the development of Bioinformatics tools. New approaches include RNAseq techniques on different tissues, where the genes of interest are being expressed, getting the whole transcriptome of the prevalent cell types of the organ, and sequencing of the genomes of the species of interest, followed by the prediction of the different coding regions with the use of computational tools. Many research groups around the world have used this last approach in different cultivated, exploited or aquarium species, making the gene databases full of predicted sequences and their corresponding proteins. Although this methodology has the support of previously obtained data by classical sequencing of Expressed Sequence Tags (ESTs) in some of the species, not all them, neither all the sequences, have an EST database behind or an EST counterpart to support the findings, making the annotation of the new cDNAs very complicated. Under this situation, searching the databases for a given type of gene product using algorithms like BLAST will provide numerous results of the type *gene-like* as well as many different type of isoforms not previously found for other species analysed in a more classical approach. This increases the difficulty to assign a given classically cloned sequence to a particular isoform type, given the new sequences arisen from genome analyses do not have a real transcriptomic body of data to support the annotations, then adding confusion at a high level to the complexity of the molecular analyses.

As an example to this statement, the reader can search the National Center for Biotechnology Information (NCBI) nucleotide (nt) database (GenBank) and discover for the genes of interest related to the gastrointestinal tract (GIT) the existence of several whole genome shotgun (WGS) sequence databases for different fish species, like the ones shown in Table 3.2. Another search for one of these proteins, Chymotrypsin, gave, at the date this book chapter was being written (15/03/2017), 616 hits for mRNA (157 hits in total) and genomic (362 hits in total) sequences, and 251 ESTs additionally. From the 616 hits, 259 were “predicted”, and all belonged to 22 bony fish Orders and 51 fish species.

The mentioned difficulty is clear when analyzing phylogenies using protein sequences, given their abundant number available in GenBank database coming from the predicted coding sequences (CDSs) from the bioinformatics platforms. This can be exemplified with trypsins (Try; Fig. 3.1). The amount of information related to different isoforms available in GenBank is overwhelming, and, therefore to avoid confusion, the predicted sequences can be removed. Downloading 65 protein sequences entirely obtained by classical cloning and Sanger sequencing methods, from exclusively species being cultured or having potential for aquaculture, and using free software downloaded from Molecular Evolutionary Genetics Analysis (MEGA) and version 7 for Windows (Kumar et al. 2016), it is possible to infer the molecular evolutionary relationship of taxa using the Neighbor-Joining (Saitou and Nei 1987), with Poisson corrections methods (Zuckerkandl and Pauling 1965) and 1000 bootstraps to enhance the reliability of the obtained tree (Felsenstein 1985). The result shows how the different described and known trypsin isoforms clusters together. Moreover, an additional analysis using EMBOSS explorer to calculate the isoelectric points for each of the proteins and their charge at pH 7.0 ensures the pertaining of each type of isoform to its corresponding

Table 3.2 Examples of bioprojects for genome sequencing from GenBank for commercial fish species

Bioproject	Assembly	Level	WGS	Chrs	Biosample	Taxonomy	Uses	References
PRJEB5099	GCA_000689215.1	Scaffold	CBXY000000000		SAMEA3271076	*Dicentrarchus labrax*	F, A, G	1
PRJEB9465	GCA_001403095.1	Contig	CVRK00000000		SAMEA3416743	*Squalius pyrenaicus*	G	
PRJNA11776	GCA_000767325.1	Scaffold	CABZ00000000		SAMEA3146315	*Danio rerio*	a, R	2
	GCA_000002035.3	Chromosome		25	SAMN03020626			
PRJNA221548	GCA_000721915.3	Chromosome	AZJR00000000	25	SAMN02439989	*Esox lucius*	F, A, G, a	3, 4
PRJNA245366	GCA_000972845.1	Scaffold	JRPU00000000		SAMN03092871	*Larimichthys crocea*	F, A	5, 6
PRJNA62009	GCA_000475215.1	Scaffold	AUPQ00000000		SAMN02981552	*Sebastes rubrivinctus*	G	
PRJNA72713	GCA_000233375.4	Chromosome	AGKD00000000	29	SAMN02749551	*Salmo salar*	F, A, G	7, 8, 9

F fisheries: *A* aquaculture; *G* gamefish; *a* aquarium; *R* research. *References* 1—Tine et al. (2014); 2—Howe et al. (2013); 3—Rondeau et al. (2014); 4—Ishiguro et al. (2003); 5—Ao et al. (2015); 6—Cui et al. (2008); 7—Lien et al. (2016); 8—Davidson et al. (2010); 9—Hurst et al. (1999)

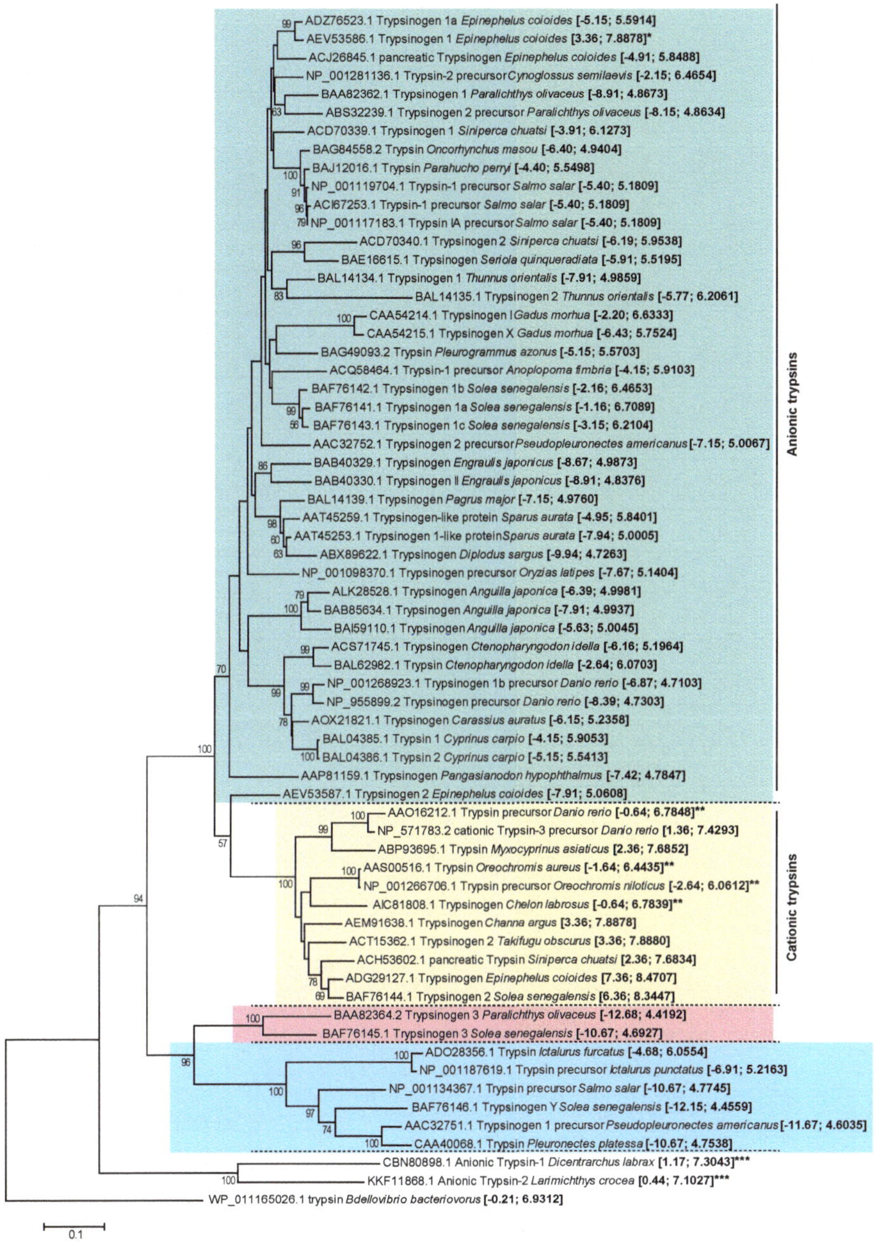

Fig. 3.1 Phylogenetic tree reconstruction for 65 different Trypsin proteins. Values for the charge of each protein at pH 7.0 and its isoelectric point are between brackets next to the species name. Anionic Trypsins or Trypsins type 1 are highlighted with green background (single asterisk, indicates exceptions to the rule); Cationic Trypsins or Trypsins type 2 are in yellow background (double asterisk, indicates exceptions to the rule); Trypsins type 3 are highlighted with pink; and finally, Trypsins type Y are with light blue background. Two trypsins, clustered in a separated branch, are labeled with triple asterisk. The tree was rooted using a bacterial trypsin

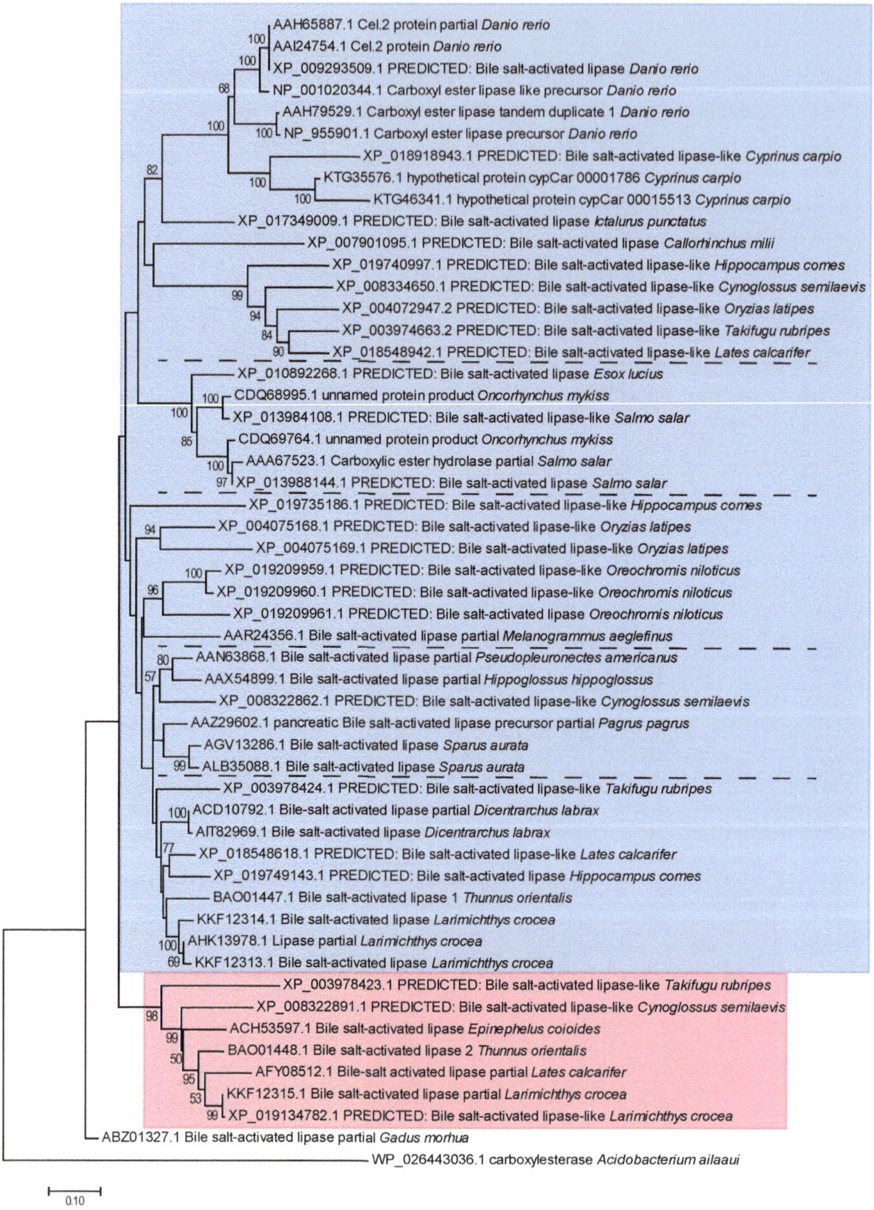

Fig. 3.2 Phylogenetic tree reconstruction. For 53 different bile-salt activated lipases (Cel) proteins. The two main clusters are highlighted with blue and pink backgrounds, one corresponding to Cel type 1 isoforms and the other to Cel type 2. The tree was rooted using a bacterial carboxylesterase

isoform group, with some exceptions (labeled with asterisks). To avoid interference, only full proteins were used in the analyses. The resulting tree clearly shows a main cluster conformed by anionic trypsins or trypsins type 1, a second cluster with cationic trypsins or trypsins type 2, and two more clusters for trypsins type 3 and type Y.

On the other side, we have the example for bile-salt activated lipase (Bal), more correctly named carboxyl-ester lipase (Cel; Fig. 3.2). The number of sequences in GenBank derived from classical cloning and sequencing is so scarce that will be almost impossible to infer a phylogeny without maintaining the predicted sequences from mass sequencing projects (NGS). For this case, two main clusters can be inferred, one belonging to type 1 and the other one to type 2. At the same time type 1 can be divided in several subtypes. It can be observed a number of sequences Cel or Bal-like for the same species but with different accession numbers, which reinforces the drawbacks already mentioned above.

In conclusion, although the development of new sequencing genome projects is giving a great number of new sequences in the gene databases, the fact of them not having the corresponding transcriptomic counterparts makes difficult the assignment to real mRNAs (are them really transcribed?). In a close future, after filling the gap in knowledge to deal with the complexity of the system, it will be a powerful tool for the discovery of new real transcripts regulated under certain physiological circumstances, which, otherwise, because of their scarcity, would have been undiscovered.

3.4 Ontogenetic Patterns of Activity and Molecular Expression

As mentioned in the Introduction, there is a great number of published works describing ontogeny of digestive enzymes in larval fish, but performing a comparative study is a quite difficult task considering all the above-mentioned variability in methodologies used by the different authors. Nevertheless, such kind of study should be of a great interest in order to describe general patterns of appearance of the enzymes related to species-specific features, feeding habits, etc. Having this in mind and only as an example, we have used the data reported for trypsin, pepsin and amylase activities expressed as mg^{-1} soluble protein, measured in several studies detailed in Table 3.3. These studies were selected since the authors used quite similar methodologies, although even in these cases the possibility of comparing absolute values was rather limited. The main conclusions obtained were:

Variations in Total Activities The average activity of trypsin measured during the initial stages in these species showed a great variability, ranging from 0.005 to 0.255 Units mg^{-1} soluble protein (Fig. 3.3). No specific pattern related to the taxonomic group could be observed; while several Sparidae (sharpsnout seabream

Table 3.3 List of species used for comparison of the ontogeny of several digestive enzymes

Species	Family	Authors
Huso huso	Acipenseridae	Asgari et al. (2013)
Acipenser persicus	Acipenseridae	Babaei et al. (2011)
Centropomus undecimalis	Centropomidae	Jiménez-Martínez et al. (2012)
Cyprinus carpio	Ciprinidae	Farhoudi et al. (2013)
Cichlasoma urophthalmus	Cichlidae	López-Ramírez et al. (2011)
Melanogrammus aeglefinus	Gadidae	Pérez-Casanova et al. (2006)
Dicentrarchus labrax	Moronidae	Zambonino-Infante and Cahu (1994)
Oplegnathus fasciatus	Oplegnathidae	He et al. (2012)
Paralichthys californicus	Paralichthyidae	Alvarez-González et al. (2006)
Miichthys miiuy	Scienidae	Shan et al. (2009)
Atractoscion nobilis	Scienidae	Galaviz et al. (2011)
Dentex dentex	Sparidae	Gisbert et al. (2009)
Diplodus puntazzo	Sparidae	Suzer et al. (2007a)
Pagrus pagrus	Sparidae	Suzer et al. (2007b)
Pagellus erytrhinus	Sparidae	Suzer et al. (2006)

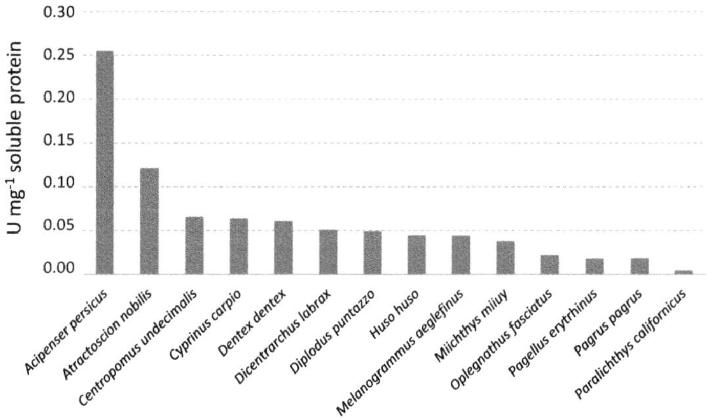

Fig. 3.3 Maximum values of trypsin activity (in U mg^{-1} soluble protein) measured during the initial stages of development in different fish species

Diplodus puntazzo, red porgy *Pagrus pagrus*, common pandora *Pagellus erythrinus*) presented quite similar values, common dentex presented a much higher activity. The two species of Scienidae, miiuy croaker *Miichthys miiuy* and white seabass, also offered quite different activities.

In the case of pepsin, most of the values measured in the different species were in a much narrower range; from 3.5 mg^{-1} soluble protein in white seabass to 28.3 mg^{-1} soluble protein in common snook. The comparison of amylase activities was not possible provided the huge variations in the units used to measure this enzyme by the different authors.

Patterns of Appearance of the Enzymes To perform this comparison accurately a previous normalization of the data was performed: in one hand, as indicated before, the age of the larvae was expressed in degree-days (DD) considering rearing temperature used in each case. On the other hand, the values of activity were recalculated and expressed as a percentage of the maximum value measured in each experiment. This way, the different profiles could be easily compared, paying attention to two main features; the shape of the profiles and the moment of appearance of the maximum activity within the studied period.

In general terms the shapes of the profiles could be classified as follows:

- Trypsin; with the only exceptions found in common dentex (which showed a progressive decrease of the specific activity with age) and in major carp *Catla catla* and common carp *Cyprinus carpio*, (which showed just the opposite trend), the profiles were always characterized by the presence of a single sharp peak of maximum activity, reached at different ages depending on the species, followed by a further decrease. Since it should be expected that once a given activity reaches a high production it should be maintained from that moment onwards, the observed decreases suggest that the onset of this enzyme may be characterized by a genetically driven over production (a sort of "switch on" process) followed by a further regulation dependent on the real metabolical needs of the individuals (Fig. 3.4).
- Amylase; the profiles are somewhat more complex, showing more than one peak during the ontogenetic development. A more complex regulation of the activity of this enzyme is suggested, mainly considering that in some carnivorous species, as in common dentex or European sea bass *Dicentrarchus labrax*, the maximum activity is detected very early in the development, being this no connected to the composition of the diet (Fig. 3.5).
- Pepsin; the profiles are characterized by the absence of sharp peaks. It seems that the activity is maintained once reached the maximum, irrespective if this maximum appears progressively from an early age (as in Beluga sturgeon *Huso huso* and Persian sturgeon) or at a given moment during the larval development (e.g. sharpsnout seabream, white seabass, common snook). This pattern seems to be related to the onset of a functional stomach that does not require a further fitting as in the case of trypsin (Fig. 3.6).

On the other hand, important variations in the time required to reach maximum activities can be observed among species, being possible to classify them into "precocious" (if the maximum activity is detected prior 300 DD), "medium" (between 300 and 600 DD) or "late" (after 600 DD). For trypsin, as detailed in Fig. 3.4, it is worthwhile to mention that a certain similitude between closely

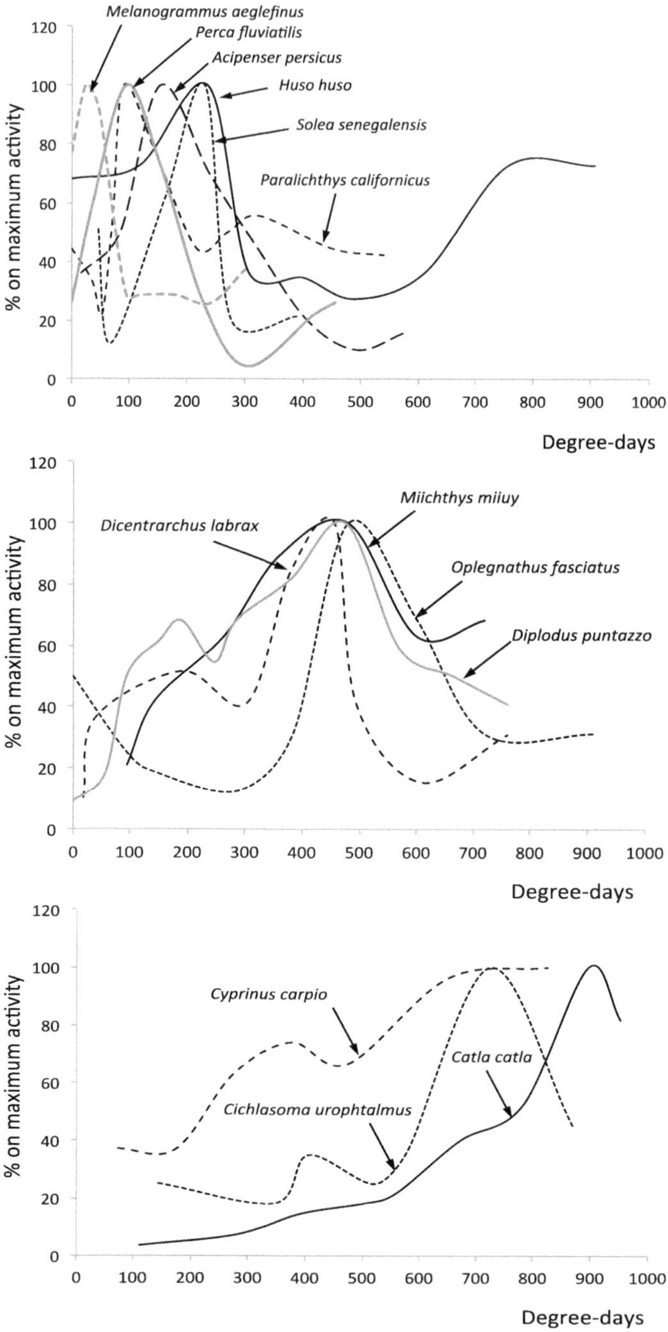

Fig. 3.4 Differences in the moment of appearance of maximum trypsin activity during early development in different fish species (made from data of authors cited in the text)

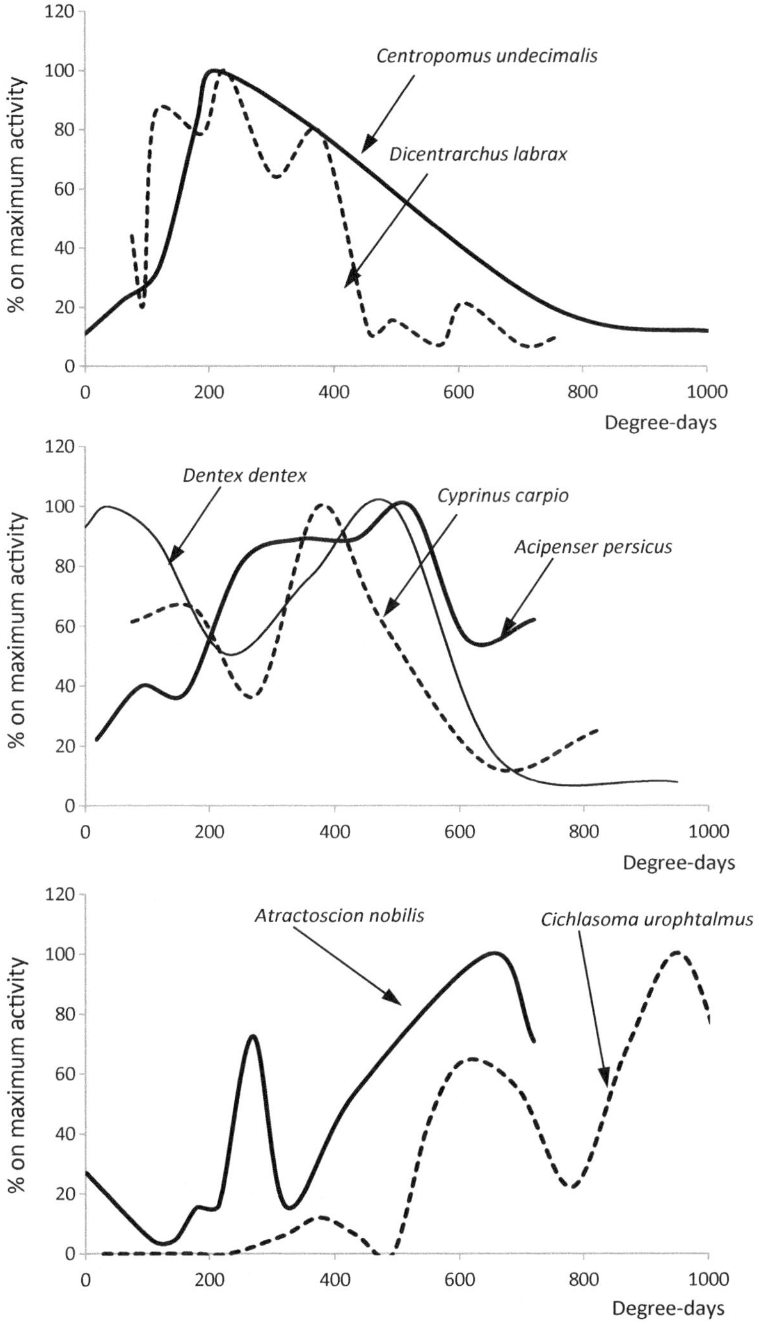

Fig. 3.5 Differences in the moment of appearance of maximum amylase activity during early development in different fish species (made from data of authors cited in the text)

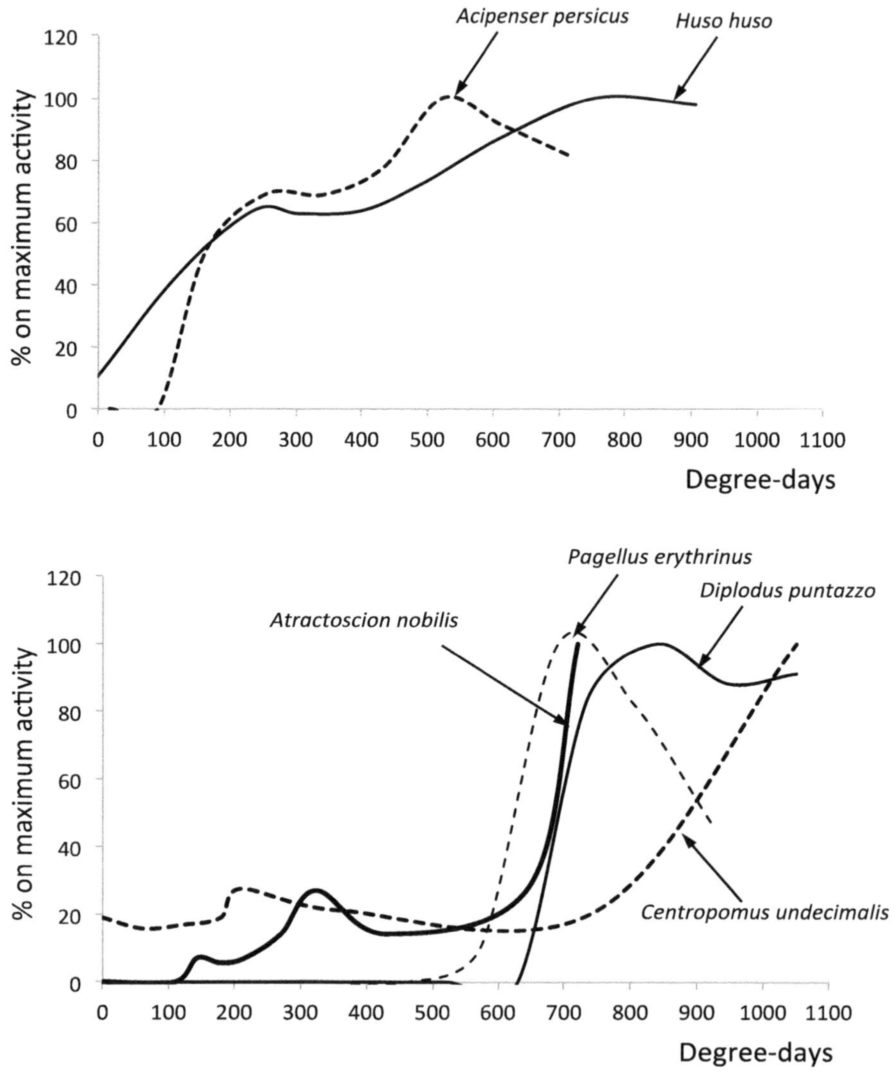

Fig. 3.6 Differences in the moment of appearance of maximum pepsin activity during early development in different fish species (made from data of authors cited in the text)

related species was observed, since the two Acipenseridae (Persian sturgeon and beluga) as well as the two Pleuronectiformes (Senegal sole and California flounder *P. californicus*) were classified as "precocious". The patterns observed for amylase activity resembled those obtained from trypsin but, of course, no coincidence between categories could be expected for the enzymes measured in the same species; as an example, European sea bass was precocious in the developmental pattern of amylase, but medium in that of trypsin, while Persian sturgeon was just the opposite (Fig. 3.5). Two different patterns were observed in the development of pepsin activity (Fig. 3.6). In one hand, the profile showed by the two Acipensaridae was characterized by the presence of high values of activity at early ages, while in the other species significant pepsin activity was only detected from 600 DD onwards. It is supposed that precocious species are better prepared to develop an effective digestion of protein at early stages of development and this should be paced to a similar behavior of other enzymes also involved in the digestion process.

The kind of comprehensive approach exemplified here with only a few species, if extended to the high number of already published works in this field, could provide a highly useful insight into species-specific differences, both in the ability to develop a fully functional digestive system and to the efficient use of the main types of nutrients. The different points to consider in such global evaluation should be: firstly, the above mentioned differences in the time required to reach maximum activity for the main digestive enzymes acting in the lumen of the digestive tract (proteases, lipases, amylase); secondly, a similar estimation made on the enzymes involved in digestion and transport in the intestinal epithelium (amino peptidases, phosphatase); and thirdly, the degree of concordance along the development for all the enzymes; to be precise, the presence of a more or less homogeneous pattern for all of them or the appearance of some enzymes abnormally late in the development compared to the rest.

At molecular level, studies on the ontogeny of the digestive function have been more erratic and still remain very incomplete. The first molecular expression studies on fish larvae using PCR were not strictly quantitative, allowing only the visualization of not very well defined increasing or decreasing patterns throughout the development (Darias et al. 2006, 2007; Douglas et al. 1999, 2000; Murray et al. 2003, 2006). Nevertheless, these studies helped to determine when the different digestive enzymes started to be available and functional. Furthermore, the in situ hybridization (ISH) techniques complemented these early analyses by locating the increasing expression of the enzyme precursors in the corresponding organs and tissues. Later, more precise quantitative reverse transcription real time PCR analyses (qRT-PCR or qPCR) provided good expression profiles based in one sample of individuals taken in different days throughout the development, as for enzyme activities, allowing comparing both molecular expression and activity patterns. Thus, although profiles showing the ontogenetic expression of several digestive enzymes—mainly proteases—have been published for some species (Gao et al. 2013; Hansen et al. 2013; Kortner et al. 2011b; Mata-Sotres et al. 2016), studies aimed to obtain more complete screenings, particularly those on which activities and molecular expression are analysed simultaneously at the same time, are scarce (Galaviz et al. 2012, 2015; Murashita et al. 2013, 2014; Peres et al. 1998; Ruan et al. 2010; Sahlmann et al. 2015; Srichanun et al. 2013). These latter studies showed that, in spite of showing some differences, the ontogenetic profiles of activity and gene expression of proteases, lipases and carbohydrases, also present clear similarities. Time patterns of increase and decrease as well as the peaks of maximum values for both biochemical and gene expression of the enzymes are overall coincident, being the net result, from combining the sequential development of specific digestion and the variations in diet composition, linked to changes in prey items or weaning onto inert diets. On the other hand, the observed differences could be explained by several reasons; e.g. while biochemical determinations involve activities that may correspond to a group of enzyme isoforms, the expression of a transcript corresponds to only one specific isoform. In addition, it should be remarked that it is common to observe a delay between the detection of the molecular expression of a given enzyme and of its corresponding activity. Anyway, considering the available data, it could be deduced that time course profiles of gene expression during early larval development for the main digestive enzymes seem to be genetically programmed and scarcely influenced by dietary changes.

3.5 The Digestion as a 24 h Work

As indicated previously, most studies oriented to assess the general patterns of ontogeny for the activity and molecular expression of digestive enzymes in fish larvae have been performed on the basis of one single daily sample, in many cases after several hours of starvation intending to prevent the potential interference of the gut content. A similar one-time sampling schedule is being used for comparing the digestive capacity during or at the end of the experimental period in nutritional studies testing different feeding conditions in juvenile fish. Nevertheless, the information obtained in such a way is very incomplete, because it assumes that the enzymatic activities remain relatively constant for long periods under given nutritional conditions. However, the activity of digestive enzymes may change as a function of the amount and quality of the digestive content, which on turn depends on feeding behavior and external conditions. Living organisms have developed an adaptation to the predictable changes of environmental conditions by means of a biological rhythmicity that follows daily, lunar and yearly cycles. For fish larvae, and considering their relatively short developmental time span, the most relevant is the circadian cycle involving day/night alternation. At organismal level there is a group of clock-genes whose expression changes following a circadian period by means of an autonomous feedback loop of positive and negative transcriptional-translational factors. This self-sustainable molecular clock plays a determinant role in synchronizing many physiological processes, particularly those related to the feeding activity and digestion (see Chap. 2). These clock-genes have been described in embryos and larvae of several fish species such as zebrafish *Danio rerio* (Dekens and Whitmore 2008), Senegal sole (Martín-Robles et al. 2013), rainbow trout *Oncorhynchus mykiss* (Davie et al. 2011), medaka *Oryzias latipes* (Cuesta et al. 2014) and gilthead seabream (Mata-Sotres et al. 2015). These studies showed that in general terms in diurnal fish, the expression of positive elements (*clock* and *bmal* transcript families) reach the acrophase by the end of the light period and the negative elements (*per1 per3* and *cry* transcripts) during the beginning of the dark period. Accordingly, several studies have demonstrated that, as a consequence of the evolution and the adaptation to environment, each fish species has a particular feeding behavior. This larval feeding behavior changes along the day following a specific pattern determined primarily by the density and accessibility of prey items. Moreover, studies performed in wild larvae evidenced hourly preferences aimed to maximize food intake and minimize the feeding effort (MacKenzie et al. 1999; Østergaard et al. 2005; Shoji et al. 1999). Accordingly, while a continuous feeding not related to day/night sequence has been described in some species, in others the feeding activity shifts from mainly diurnal to nocturnal with the development. In most cases, the ingestion rate (usually estimated from the gut contents) tends to be higher at twilight, either at sunrise, sunset or both. The same has been observed at the laboratory in species such as Japanese Spanish mackerel *Scomberomorus niphonius* (Shoji et al. 2001), tongue sole, *Cynoglossus semilaevis* (Ma et al. 2006), loach *Misgurnus anguillicaudatus* (Wang et al. 2008), Japanese flounder *Paralichthys olivaceus* (Kotani and Fushimi 2011), gilthead seabream (Mata-Sotres et al. 2015) and Senegal sole (Navarro-Guillén et al. 2015). Therefore, a synchronization of the digestive function to these time patterns of food uptake, and hence the appearance of a daily rhythm, may be expected. Most of current knowledge on the dynamics of the digestive function in teleosts has been obtained in experiments with juvenile fish and later extrapolated to larvae. The studies on the postprandial response in juveniles point to a relation between digestive activity, the amount of gut content, and the transit time, although the relation with this last factor remains practically unexplored. Once the digestive tract is completely developed and the stomach becomes fully functional (except in agastric fish), a better control on the food transit is reached and the patterns of activity and molecular expression of enzymes are adapted to feeding times and frequencies (Montoya et al. 2010; Montoya-Mejía et al. 2016; Santigosa et al. 2008; Yúfera et al. 2012, 2014; Zeng et al. 2014). Nevertheless, the feeding process is somewhat different in larvae than in juveniles. Juvenile fish usually eat at discrete time intervals, being the food retained for a while in the stomach to complete the acidic digestion before being transferred to the intestine, a step that determines further transit throughout the rest of gut segments. In contrast, fish larvae eat relatively large amounts of food continuously, this resulting in very short residence times of the digesta within the gut. Rearing protocols of fish larvae are usually characterized by maintaining a permanent food availability in the surrounding water, but these food items are not always detectable by the sensory organs, since larvae of most fish species are visual predators highly dependent on the presence of suitable light intensity and duration (Villamizar et al. 2011; Yoseda et al. 2008). This makes the feeding of larval fish highly dependent on daily light/dark cycles, being photoperiod and feeding time considered as key regulators of their daily rhythms, as in other vertebrates (López-Olmeda 2017; Vera et al. 2013; Whitmore et al. 2000).

While these daily dynamics in the feeding activity of larval fish was already observed some years ago, with the exception of some punctual studies (Fujii et al. 2007; MacKenzie et al. 1999), the potential rhythmicity of the digestive function has been studied only recently in depth (Mata-Sotres et al. 2016; Navarro-Guillén et al. 2015, 2017; Zeytin et al. 2016). Although the number of these studies is still very scarce, it is possible to envisage clearly that the digestive function in larvae also exhibits daily patterns, but with different nuances. Variations along the 24 h cycle can be observed in both enzymatic activities and molecular expression of their corresponding mRNA transcripts. However, strong differences exist in these daily patterns among digestive enzymes, as well as between the patterns of molecular expression and of activity for a given enzyme. This may be explained to a certain extent considering that in fish larvae, as occurs in juveniles and adults, pancreatic and intestinal digestive enzymes involved in the hydrolysis of the ingested macronutrients are not produced and activated simultaneously and in the same manner, but following a serial schedule along the digestive tract.

In these recent studies, daily patterns of food intake and pancreatic enzyme production under a light/dark photoperiod have partially been assessed in larvae of two species with different life styles and feeding behavior, the gilthead seabream and the Senegal sole. Under permanent prey availability in the water column, seabream larvae exhibit a net diurnal feeding with maximum ingestion at the end of the light period (Mata-Sotres et al. 2016). This means that gut contents increases progressively during the daytime and evacuation takes place in few hours during the dark period, being this pattern maintained during the whole larval stage. In addition, these circadian rhythms and the parallelism between trypsin activity and feeding activity were maintained even when larvae are reared under permanent illumination, although with asynchrony with the theoretical day/night hours (Yúfera 2016).

The flatfish Senegal sole presents a similar feeding profile to that of gilthead sea bream during the pelagic stage, but after the eye migration, that takes place during the second to third week of life, post-larvae switch to a continuous feeding that is maintained during day and night (Navarro-Guillén et al. 2015). In both species, trypsin activity follows a similar pattern to that exhibited by the feed intake, characterized by a very quick release of the enzyme as a response to food intake, this ensuring an optimum hydrolysis of protein and polypeptides and a maximum bioavailability of nitrogen compounds present in food (Fig. 3.7). This is not surprising because trypsin is considered the key digestive enzyme in fish larvae (Ueberschär 1993), mainly taking into account the high amount of amino acids required to build new tissues and undergo such as fast growth. Furthermore, as the residence time of food in the gut is very short in sparids (Andrade et al. 2012; Yúfera et al. 1995), digestion must take place as soon as food protein is ingested in order to maximize bioavailability. Likewise, Morais et al. (2004a) found that this process is also very fast in Senegal sole post-larvae, being over 3 h the time necessary to digest and incorporate the ingested amino acids into the body tissues.

Contrarily to trypsin, the activities of lipase and α-amylase seem to be more independent of the feeding pattern. Although also a parallelism for lipase and food content was observed at certain ages during most of the larval period, they present irregular variations not following a clear daily rhythm but only a certain trend to decrease during the transition from dark to light. The absence of an evident temporal correlation between the activities of these enzymes and the presence of food within the gut may be due to a different secretion mechanism to that described for

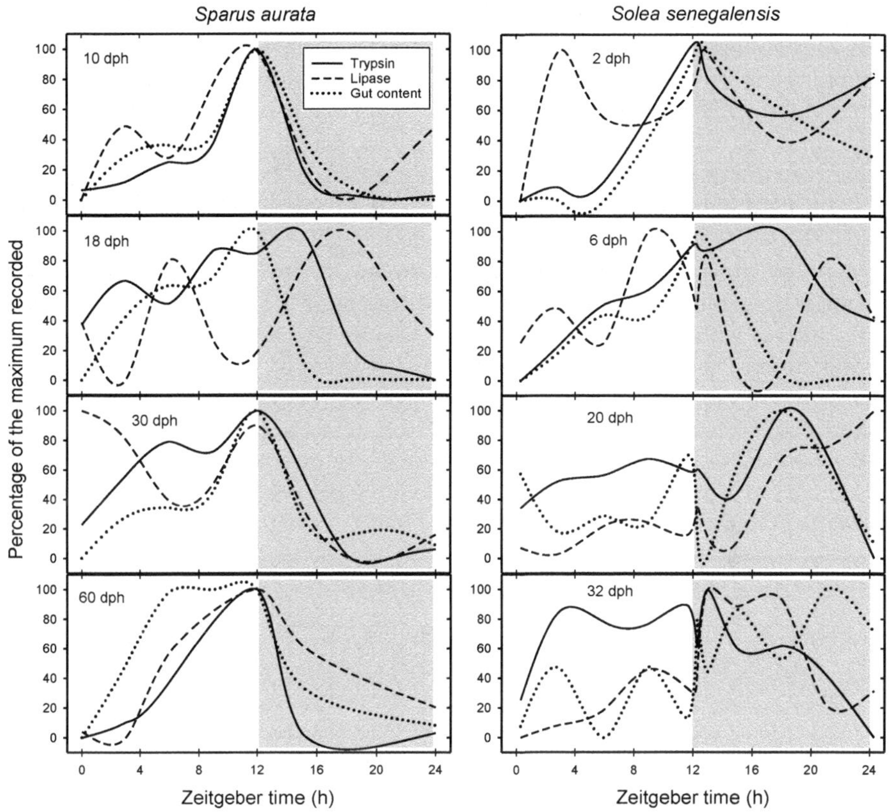

Fig. 3.7 Trypsin and lipase activities during the daily cycle in gilthead seabream and Senegal sole developing larvae compared to the gut fullness. To enhance the daily pattern activity, values have been recalculated in the range from 0 to 100 for the minimum and maximum detected enzyme activities. Grey area indicates the dark period. Modified from Navarro-Guillén et al. (2015) and Mata-Sotres et al. (2016)

proteases together with other factors (e.g. longer time required to complete lipid digestion).

On the other hand, daily changes of the mRNA expression for the main pancreatic digestive proenzymes (*trypsinogen*, *try*; *chymotrypsinogen b*, *ctrb*; *phospholipase a2*, *pla2*; *bile-salt activated lipase 1b* or *carboxyl ester lipase 1b*, *cel1b;* and *pancreatic α-amylase*, *amy2a*) have been examined, and they differ from those of the corresponding activities, at least in developing larvae of gilthead seabream (Mata-Sotres et al. 2016). Such discordance is not unexpected for different reasons. Firstly, both the molecular expression and the biochemical activity are regulated and activated in a different manner and they do not necessarily have to exhibit similar patterns. Secondly, there is not a direct correspondence between the enzyme isoforms tested at activity and molecular levels. As explained in Sect. 3.3, qPCR analyses are performed on a specific isoform, while biochemically analyzed activity may refer to a group of isoforms or even a group of enzymes hydrolyzing a given macronutrient. In any case and perhaps in a simplistic way, taking together the daily pattern of the different enzymes, it is worth noting that the lowest expression values were observed within the first hours of the dark period while highest expression values occur at the transition from dark to light (Fig. 3.8), this being in agreement to the pattern observed in the negative elements of the molecular clock (*per3* and *cry1*). This resemblance may indicate their regulation by these clock genes to prepare the enzymatic machinery for the coming feeding cycle. This anticipatory response to feeding could allow a better utilization of the nutrients.

In species with stomach, pepsin activity usually appears at the end of the larval period during the transition to juvenile. The activation of pepsin from pepsinogen, and therefore the level of pepsin activity, depends on an adequate gastric acidification. For this reason, as explained in Sect. 3.2, actual pepsin activity in fish larvae and juveniles should be measured at the real pH measured in their gut lumen. Gastric pH in older larvae can be measured with a pH microelectrode (Yúfera et al. 2004) or by injecting pH indicator solutions within the gut (Rønnestad et al. 2000).

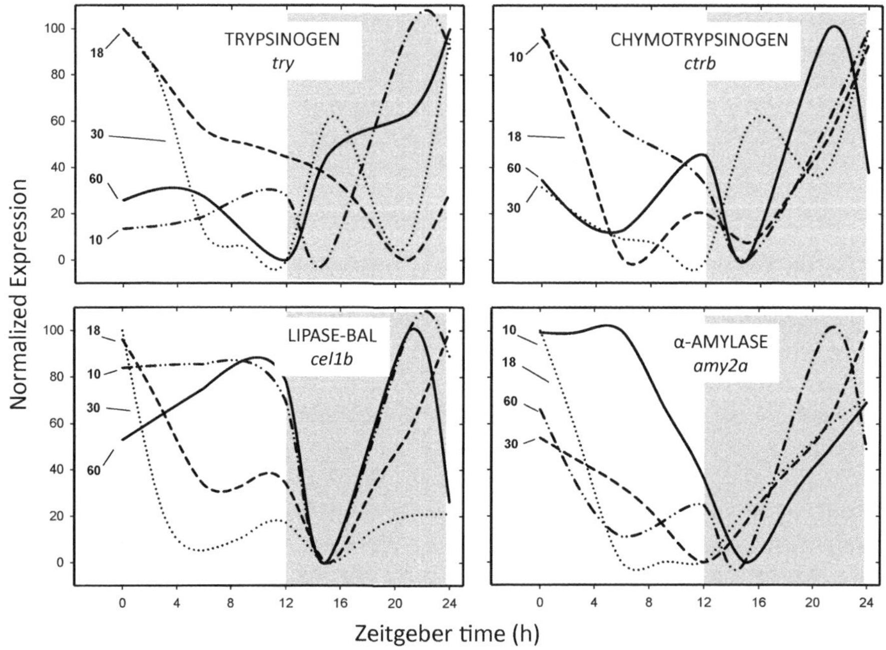

Fig. 3.8 Daily expression of *try*, *ctrb*, *cel1b* and *amy2a* in developing gilthead seabream larvae. To enhance the daily expression patterns, values have been recalculated in the range from 0 to 100 for the minimum and maximum expressions detected for each gene. Numbers inside the graphs indicate the larval age (dph). Grey area indicates the dark period. Modified from Mata-Sotres et al. (2016)

However, due to methodological constrains derived of their small gut size, it is hard to perform routinely this kind of measurements with accuracy. Consequently, daily cycles of gastric pH and pepsin activity have been determined in early juveniles over 1 g size. Empty stomach in most teleostean species presents a neutral pH and pepsin activity is triggered only with the pH decrease produced by the release of acid associated with feeding. Initial values are recovered when the stomach is emptied, being this pattern highly dependent on the feeding regime of the species (Hlophe et al. 2014; Yúfera et al. 2012, 2014). In a general way, the period during which the luminal pH is low enough to allow the pepsin activation from pepsinogen increases with feeding frequency. If a similar response is supposed to take place in weaning larvae, the total cumulative pepsin activity along a day could increase with frequent food supplies. Nevertheless, the development of acidification capacity is a process that may last from few days to several weeks (Yúfera et al. 2004; Yúfera and Darias 2007) as well as the appearance of measurable levels of active pepsin. On the other hand, not all fish species have the same acidification strategy. Some species, as the rainbow trout (Bucking and Wood 2009) or cobia *Rachycentron canadum* (Yúfera et al. 2016) exhibit a permanent gastric acidification irrespectively of the fed or fasting status, although the gastric pH may increase after a meal due to the buffering effect of the food. In these species, the appearance of the stomach functionality and postprandial peptic hydrolysis of proteins have not been examined yet. Therefore, in fish with this mode of gastric digestion, whether or not the actual pepsin activity is subjected to daily cycles in late larvae/early juveniles remains nowadays an open question, though it can be assumed that it would be related to feeding rhythm.

At molecular level, the expression of pepsinogen and gastric proton pump (H^+/K^+-ATPase) shows different patterns in the studied species. In juveniles of white bream *Diplodus sargus* (1–2.5 g) fed a single meal in the morning, both pepsinogen (*pep*) and gastric proton pump (*atp4a*) transcripts showed parallel patterns and were clearly over expressed during the night (Yúfera et al. 2012). The mRNAs are produced and accumulated during the resting period, when the stomach content and the acidic digestion decreased. The expression was lowered to initial values just before the next meal, this indicating that the transcripts have been used to synthesize the pepsinogen required for the forthcoming meal. In gilthead seabream juveniles (5–7 g) not so well established expression patterns were detected (Yúfera et al. 2014), which were dependent on the different feeding protocols. Anyway, although not as evident as in white bream, a progressive increase of pepsinogen expression was observed during the night in fish fed one single meal in the morning. It is obvious that more research is required to elucidate the postprandial pattern of expression of gastric enzymes in larval and juvenile fish. The interpretation of the potential relation between the precursor transcript and the final pepsin activity is even more complicated. Interestingly, in white bream total available pepsinogen (activity determined at pH 2) remains constant during the whole day. This profile would indicate that pepsinogen mRNA is being translated permanently to the corresponding protein in order to maintain a stable amount of the inactive precursor. Therefore, the acidic digestion would be regulated by the activation of the proton pump. Furthermore, more research is still necessary to understand the internal regulation mechanisms of this key process.

These recent findings confirm that feeding and digestion in fish larvae are organized in daily rhythms as a 24 h work, mainly synchronized by the daily illumination cycle but also by internal signaling. In fact, a recent study in gilthead seabream demonstrates that the transcriptome of growing larvae is expressed in four daily successive waves with a very high synchronicity with the cell division cycle and multicellular processes (Yúfera et al. 2017). The larvae seem to have certain ability to adapt to the prevalent feeding conditions in order to optimize their digestive process. These results also suggest that food intake and further digestion by fish larvae requires a resting period even under an artificial environment with continuous feeding and illumination. This is probably necessary to restore the supply of enzymes and other metabolic compounds required for a proper digestion, as well as to synthesize new tissues. In the practice, a dark/light alternation would benefit this process. Some studies on European sea bass and Senegal sole have demonstrated that the best ingestion, best quality of larvae, fastest development and lowest degree of deformities were achieved under the illumination conditions resembling those of their natural aquatic environment (Blanco-Vives et al. 2010, 2012; Villamizar et al. 2009).

Considering all the aforementioned, it is clear that an accurate assessment of digestive enzyme activity and gene expression in larvae along a day cycle cannot be achieved through one single sampling and that more suitable sampling schedules should be designed considering these daily patterns. Clearly, different results can be obtained if we consider the value obtained either from one single sample taken always at the same hour (the most common method), the highest value recorded along the whole day, the average of all daily samples or the cumulative activity during the day. A different value would be obtained for the activity (and gene expression) with each estimation criterion (Fig. 3.9). This situation could be particularly critical for comparing result obtained in different species using different sampling schedules or in a species in which the feeding behavior changes with the progress of the development, as it is the case of some flatfishes. Therefore, a better consensus would be required for future comparisons in a more realistic way than that developed in most studies performed to date.

3.6 Effect of Diet Quality

The limitations of a single daily sample take special relevance when exploring potential effects of food composition on the digestive process. If there is few information of daily patterns, still less information is available on the regulatory mechanisms at molecular level related to diet quality. Contradictory results have been reported in different studies with different species. Cahu et al. (2004), working with European sea bass larvae fed microdiets, found that trypsin activity was modulated at translational level by the protein content of the diet but in older larvae

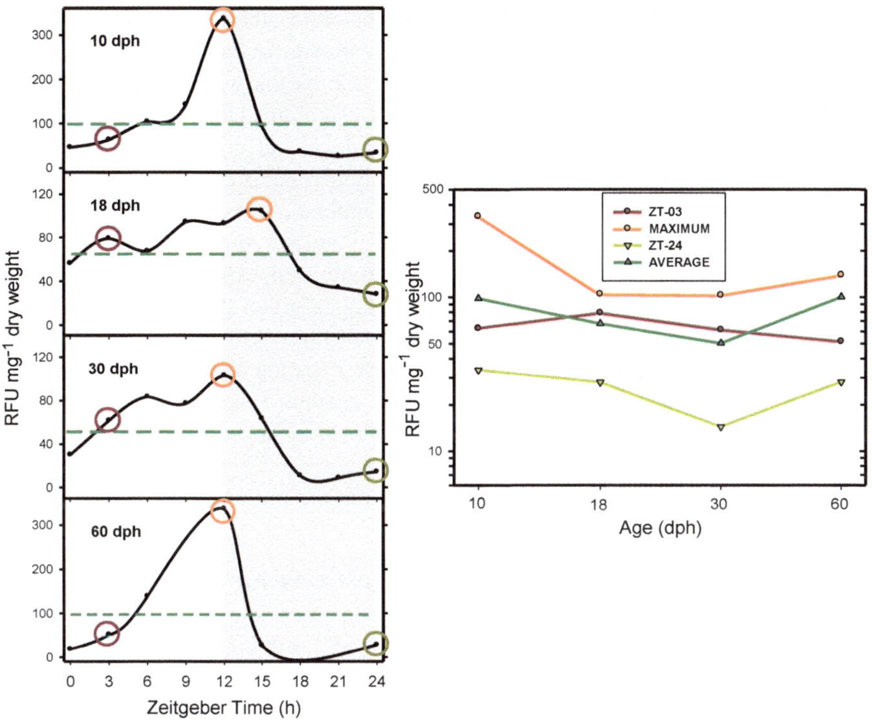

Fig. 3.9 Daily pattern of trypsin activity (left) and comparison of the trypsin activity at different ages using different criteria (right) in gilthead seabream *Sparus aurata* larvae. Four criteria have been compared: **a** one single sample three hours after the feeding start (ZT-03; **b** the maximum activity value recorded during the whole day; **c** one single sample at the end of the night when the gut is empty and minimum activities are recorded (ZT-24); **d** the average of the all daily samples (dashed line). Grey area indicates the dark period

the trypsin activity appeared regulated at transcriptional level. In this species, lipase activity was different depending on the dietary oil sources and larval age but no significant differences were found in the lipase/glyceraldehyde-3-phosphate dehydrogenase mRNA expression (Morais et al. 2004b). Dietary protein level also affected activity and mRNA level of pepsin and trypsin in yellow catfish *Pelteobagrus fulvidraco* larvae but without consistent trends (Wang et al. 2006). On the other hand, Senegal sole early larvae exhibited similar trypsin and chymotrypsin activities when fed different live prey and a microdiet although the expression of *ssetryp1*, one of the six trypsinogen isoforms described for this species, differed among diets (Gamboa-Delgado et al. 2011). In common sole *Solea solea* larvae, mRNA profiles of three trypsinogens, chymotrypsinogen, amylase and hepatic lipase, remained poorly affected by different early weaning protocols onto a commercial microdiet (Parma et al. 2013). In cod larvae, the genes involved in appetite regulation and digestion showed differential expression profiles when fed on different live prey (Kortner et al. 2011a); in this study the differential expression was also observed in older larvae. In gilthead seabream larvae no differences in the trypsin activity and the trypsinogen molecular expression was observed when fed on either live prey or an experimental microdiet, but in this latter case lipase activity was much lower although the gene expression level did not change (Mata et al. 2014). Besides a possible negative effect of this particular experimental microdiet on lipase activity, these results suggest that its regulation or potential inhibition took place at post-transcriptional level. Comparison among these results is difficult because only this last study evaluated the activity and expression over 24 h. In addition to the mentioned limitations for relating a given transcript and the biochemically determined activity of an enzyme, all this variety of results suggest a species-specific component and probably an age-dependent effect in the transcriptional or post-transcriptional regulation of the enzyme function. Only with more complete studies at molecular and biochemical level the actual effect of diet quality will be elucidated.

3.7 Concluding Remarks and Suggestions for Future Research

We have showed here the main achievements of recent research for progressing in our understanding on the digestive function in developing fish larvae. However, there are still many questions to solve and gaps to fill in order to have a complete picture on the complex cascade of events taking place from the first stimuli triggering food intake, followed by ingestion and digestion, until nutrients are absorbed and retained into the larval tissues. All of them are linked processes whose regulation mechanisms are closely interdependent and driven by biotic (food availability) and abiotic (temperature, illumination cycle) environmental factors, but also by internal signaling. New tools and analytical methodologies are contributing to advance in obtaining more realistic and comparable information, but considering the characteristic species-specific variability in fish it is also necessary to expand these studies to other species and taxonomic groups.

Different ontogenetic patterns, precocious, medium and late, have been evidenced in relation to the temporal sequence of enzyme appearance, but it is expected that a more complete picture will be constructed when obtaining information from more species with different feeding habits. In this sense, besides the observed differences in the production of α-amylase, it would be interesting to assess and compare the developmental profiles of expression and activity in carnivorous, omnivorous or herbivorous fish, as well as among those living in marine, estuarine or freshwater environments. It will be particularly attractive to obtain more information from fast growing large pelagic species and those new species that are being incorporated to the list of farmed fish species in the different continents.

Current knowledge of the daily production patterns of enzymes in fish larvae is practically restricted to some pancreatic enzymes, but a similar information on cytosolic and intestinal brush border enzymes is still required. To date, only two species have been studied, being consequently difficult to extrapolate general trends to other species. Moreover, the connection between molecular expression and actual activity is not well understood. During the ontogeny, peaks in molecular expression are sometimes detected a few days prior to those of the corresponding activity, although the opposite has also been described. This may be due either to a greater sensitivity of the molecular techniques or to a different time sequence in the development of the compounds involved in the regulation and activation mechanisms. As explained in Sect. 3.2, probably these discrepancies are a consequence of the partial information available, in addition to the differences in analytical methodologies.

In some species, the activity of enzymes extracts has been characterized in relation to environmental variables, mainly temperature and pH, but the actual response of living larvae faced to changes in water physical-chemical conditions, like temperature and salinity, remains practically unknown. Likewise, it is necessary to advance in the poorly understood effect of diet quality on the regulatory mechanisms and final enzymatic capacity. The study of potential consequences of early nutritional and environmental challenges on later digestive capacity is also still in its early steps. These are critical issues from an applied point of view. One of the main current interests is to elucidate the digestive response when fish larvae are fed on microdiets. To know to what extent fish larvae of different species and at different developmental stages are able to digest formulated inert diets is the basis for improving microparticulation technologies, and for designing suitable species-specific tailored formulations to achieve the best utilization of nutrients.

Acknowledgements This work received national funds through the Spanish Ministry of Economic Affairs and Competitiveness (MINECO) by projects RIDIGEST (AGL2011-23722) and EFISHDIGEST (AGL2014-52888-R) with European Regional Development Fund (FEDER/ERDF) contributions granted to M. Yúfera.

References

Alliot E, Pastoureaud A, Trellu J (1980) Evolution des activities enzymatiques dans le tractus digestif au cour de la vie larvaire de la sole. Variations des proteinogrammes et des zymogrammes. Bioch Syst Ecol 8:441–445

Alvarez-González CA, Cervantes-Trujano M, Tovar-Ramírez D et al (2006) Development of digestive enzymes in California halibut *Paralichthys californicus* larvae. Fish Physiol Biochem 31:83–93

Andrade CAP, Nascimento F, Conceição LEC et al (2012) Red porgy, *Pagrus pagrus*, larvae performance and nutritional condition in response to different weaning regimes. J World Aquacult Soc 43:321–334

Anson ML (1938) The estimation of pepsin, trypsin, papain and cathepsin with hemoglobin. J Gen Physiol 22:79–89

Ao J, Mu Y, Xiang LX et al (2015) Genome sequencing of the perciform fish *Larimichthys crocea* provides insights into molecular and genetic mechanisms of stress adaptation. PLoS Genet 11 (4):e1005118

Asgari R, Rafiee G, Eagderi S et al (2013) Ontogeny of the digestive enzyme activities in hatchery produced Beluga (*Huso huso*). Aquaculture 416–417:33–40

Babaei S, Kenari A, Nazari RE et al (2011) Developmental changes of digestive enzymes in Persian sturgeon (*Acipenser persicus*) during larval ontogeny. Aquaculture 318:138–144

Baragi V, Lovell RT (1986) Digestive enzyme activities in striped bass from first feeding through larva development. Trans Am Fish Soc 115(3):478–484

Blanco-Vives B, Villamizar N, Ramos J et al (2010) Effect of daily thermo- and photo-cycles of different light spectrum on the development of Senegal sole (*Solea senegalensis*) larvae. Aquaculture 306:137–145

Blanco-Vives B, Aliaga-Guerrero M, Cañavate JP et al (2012) Metamorphosis induces a light-dependent switch in Senegalese sole (*Solea senegalensis*) from diurnal to nocturnal behavior. J Biol Rhythm 27:135–144

Blanco A, Planas M, Moyano FJ (2016) Ontogeny of digestive enzymatic capacities in juvenile seahorses *Hippocampus guttulatus* fed on different live diets. Aquac Res 47:3558–3569

Bolasina S, Pérez A, Yamashita Y (2006) Digestive enzymes activity during ontogenetic development and effect of starvation in Japanese flounder, *Paralichthys olivaceus*. Aquaculture 252:503–515

Bucking C, Wood CM (2009) The effect of postprandial changes in pH along the gastrointestinal tract on the distribution of ions between the solid and fluid phases of chime in rainbow trout. Aquacult Nutr 15:282–296

Cahu CL, Zambonino-Infante JL (1994) Early weaning of sea bass (*Dicentrarchus labrax*) larvae with a compound diet: effect on digestive enzyme. Comp Biochem Physiol A 109:213–222

Cahu CL, Zambonino-Infante JL (1995) Maturation of the pancreatic and intestinal digestive functions in sea bass (*Dicentrarchus labrax*): effect of weaning with different protein sources. Fish Physiol Biochem 14:431–437

Cahu C, Rønnestad I, Grangier V et al (2004) Expression and activities of pancreatic enzymes in developing sea bass larvae (*Dicentrarchus labrax*) in relation to intact and hydrolyzed dietary protein; involvement of cholecystokinin. Aquaculture 238:295–308

Cara JB, Moyano FJ, Fauvel C et al (2007) Trypsin and chymotrypsin activities as nutritional indicators of quality in cultured fish larvae. J Fish Biol 70:1798–1808

Conceição LEC, van der Meeren T, Verreth JAJ et al (1997) Amino acid metabolism and protein turnover in larval turbot (*Scophthalmus maximus*) fed natural zooplankton or *Artemia*. Mar Biol 129:255–265

Cousin JCB, Baudin-Laurencin F, Gabaudan J (1987) Ontogeny of enzymatic enzyme activities in fed and fasting turbot, *Scophthalmus maximus* L. J Fish Biol 30:15–33

Cuesta IH, Lahiri K, Lopez-Olmeda JF et al (2014) Differential maturation of rhythmic clock gene expression during early development in medaka (*Oryzias latipes*). Chronobiol Int 31:468–478

Cui Z, Liu Y, Li CP et al (2008) The complete mitochondrial genome of the large yellow croaker, *Larimichthys crocea* (Perciformes, Sciaenidae): unusual features of its control region and the phylogenetic position of the Sciaenidae. Gene 432:33–43

Darias MJ, Murray HM, Gallant JW et al (2006) Characterization of a partial alpha-amylase clone from red porgy (*Pagrus pagrus*): expression during larval development. Comp Biochem Physiol B 143:209–218

Darias MJ, Murray HM, Gallant JW et al (2007) The spatiotemporal expression pattern of trypsinogen and bile salt activated lipase during the larval development of red porgy (*Pagrus pagrus*, Pisces, Sparidae). Mar Biol 152:109–118

Davie A, Sánchez JA, Vera LM et al (2011) Ontogeny of the circadian system during embryogenesis in rainbow trout (*Oncorhynchus Mykyss*) and the effect of prolonged exposure to continuous illumination on daily rhythms of *per1*, *clock*, and *aanat2* Expression. Chronobiol Int 28:177–186

Davidson WS, Koop BF, Jones S et al (2010) Sequencing the genome of the Atlantic salmon (*Salmo salar*). Genome Biol 11(9):403

Dekens MPS, Whitmore D (2008) Autonomous onset of the circadian clock in the zebrafish embryo. EMBO J 27:2757–2765

Douglas SE, Gawlicka A, Mandla S et al (1999) Ontogeny of the stomach in winter flounder: characterization and expression of the pepsinogen and proton pump genes and determination of pepsin activity. J Fish Biol 55:897–915

Douglas SE, Mandla S, Gallant JW (2000) Molecular analysis of the amylase gene and its expression during development in the winter flounder, *Pleuronectes americanus*. Aquaculture 190:247–260

Engrola S, Conceição LEC, Dias L et al (2007) Improving weaning strategies for Senegalese sole: effects of body weight and digestive capacity. Aquac Res 38:696–707

Engrola S, Figueira L, Conceição LEC et al (2009) Co-feeding in Senegalese sole larvae with inert diet from mouth opening promotes growth at weaning. Aquaculture 288:264–272

Farhoudi A, Abedian Kenari AM, Nazari RM et al (2013) Changes of digestive enzymes activity in common carp (*Cyprinus carpio*) during larval ontogeny. Iran J Fish Sci 12(2):320–334

Faulk CK, Holt GJ (2009) Early weaning of southern flounder, *Paralichthys lethostigma*, larvae and ontogeny of selected digestive enzymes. Aquaculture 296:213–218

Felsenstein J (1985) Confidence limits on phylogenies: an approach using the bootstrap. Evolution 39:783–791

Fujii A, Kurokawa Y, Kawai S et al (2007) Diurnal variation of tryptic activity in larval stage and development of proteolytic enzyme activities of malabar grouper (*Epinephelus malabaricus*) after hatching. Aquaculture 270:68–76

Galaviz MA, García-Gasca A, Drawbridge M, Álvarez-González CA, López LM (2011) Ontogeny of the digestive tract and enzymatic activity in white seabass, *Atractoscion nobilis*, larvae. Aquaculture 318:162–168

Galaviz MA, García-Ortega A, Gisbert E et al (2012) Expression and activity of trypsin and pepsin during larval development of the spotted rose snapper *Lutjanus guttatus*. Comp Biochem Physiol Part B 161:9–16

Galaviz MA, López LM, García Gasca A et al (2015) Digestive system development and study of acid and alkaline protease digestive capacities using biochemical and molecular approaches in totoaba (*Totoaba macdonaldi*) larvae. Fish Physiol Biochem 41:1117–1130

Gamboa-Delgado J, Le Vay L, Fernández-Díaz C et al (2011) Effect of different diets on proteolytic enzyme activity, trypsinogen gene expression and dietary carbon assimilation in Senegalese sole (*Solea senegalensis*) larvae. Comp Biochem Physiol B 158:251–258

Gao C, Yang RB, Hu WB et al (2013) Ontogeny of the stomach in yellow catfish (*Pelteobagrus fulvidraco*): detection and quantification of pepsinogen and H+/K+-ATPase gene expression. J Anim Physiol Anim Nutr 97:20–26

Gisbert E, Giménez G, Fernandez I et al (2009) Development of digestive enzymes in common dentex, *Dentex dentex*, during early ontogeny. Aquaculture 287:381–387

Gisbert E, Morais S, Moyano FJ (2013) Feeding and digestion. In: Qin JG (ed) Larval fish culture. Nova Publisher, New York, pp 73–124

Guerreiro I, de Vareilles M, Pousao-Ferreira P et al (2010) Effect of age-at-weaning on digestive capacity of white seabream (*Diplodus sargus*). Aquaculture 30:194–205

Hansen TW, Folkvord A, Grøtan E et al (2013) Genetic ontogeny of pancreatic enzymes in *Labrus bergylta* larvae and the effect of feed type on enzyme activities and gene expression. Comp Biochem Physiol Part B 164:176–184

Hauville MR, Zambonino-Infante JL, Bell G et al (2014) Impacts of three different microdiets on Florida Pompano, *Trachinotus carolinus*, weaning success, growth, fatty acid incorporation and enzyme activity. Aquaculture 422–423:268–276

He T, Xiao Z, Liu Q et al (2012) Ontogeny of the digestive tract and enzymes in rock bream (*Oplegnathus fasciatus*) (Temminck et Schlegel 1844) larvae. Fish Physiol Biochem 38:297–308

Hjelmeland K, Pedersen BH, Nilssen EM (1988) Trypsin content in intestines of herring larvae, *Clupea harengus*, ingesting inert polystyrene spheres or live crustacea prey. J Mar Biol 98:331–335

Hlophe SN, Moyo NAG, Ncube I (2014) Postprandial changes in pH and enzyme activity from the stomach and intestines of *Tilapia rendalli* (Boulenger, 1897), *Oreochromis mossambicus* (Peters, 1852) and *Clarias gariepinus* (Burchell, 1822). Appl Ichthyol 30:35–41

Houde ED (1989) Comparative growth, mortality, and energetics of marine fish larvae: temperature and implied latitudinal effects. Fish Bull US 87:471–495

Howe K, Clark MD, Torroja CF et al (2013) The zebrafish reference genome sequence and its relationship to the human genome. Nature 496(7446):498–503

Hurst CD, Bartlett SE, Davidson WS et al (1999) The complete mitochondrial DNA sequence of the Atlantic salmon, *Salmo salar*. Gene 239:237–242

Ishiguro NB, Miya M, Nishida M (2003) Basal euteleostean relationships: a mitogenomic perspective on the phylogenetic reality of the "Protacanthopterygii". Mol Phylogenet Evol 27:476–488

Jiménez-Martínez LD, Álvarez-González CA, Tovar-Ramírez D et al (2012) Digestive enzyme activities during early ontogeny in Common snook (*Centropomus undecimalis*). Fish Physiol Biochem 38:441–454

Kortner TM, Overrein I, Øie G et al (2011a) The influence of dietary constituents on the molecular ontogeny of digestive capability and effects on growth and appetite in Atlantic cod larvae (*Gadus morhua*). Aquaculture 315:114–120

Kortner TM, Overrein I, Øie G et al (2011b) Molecular ontogenesis of digestive capability and associated endocrine control in Atlantic cod (*Gadus morhua*) larvae. Comp Biochem Physiol A 160:190–199

Kotani T, Fushimi H (2011) Determination of appropriate feeding schedules from diel feeding rhythms in finfish larviculture. Aquaculture 315:104–113

Kumar S, Stecher G, Tamura K (2016) MEGA7: molecular evolutionary genetics analysis version 7.0 for bigger datasets. Mol Biol Evol 33:1870–1874

Lazo JP, Darias MJ, Gisbert E (2011) Ontogeny of the digestive tract. In: Holt GJ (ed) Larval fish nutrition. Wiley-Blackwell, West Sussex, pp 3–46

Lien S, Koop BF, Sandve SR et al (2016) The Atlantic salmon genome provides insights into rediploidization. Nature 533(7602):200–205

López-Olmeda JF (2017) Nonphotic entrainment in fish. Comp Bioch Physiol Part A 203:133–143

López-Ramírez G, Cuenca-Soria CA, Álvarez-González CA et al (2011) Development of digestive enzymes in larvae of Mayancichlid *Cichlasoma urophthalmus*. Fish Physiol Biochem 37:197–208

Ma AJ, Liu XZ, Xu YJ et al (2006) Feeding rhythm and growth in the tongue sole, *Cynoglossus semilaevis* Gunther, during its early life stages. Aquac Res 37:586–593

MacKenzie BR, Ueberschär B, Basford D et al (1999) Diel variability of feeding activity in haddock (*Melanogrammus aeglifinus*) larvae in the East Shetland area, North Sea. Mar Biol 135:361–368

Martín-Robles AJ, Whitmore D, Pendón C et al (2013) Differential effects of transient constant light-dark conditions on daily rhythms of period and clock transcripts during Senegalese sole metamorphosis. Chronobiol Int 30:699–710

Marx MC, Wood M, Jarvis SC (2001) A microplate fluorimetric assay for the study of enzyme diversity in soils. Soil Biol Biochem 33:1633–1640

Mata JA, Moyano FJ, Martínez-Rodríguez G et al (2014) Effect of feeding frequency on digestive function of gilthead seabream (*Sparus aurata*) larvae fed on microdiet. In: Aquaculture Europe 2014, San Sebastián, Spain, 14–17 Oct 2014, pp 794–795 (book of abstracts)

Mata-Sotres JA, Moyano FJ, Martínez-Rodríguez G et al (2016) Daily rhythms of digestive enzyme activity and gene expression in gilthead seabream (*Sparus aurata*) during ontogeny. Comp Biochem Physiol Part A 197:43–51

Mata-Sotres JA, Martínez-Rodríguez G, Pérez-Sánchez J et al (2015) Daily rhythms of clock gene expression and feeding behaviour during the development in gilthead seabream, *Sparus aurata*. Chronobiol Int 32:1061–1074

Martínez I, Moyano FJ, Fernández-Díaz C et al (1999) Digestive enzyme activity during larval development of Senegal sole (*Solea senegalensis*). Fish Physiol Biochem 21:317–323

Mazurais D, Coves D, Papandroulakis N et al (2015) Gene expression pattern of digestive and antioxidant enzymes during the larval development of reared Atlantic bluefin tuna (ABFT), *Thunnus thynnus* L. Aquac Res 46:2323–2331

Moguel-Hernández I, Peña R, Andree KB et al (2016) Ontogeny changes and weaning effects in gene expression patterns of digestive enzymes and regulatory digestive factors in spotted rose snapper (*Lutjanus guttatus*) larvae. Fish Physiol Biochem 42:1319–1334

Montoya A, López-Olmeda JF, Yúfera M et al (2010) Feeding time synchronizes daily rhythms of behavior and digestive physiology in gilthead seabream (*Sparus aurata*). Aquaculture 306:315–321

Montoya-Mejía M, Rodríguez-González H, Nolasco-Soria H (2016) Circadian cycle of digestive enzyme production at fasting and feeding conditions in Nile tilapia, *Oreochromis niloticus* (Actinopterygii: Perciformes: Cichlidae). Acta Ichthyol Piscat 46:163–170

Morais S, Lacuisse M, Conceição LEC et al (2004a) Ontogeny of the digestive capacity of Senegalese sole (*Solea senegalensis*), with respect to digestion, absorption and metabolism of amino acids from *Artemia*. Mar Biol 145:243–250

Morais S, Cahu C, Zambonino-Infante JL et al (2004b) Dietary TAG source and level affect performance and lipase expression in larval sea bass (*Dicentrarchus labrax*). Lipids 39:449–458

Moyano FJ, Díaz M, Alarcón FJ et al (1996) Characterization of digestive enzyme activity during larval development of gilthead seabream (*Sparus aurata*). Fish Physiol Biochem 15:121–130

Murashita K, Furuita H, Matsunari H et al (2013) Partial characterization and ontogenetic development of pancreatic digestive enzymes in Japanese eel *Anguilla japonica* larvae. Fish Physiol Biochem 39:895–905

Murashita K, Matsunari H, Kumon K et al (2014) Characterization and ontogenetic development of digestive enzymes in Pacific bluefin tuna *Thunnus orientalis* larvae. Fish Physiol Biochem 40:1741–1755

Murray HM, Gallant JW, Perez-Casanova JC et al (2003) Ontogeny of lipase expression in winter flounder. J Fish Biol 62:816–833

Murray HM, Gallant JW, Johnson SC et al (2006) Cloning and expression analysis of three digestive enzymes from Atlantic halibut (*Hippoglossus hippoglossus*) during early development: predicting gastrointestinal functionality. Aquaculture 252:394–408

Navarro-Guillén C, Moyano FJ, Yúfera M (2015) Diel food intake and digestive enzyme production patterns in *Solea senegalensis* larvae. Aquaculture 435:33–42

Navarro-Guillén C, Rønnestad I, Jordal A-EO et al (2017) Involvement of cholecystokinin (CCK) in the daily pattern of gastrointestinal regulation of Senegalese sole (*Solea senegalensis*) larvae reared under different feeding regimes. Comp Biochem Physiol Part A 203:126–132

Nguyen HQ, Reinertsen H, Wold P-A et al (2011) Effects of early weaning strategies on growth, survival and digestive enzyme activities in cobia (*Rachycentron canadum* L.) larvae. Aquacult Int 19:63–78

Østergaard P, Munk P, Janekarn V (2005) Contrasting feeding patterns among species of fish larvae from the tropical Andaman Sea. Mar Biol 146:595–606

Parma L, Bonaldo A, Massi P et al (2013) Different early weaning protocols in common sole (*Solea solea* L.) larvae: implications on performances and molecular ontogeny of digestive enzyme precursors. Aquaculture 414–415:26–35

Peres A, Zambonino Infante JL, Cahu CL (1998) Dietary regulation of activities and mRNA levels of trypsin and amylase in seabass (*Dicentrarchus labrax*) larvae. Fish Physiol Biochem 19:145–152

Pedersen BH, Hjelmeland K (1988) Fate of trypsin and assimilation efficiency in larval herring (*Clupea harengus*) following digestion of copepods. Mar Biol 97:467–476

Pérez-Casanova JC, Murray HM, Gallant JW et al (2006) Development of the digestive capacity in larvae of haddock (*Melanogrammus aeglefinus*) and Atlantic cod (*Gadus morhua*). Aquaculture 251:377–401

Ribeiro L, Zambonino-Infante JL, Cahu C et al (1999) Development of digestive enzymes in larvae of *Solea senegalensis*, Kaup 1858. Aquaculture 179:465–473

Rodrigues JB, Leitão NJ, Chaves KS et al (2014) High protein microparticles produced by ionic gelation containing *Lactobacillus acidophilus* for feeding pacu larvae. Food Res Int 63:25–32

Rondeau EB, Minkley DR, Leong JS et al (2014) The genome and linkage map of the northern pike (*Esox lucius*): conserved synteny revealed between the salmonid sister group and the Neoteleostei. PLoS ONE 9(7):e102089

Rønnestad I, Pérez Domínguez R, Tanaka M (2000) Ontogeny of digestive tract functionality in Japanese flounder, *Paralichthys olivaceus* studied by in vivo microinjection: pH and assimilation of free amino acids. Fish Physiol Biochem 22:225–235

Rønnestad I, Yúfera M, Ueberschär B et al (2013) Feeding behaviour and digestion physiology in larval fish—current knowledge, and gaps and bottlenecks in research. Rev Aquacult 5(Suppl. 1):S59–S98

Ruan GL, Li Y, Gao ZX, Wang HL et al (2010) Molecular characterization of trypsinogens and development of trypsinogen gene expression and tryptic activities in grass carp (*Ctenopharyngodon idellus*) and topmouth culter (*Culter alburnus*). Comp Biochem Physiol Part B 155:77–85

Sahlmann C, Gu J, Kortner TM et al (2015) Ontogeny of the digestive system of Atlantic salmon (*Salmo salar* L.) and effects of soybean meal from start-feeding. PLoS ONE 10(4):e0124179

Saitou N, Nei M (1987) The neighbor-joining method: a new method for reconstructing phylogenetic trees. Mol Biol Evol 4:406–425

Saleh R, Betancor MB, Roo R et al (2013) Optimum soybean lecithin contents in microdiets for gilthead seabream (*Sparus aurata*) larvae. Aquacult Nutr 19:585–597

Santigosa E, Sánchez J, Médale F et al (2008) Modifications of digestive enzymes in trout (*Oncorhynchus mykiss*) and sea bream (*Sparus aurata*) in response to dietary fish meal replacement by plant protein sources. Aquaculture 282:68–74

Schaefer FJ, Flues S, Meyer S et al (2017) Inter- and intra-individual variability in growth and food consumption in pikeperch, *Sander lucioperca* L., larvae revealed by individual rearing. Aquac Res 48:800–808

Shan XJ, Huang W, Cao L et al (2009) Ontogenetic development of digestive enzymes and effect of starvation in miiuy croaker *Miichthys miiuy* larvae. Fish Physiol Biochem 35:385–398

Shoji J, Maehara T, Aoyama M et al (2001) Daily ration of Japanese Spanish mackerel *Scomberomorus niphonius* larvae. Fish Sci 67:238–245

Shoji J, Maehara T, Tanaka M (1999) Diel vertical movement and feeding rhythm of Japanese mackerel larvae in the central Seto inland sea. Fish Sci 65:726–730

Sinsabaugh RL, Findlay S, Franchini P et al (1997) Enzymatic analysis of riverine bacterioplankton production. Limnol Oceanogr 42(1):29–38

Skoog DA, West DM, Holler FJ (1995) Química Analítica. McGraw Hill Interamericana ed., México

Solovyev M, Campoverde C, Öztürkd C et al (2016) Morphological and functional description of the development of the digestive system in meagre (*Argyrosomus regius*): an integrative approach. Aquaculture 464:381–391

Srichanun M, Tantikitti C, Utarabhand P et al (2013) Gene expression and activity of digestive enzymes during the larval development of Asian seabass (*Lates calcarifer*). Comp Biochem Physiol Part B 165:1–9

Suzer C, Fırat K, Saka Ş (2006) Ontogenic development of the digestive enzymes in common pandora, *Pagellus erythrinus* L. larvae. Aquac Res 37:1565–1571

Suzer C, Aktülün S, Çoban D et al (2007a) Digestive enzyme activities in larvae of sharpsnout seabream (*Diplodus puntazzo*). Comp Biochem Physiol A 148:470–477

Suzer C, Kamacı HO, Çoban D et al (2007b) Digestive enzyme activity of the red porgy (*Pagrus pagrus*, L.) during larval development under culture conditions. Aquac Res 38:1178–1785

Tine M, Kuhl H, Gagnaire PA et al (2014) European sea bass genome and its variation provide insights into adaptation to euryhalinity and speciation. Nat Commun 5:5770
Toledo-Cuevas M, Moyano FJ, Tovar D et al (2011) Development of digestive biochemistry in the initial stages of three cultured Atherinopsids. Aquac Res 42:776–786
Ueberschär BFR (1988) Determination of the nutritional condition of individual marine fish larvae by analyzing their proteolytic enzyme activities with a highly sensitive fluorescence technique. Meeresforsch 32:144–154
Ueberschär B (1993) Measurement of proteolytic enzyme activity: significance and application in larval fish research. In: Walther BT, Fhyn HJ (eds) Physiological and biochemical aspects of fish development. University of Bergen, Norway, pp 233–239
Ueberschär B, Pedersen BH, Hjelmeland K (1992) Quantification of trypsin with a radioimmunoassay in herring larvae (*Clupea harengus*) compared with a highly sensitive fluorescence technique to determine tryptic enzyme activity. Mar Biol 13:468–473
Vera LM, Negrini P, Zagatti C et al (2013) Light and feeding entrainment of the molecular circadian clock in a marine teleost (*Sparus aurata*). Chronobiol Int 30:649–661
Villamizar N, Blanco-Vives B, Migaud H et al (2011) Effects of light during early larval development of some aquacultured teleosts: a review. Aquaculture 315:86–94
Villamizar N, García-Alcazar A, Sánchez-Vázquez FJ (2009) Effect of light spectrum and photoperiod on the growth, development and survival of European sea bass (*Dicentrarchus labrax*) larvae. Aquaculture 292:80–86
Wang Y, Hu M, Wang W et al (2008) Transpositional feeding rhythm of loach *Misgurnus anguillicaudatus* from larvae to juveniles and its ontogeny under artificial rearing conditions. Aquacult Int 16:539–549
Wang C, Xie S, Zhu X, Lei W, Yang Y, Liu J (2006) Effects of age and dietary protein level on digestive enzyme activity and gene expression of Pelteobagrus fulvidraco larvae. Aquac 254 (1–4):554–562
Whitmore D, Foulkes NS, Sassone-Corsi P (2000) Light acts directly on organs and cells in culture to set the vertebrate circadian clock. Nature 404:87–91
Yoseda K, Yamamoto K, Asami K et al (2008) Influence of light intensity on feeding, growth, and early survival of leopard coral grouper (*Plectropomus leopardus*) larvae under mass-scale rearing conditions. Aquaculture 279:55–62
Yúfera M (2016) Intake and digestion of food by fish larvae and juveniles: a strategy planned over twenty-four hours. World Aquacult Mag 47(2):28–31
Yúfera M, Darias MJ (2007) Changes in the gastrointestinal pH from larvae to adult in Senegal sole (*Solea senegalensis*). Aquaculture 267:94–99
Yúfera M, Fernández-Díaz C, Vidaurreta A et al (2004) Gastrointestinal pH and development of the acid digestion in larvae and early juveniles of *Sparus aurata* L. (Pisces: teleostei). Mar Biol 144:863–869
Yúfera M, Fernández-Díaz C, Pascual E (1995) Feeding rates of gilthead seabream, *Sparus aurata*, larvae on microcapsules. Aquaculture 134:257–268
Yúfera M, Moyano FJ, Astola A et al (2012) Acidic digestion in a teleost: postprandial and circadian pattern of gastric pH, pepsin activity, and pepsinogen and proton pump mRNAs expression. PLoS ONE 7(3):e33687
Yúfera M, Romero MJ, Pujante IM et al (2014) Effect of feeding frequency on the daily rhythms of acidic digestion in a teleost fish (gilthead seabream). Chronobiol Int 31:1024–1033
Yúfera M, Nguyen MV, Engrola S et al (2016) Cobia exhibits a permanent gastric acidity as digestion strategy. In: Aquaculture Europe 2016, Edinburgh, Scotland, UK, 20–23 Sept 2016
Yúfera M, Perera E, Mata-Sotres JA et al (2017) The circadian transcriptome of marine fish (*Sparus aurata*) larvae reveals highly synchronized biological processes at the whole organism level. Sci Rep 7:12943
Zambonino-Infante JL, Cahu C (1994) Development and response to a diet change of some digestive enzymes in sea bass (*Dicentrarchus labrax*) larvae. Fish Physiol Biochem 12:399–408

Zeng LQ, Fu SJ, Li XM et al (2014) Physiological and morphological responses to the first bout of refeeding in southern catfish (*Silurus meridionalis*). J Comp Physiol B 184:329–346
Zeytin S, Schulz C, Ueberschär B (2016) Diurnal patterns of tryptic enzyme activity under different feeding regimes in gilthead sea bream (*Sparus aurata*) larvae. Aquaculture 457:85–90
Zuckerkandl E, Pauling L (1965) Evolutionary divergence and convergence in proteins. In: Bryson V, Vogel HJ (eds) Evolving genes and proteins. Academic Press, New York, pp 97–166

Chapter 4
Variability in Digestive Enzyme Capacity in Early Stages of Marine Fish Larvae: Ontogenetic Variations, Biorhythms, Hormonal Control and Nutrient Sensing Mechanisms

Bernd Ueberschär, Carmen Navarro-Guillén, Ana Gomes, Ivar Rønnestad, Carlos Rojas-Garcia, Inken Hanke, Dagh Sommerfeld and Robert Tillner

Abstract In recent years, no substantial progress has been achieved to significantly improve survival rates of cultured marine fish larvae and make the viability in offspring production more predictable. Improving survival and quality of marine larvae in aquaculture industry beyond the present status requires consideration of nutritional quality of the feed but also an in-depth knowledge of larval physiology to account for biorhythms in digestion capacity, appetite and regulation of food ingestion and endocrine control of digestion processes associated to nutrient sensing and feeding activity. Since gut hormones are most likely stimulated and released in response to specific nutrients it is crucial to study the nutrient sensing mechanisms also in the gastrointestinal tract of fish larvae and the importance of nutrients as they are not only substrates, but active regulators of physiological processes. This chapter focuses on the importance of the daily and circadian rhythms in the digestion capacity in larval stages of fish, emphasising the potential variability of enzyme secretion during the day, addresses the specific role of the gut hormones cholecystokinin and ghrelin on appetite and digestive functions in fish larvae and provides a comprehensive consideration of the food sensing mechanisms in the gastrointestinal tract of fish larvae associated to specific nutrients. Knowledge

B. Ueberschär (✉) · D. Sommerfeld · R. Tillner
Gesellschaft für marine Aquakultur, MBH, (GMA), Büsum, Germany
e-mail: ueberschaer@gma-buesum.de

C. Navarro-Guillén
Instituto de Ciencias Marinas de Andalucía (ICMAN-CSIC), Cádiz, Spain

A. Gomes · I. Rønnestad · C. Rojas-Garcia
Department of Biology, University of Bergen, Bergen, Norway

I. Hanke
Helmholtz Centre for Polar and Marine Research (AWI), Alfred Wegener Institute, Bremerhaven, Germany

M. Yúfera (ed.), *Emerging Issues in Fish Larvae Research*,
https://doi.org/10.1007/978-3-319-73244-2_4

in these research areas, through larval ontogeny, will be specifically beneficial for the development of stage-specific micro diets and replacement of live feed with formulated feed. Recommendations on future research aspects and approaches within these themes are discussed and suggested in the specific sections.

Keywords Digestion · Nutrient sensing · Hormonal control · Enzyme capacity Circadian rhythms · Appetite and ingestion

4.1 Assessment of the Circadian Rhythm in Digestive Capacity of Larval Fish

4.1.1 Introduction

With only few exceptions, all organisms have evolved under predictable daily cycles connected to the natural-day-and-night-rhythm. The advantage for the organisms that anticipate such environmental cycles has driven the evolution of endogenous circadian rhythms that tune internal physiology to external conditions.

A circadian rhythm is any biological process that displays an endogenous, untrainable oscillation of about 24 h. These 24-h rhythms are driven by circadian clocks, and have been widely observed in plants, animals, and even fungi and cyanobacteria (Edgar et al. 2012). The term circadian comes from the Latin "circa", meaning "around" (or "approximately"), and diēm, meaning "day". Processes with 24-h oscillations are generally called daily or diel rhythms; strictly speaking, they should not be called circadian rhythms unless their endogenous nature is confirmed (Vitaterna et al. 2001).

Biological clocks are genetically encoded oscillators that allow organisms to anticipate predictable changes in the environmental conditions that are tied to the rotation of Earth. Biological clocks in general enhance fitness and growth and they are expressed throughout the central nervous system (CNS) and peripheral tissues of multicellular organisms in which they mainly influence behaviour, feeding and metabolism. Although circadian rhythms are endogenous ("built-in", self-sustained), they can be adjusted (entrained) to the local environmental conditions by external cues called "zeitgebers" (from German, "time giver"), which include mainly light-dark cycle and temperature (Bass 2012).

In general, circadian rhythms in mammals are easily classifiable due to their stability, with active and resting phases that include distinct sleeping and feeding patterns. It is well recognized that disruption of this rhythm usually has significant adverse health effects, particularly in the long term. Most physiological and behavioural functions show rhythms orchestrated by a biological clock, which enables animals to anticipate cyclic events in their environment and mount an appropriate response (Pittendrigh 1993).

Concluding from the introduction above, it is well accepted that circadian (or daily) rhythms widely exists and this is also confirmed in adult fish. It is clear that most fish do not feed continuously during 24 h, but display particular day/night rhythmic patterns. Although a few species that consume low-energy food, such as grass carp *Ctenopharyngodon idella*, may need to forage almost continuously to obtain the energy they need, distinct feeding rhythms are widespread in fish. Several reviews of rhythmicity in fish reported in a number of species the existence of consistent patterns, although in some cases with diurnal fish becoming nocturnal and vice versa at a certain time (Ali 1992; Thorpe 1978). In that context, there are very few studies of resting or sleeping habits of fish throughout a long period. One rare example is a study on Mozambique tilapia *Oreochromis mossambicus* that concluded that juvenile tilapia does not show any sign of sleep, at any time. In these experiments, the young fish took 22 weeks to develop the adult sleep patterns. Yet, it is unknown if such lack of sleep in the early stages of life is commonplace among fishes (Shapiro and Hepburn 1976; Reebs 1991, 2011).

Biorhythms may have obviously significant consequences on feeding regimes applied to the various species under aquaculture conditions. Specifically, the regulation of feed intake is very complex and involves interaction among the circadian and homeostatic control systems within the CNS, the gastrointestinal tract (GIT) and the environment. The hypothalamus, which receives, integrates and transmits relevant internal and external signals, is recognized as the primary centre of regulation of feed intake. The neuroendocrine factors that originate from the hypothalamus either stimulate or inhibit feed intake. Farmed and wild fish show no differences in neurohormones that regulate feed intake, but they meet different challenges (da Silva et al. 2016; Kulczykowska and Sánchez-Vázquez 2010). Finding the best time to deliver food is not an easy task, as fish may change their feeding behaviour at certain times and age (Sánchez-Vázquez et al. 1995a). Nevertheless, the study of feeding rhythms and feeding anticipatory activity may provide useful information to design improved feeding schedules that match fish appetite.

Comparably little is known about these issues in the early stages of finfish. In finfish larviculture, feeding regimes and schedules vary with hatchery and species; however, they have mostly no biological or technological foundation and are dependent on operators and previous experiences. Inadequate feeding regimes and/or inappropriate food intake, especially during early larval stages, may result in a decline in health and/or quality and high mortalities. As described above, fish have species-specific diel feeding rhythms; and therefore, feeding schedules in larviculture should not be determined uniformly among target species. To improve survival and quality, it is important to establish feeding schedules corresponding to the diel feeding rhythms in the larval stages of various fish species (Kotani and Fushimi 2011).

In commercial larval fish rearing, specifically in marine species, it is common to apply a prolonged light period beyond natural conditions, or even 24 h of light combined with frequent feeding events. The hypothesis is that larvae can feed all day around as long as light and sufficient food is available. However, these conditions are mostly in contrast to natural conditions, where fish larvae show a pronounced circadian rhythm, with an alternation in resting and active feeding periods and it has hardly been verified whether the currently used diel feeding schedules in commercial larval production are appropriate. Moreover, growth rates under natural environmental conditions are often superior to the results under culture conditions, although feeding activities are restricted to a relatively short period per day. Previous studies have reported that larvae do not display constant food ingestion under natural or laboratory conditions but have rather diel feeding rhythms (Boujard and Leatherland 1992; Dou et al. 2000; Mata-Sotres et al. 2015; Navarro-Guillén et al. 2015: Okauchi et al. 1980; Picapedra et al. 2015; Shoji et al. 1999; Ueberschär 1995; Yamamoto 1996; Yamamoto et al. 2003, 2005). Non-natural photoperiods, especially continuous light, is considered to be rather stressful and in the long term can even cause life-threatening conditions. In fact, stress symptoms such as hyperinflation of the swim bladder in larval rearing are often related to an improperly adjusted light regime, and a common advice in this situation is to change the light regime (Büke 2002; Gisbert et al. 2014).

Apparently, 24 h availability of light and feed is not necessarily an advantage. Yamamoto et al. (2003) published results on the feeding rhythm of some common species in aquaculture, and in summary, they discovered that the different species have different feeding preferences, concerning the time of the day. In these experiments, a day and night cycle was applied including the option to feed, thus the larvae were able to choose among the different conditions. This is not the case if 24 h of light are applied. Therefore, feeding schedules should be adapted to the diel feeding rhythm and digestive capacity and changes according to larval growth.

In contrast to the juvenile and adult stages, the digestive capacity of larvae is limited because of the immature condition of the digestive tract. The low digestive capacity of the early stages of altricial larvae is mainly owing to the lack of a functional stomach (Govoni et al. 1986) which increases the efficiency in digestion tremendously and normally develops only close to the metamorphosis. This has significant consequences on the ability to digest preyed items in the early stages. Even a high density of food available to the larvae does not consequently result in a good nutritional condition. Continuous supply of feeding specifically with live feed can even have adverse effects; Werner and Blaxter (1981) noted in experiments on gut evacuation time in Atlantic herring *Clupea harengus* larvae that almost intact *Artemia* nauplii were defecated at higher prey densities, and this was confirmed by own observations in younger stages of marine larvae. From these and other observations, it is obvious that the degree of gut fullness is not necessarily a good indicator for larval digestive capacity and efficiency, but other indicators are necessary to monitor the "true" digestive capacity.

Fast growth requires high digestive and metabolic enzyme activities to allow for fast conversion of ingested prey to somatic tissue. Therefore, enzyme activity has been used as an indicator of nutritional condition for marine fish larvae in field studies and was proven to be a suitable tool to evaluate the proteolytic digestive capacity under various feeding conditions (summary on related indicators in Ferron and Leggett 1994; Rønnestad et al. 2013; Ueberschär 1995). The monitoring of the proteolytic enzyme activity, namely the tryptic activity, in the gut of individual fish larvae to investigate their digestive processes in relation to feeding activity was found to be an appropriate indicator (Lauff and Hofer 1984; Pedersen et al. 1990; Rønnestad et al. 2013). Ueberschär (1988, 1993, 1995) was the first to apply a sensitive method to monitor the "true" proteolytic digestive capacity in relation to food supply, type of food and time of the day in individual larvae. Measurement of tryptic enzyme activity in relation to food quality, quantity and time of the day individually was applied subsequently in various experiments. For example, Drossou et al. (2006) determined the effects of two test diets on tryptic activity during early stages of Nile tilapia *Oreochromis niloticus*. Tillner et al. (2013a) evaluated the diurnal pattern of tryptic enzyme activity of Atlantic cod *Gadus morhua* larvae under various feeding protocols. Navarro-Guillén et al. (2015) examined diel food intake and tryptic enzyme activity of Senegalese sole *Solea senegalensis* larvae under laboratory conditions. Mata-Sotres et al. (2016) identified the daily pattern of tryptic activity in gilthead seabream *Sparus aurata* during ontogeny.

The sections below present and compare an assemblage of results about tryptic enzyme activity measurements which were applied to monitor the daily proteolytic digestive capacity in naturally grown fish larvae and in fish larvae of various species reared under different feeding regimes.

4.1.2 Circadian Feeding Rhythms Under Natural Conditions

As previously described, it is a common phenomenon that under natural environmental conditions circadian rhythmicity in feed intake is already existing in the larval and early juvenile stages of fish (Boujard and Leatherland 1992; Ueberschär 1995). Under natural conditions, fish larvae of most species have a strong daily feeding and resting rhythm. In the marine environment, the larval stages of most species use to feed during natural daylight, dusk or dawn and rest in the darkness. This behaviour is mostly linked to a vertical migration of marine larval fish (comprehensive article from Picapedra et al. 2015; Sclafani et al. 1993) and is well documented.

The major reason for the vertical migration is probably the access to prey (during day, dusk and dawn), to gulp air (Forward et al, 1996; Hoss et al. 1989) and to avoid predators (in the night). Most fish larvae are visual feeders and stop to feed in the night, approaching upper layers in the sea in order to rest. In that context, a post-feeding thermotaxis has been discussed with larvae migrating to the warmer surface water layers to increase the efficiency of digestion processes and eventually enhancing the growth rate (Wurtsbaugh and Neverman 1988).

Trypsin is the most important proteolytic enzyme in growing fish larvae and tryptic enzyme activity measurements are appropriate to detect actual feeding activities and digestion processes (Ueberschär 1995) and were applied in some field studies to evaluate for e.g. the daily feeding rhythm of sardine *Sardina pilchardus* and sprat *Sprattus sprattus* larvae (Figs. 4.1 and 4.2). No significant differences in the mean length distribution was found between larvae sampled from the same cohort at a 48 h-drift station in the Atlantic Ocean (Sardine larvae, Fig. 4.1) and at a 24 h-drift station in the North Sea (Sprat larvae, Fig. 4.2), respectively. Diel pattern in tryptic enzyme activity is therefore most likely considered as a reaction to feed ingestion. Oscillations of tryptic enzyme activity can be considered as a consequence of periodic feed ingestion. According to these data, the main feeding time is supposed to be in the dusk. Main digestion periods are indicated by the highest enzyme activity values which were found from midnight until the early morning.

The results clearly demonstrate a rhythmicity in feeding activities under natural conditions. Moreover, numerous investigations (e.g., Houde 1987) about the growth performance of fish larvae under natural conditions have shown that growth rates are as fast as under laboratory conditions or even better (comparing the same species).

These considerations illustrate, that apparently it is not necessarily required to feed larvae under hatchery conditions for periods which exceeds natural photoperiods by far, i.e., 24 h of light and continuous feeding in the extreme. Therefore,

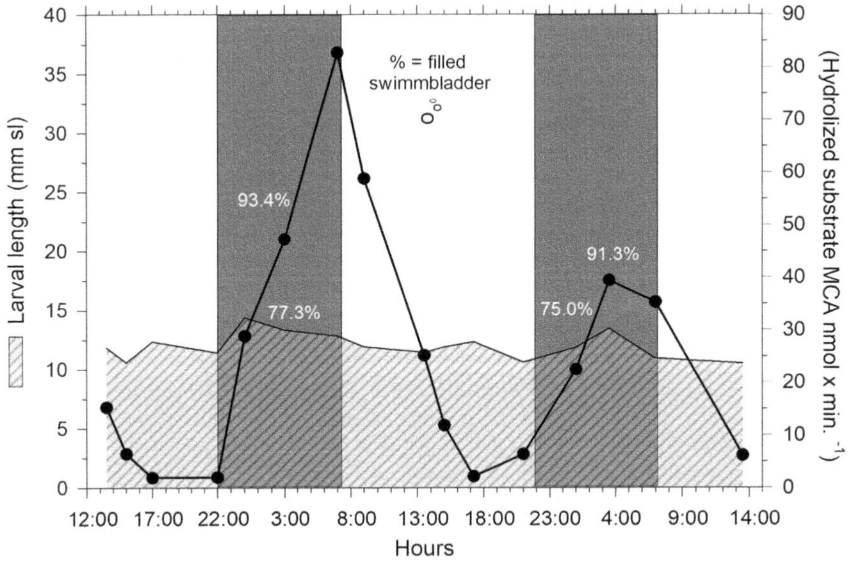

Fig. 4.1 Daily pattern of tryptic enzyme activity levels (nmol MCA $\min^{-1}$ $larva^{-1}$) for sardine larvae (*Sardina pilchardus*) sampled at a 48 h-drift station in the Atlantic Ocean off the Spanish northwest coast in May. Data points are means from 10 to 18 sardine larvae individually measured. The light grey shaded area depicts the mean size distribution of the larvae in the samples. The dark shaded areas indicate periods with no daylight. The percentage of larvae with inflated gas bladders are indicated at the corresponding sampling time. In four of the 16 samples taken on the drift station, a high percentage of larvae with inflated gas bladder were observed in the night (between 12:00 a.m. and 3:30 a.m.) when larvae were located close to the surface. The background for that periodic inflation of the gas bladder is not yet clear (Hoss et al. 1989), it is suggested that a balanced buoyancy allows the larvae to "rest" close to the surface in order to preserve energy which can be invested into digestion processes

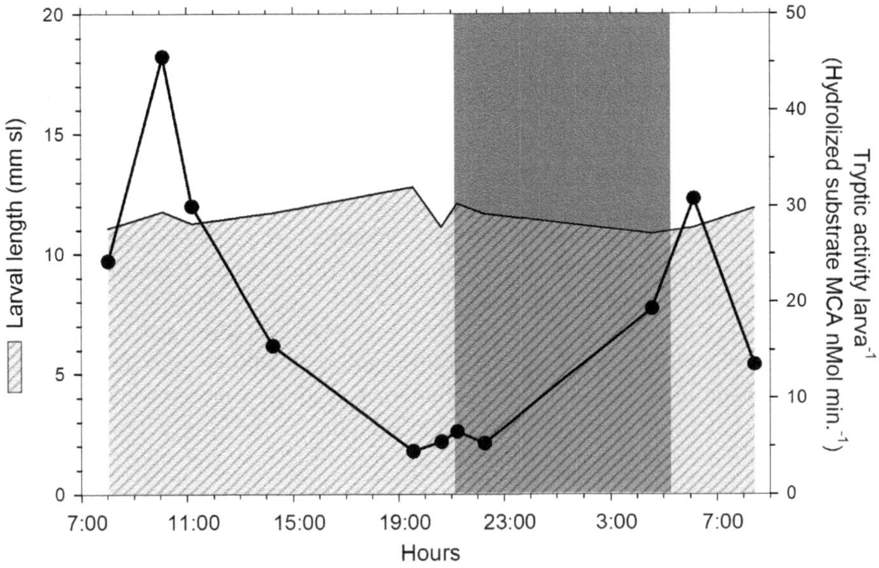

Fig. 4.2 Daily pattern of tryptic enzyme activity levels (nmol MCA min^{-1} larva^{-1}) for sprat larvae (*Sprattus sprattus*) sampled at a 24 h-drift station in the North Sea (German Bight) in May. Data points are means from 10 to 15 sprat larvae individually measured. The light-grey shaded area depicts the mean size distribution of the larvae in the samples. The dark grey-shaded area indicates periods without daylight. The background for the behaviour of this species is similar as mentioned in the caption of Fig. 4.1 about the diel behaviour of sardine larvae

it is suggested that it can be favourable to allow resting periods in order to preserve energy for catching prey which can rather be used to fuel digestion processes.

From these results, it can be concluded that fish larvae under natural conditions do not need to feed continuously to achieve fair growth rates. It can be assumed, that the larval stages have limited digestive capacity and feeding all day long up to 24 h is not necessarily beneficial for the growth rate since digestion capacity is naturally limited, to what degree depends on the species and individual ontogenetic stage. This justification was examined under laboratory conditions, as reviewed in the following paragraphs.

4.1.3 From the Nature to Observations Under Laboratory Conditions

A number of experiments about the effects of e.g., the light regime on the larval performance of cultured species has been conducted, however, the results are not yet really conclusive considering recommendations for commercial hatcheries (Barahona-Fernandes 1979; Kotani and Fushimi 2011). Nevertheless, most hatcheries use continuous light or photoperiods which are beyond the natural circadian cycle. It is supposed that under such conditions the larvae will capture prey continuously. For instance, gilthead seabream larvae fed continuously did not show apparent satiation during the availability of food (Rønnestad et al. 2013).

In order to examine the circadian digestive capacity under laboratory conditions, the diurnal rhythm in the tryptic enzyme activity has been measured in fish larvae of various species under hatchery conditions. The aim of these experiments was to examine if the natural endogenous circadian rhythms in feeding activity, digestion and resting phases changes under artificial conditions (i.e., 24 h of light) and to what extent the digestive capacity is able to deal with feeding regimes which are beyond the natural conditions.

The following data were gathered over several years and illustrates the results of various species with a range of ages, various feeding conditions and light regimes. The diurnal pattern of tryptic enzyme activity levels for herring larvae at 25, 37and 42 days post-hatch (dph) was monitored during a day-cycle from 9 a.m. to 6 p.m., with two feeding events (Rotifers and *Artemia* nauplii). The results (Fig. 4.3) reveal an obvious variation in tryptic enzyme activity as a reaction on the administration of food. Whereas the youngest stages react mainly upon the administration of the first meal of the day, the medium aged larvae show a noticeable reaction on both, the first administration of feed and on the second feeding event, although less pronounced. The oldest larvae in these experiments show an increasing tryptic enzyme activity beyond the first feeding event until the end of the sampling period. Thus, it can be concluded that the age and the related ontogenetic stage, respectively, have a significant impact on the digestive capacity; the advantage of the administration of a second meal, at least in the youngest stages, seems to be arguable.

Figure 4.4 depicts results from an experiment with 12 dph turbot *Scophthalmus maximus* larvae. The proteolytic digestive capacity was monitored from 9 a.m. to 6 p.m. following the application of three feeding events with Rotifers. A noticeable reaction on the first two feeding events was observed, but almost no reaction on the third feeding event. Apparently, the trypsinogen pool is exhausted beyond the two first meals and the larvae are obviously not able to make use of the third meal.

The proteolytic digestive capacity of Atlantic cod larvae at 21 days post first-feeding was monitored for 12 h (8 a.m.–8 p.m.) under the administration of two meals with rotifers (Fig. 4.5). In addition, cholecystokinin (CCK), supposed to trigger the secretion of trypsinogen as a consequence of food uptake (for background about CCK see part 4.2 of this chapter) was also analysed. Before and beyond the first feeding event, the larvae show a fluctuation in tryptic enzyme activity, with a trend to increasing values after the first feeding. Beyond the second feeding event, the tryptic enzyme activity decreases, even lower than compared with the initial values in the morning. At the same time, CCK values show an antagonistic behaviour. With higher tryptic enzyme activity, the CCK values are rather low. Beyond the second meal, CCK raises but this has obviously no effect on the secretion of trypsinogen and subsequent proteolytic digestive capacity. Comparable to the young stages of turbot (Fig. 4.4), the cod larvae are apparently not able to make use of the second meal, i.e., the pancreatic pool with trypsinogen seems to be temporarily exhausted.

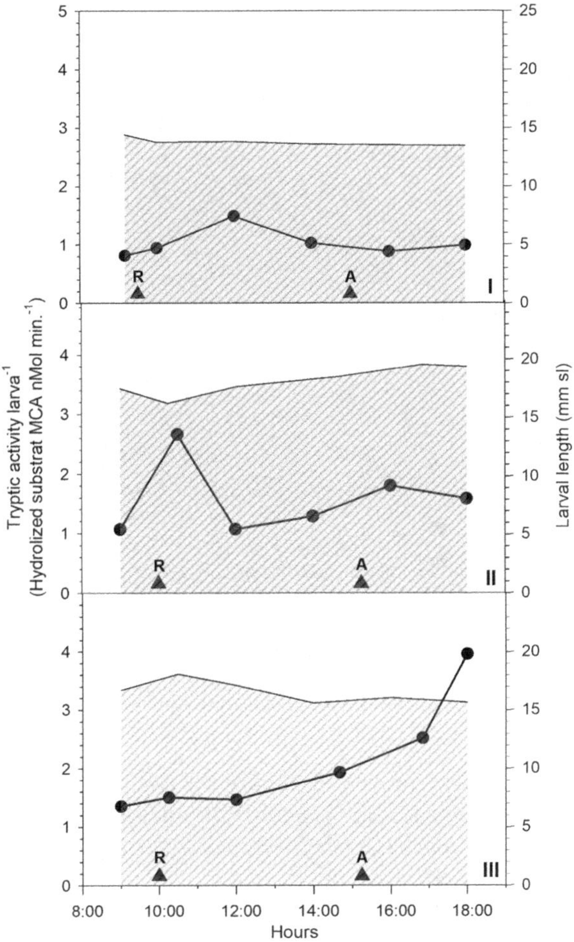

Fig. 4.3 Diurnal pattern of tryptic enzyme activity levels for 25 (I), 37 (II) and 42 (III) days-old (dph) herring larvae (*Clupea harengus*) reared under laboratory conditions. Tryptic enzyme activity (nmol MCA min^{-1} $larva^{-1}$) was related to feeding events and sampling time. The triangles indicate feeding events (two feeding events with rotifers R, *Brachionus plicatilis*, and A, *Artemia*-Nauplii). The data points represent the mean value of 10–15 individually measured larvae. The grey shaded area depicts the mean size distribution of the larvae in the samples. Light was provided 18 h a day with an intensity of 200–570 lx

Figure 4.6 depicts results from experiments with 23 dph European sea bass (*Dicentrarchus labrax*) larvae, where the proteolytic digestive enzyme capacity was monitored from 7:30 to 1:00 a.m. Two feeding regimes were applied: three events feeding *Artemia* nauplii and continuous supply (on an hourly basis) of micro diets (MD) in excess between 8 a.m. and 10 p.m. The sea bass larvae which were fed with *Artemia* nauplii reacted with a remarkable increase in proteolytic enzyme activity until the end of the sampling period, with a noticeable decrease beyond 11 p.m., the point in time when the light was switched off. The larvae which were fed with MDs did not show the same pronounced pattern compared to the *Artemia* group. Following the first feeding event at 7 a.m., the tryptic enzyme activity increases but stays beyond a lower level compared to *Artemia* feeding and more equilibrated until the end of the feeding period. This comparison clearly illustrates, that the feeding regime has a significant impact on the tryptic enzyme activity,

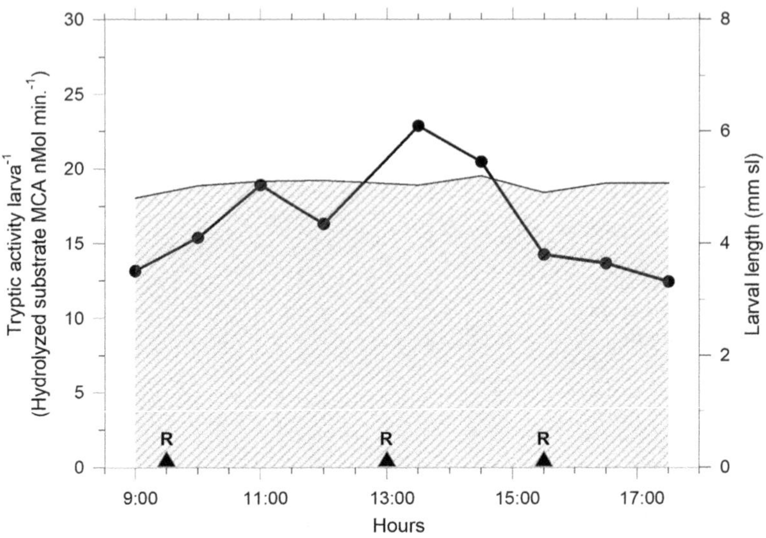

Fig. 4.4 Diurnal pattern of tryptic enzyme activity levels for 12 days-old (dph) turbot (*Scophthalmus maximus*) larvae reared under laboratory conditions. Tryptic enzyme activity (nmol MCA min^{-1} $larva^{-1}$) was related to feeding events and sampling time. The triangles indicate the 3 feeding events with rotifers (R, *Brachionus plicatilis*). The data points represent the mean value of 10–15 individually measured larvae. The grey shaded area depicts the mean size distribution of the larvae in the samples. Applied illumination period was 18/6 (light/darkness)

with live feed provoking higher activity and obviously a rather moderate reaction on a frequent feeding regime with microparticulate feed. These differences in the activity of digestive enzymes may have various reasons such as the nature of the feed: the life feed is moving around, while the microparticular feed are just slowly sinking with less physical attraction, or/and the almost continuous availability of the MDs in contrast to batch feeding events concerning the live feed, which may modify the feeding activity and digestive response.

Figure 4.7 depicts results from experiments on the diel rhythm in tryptic enzyme activity of 26 and 44 dph seabream larvae. Three feeding regimes were compared to a group of 26 dph larvae deprived of food, and two feeding regimes (*Artemia* nauplii and MD) were compared in 44 dph larvae. Samples were taken from 6:30 a. m. until 1:30 a.m. and 0:30 a.m. respectively. The experiments were conducted under 24 h light conditions to assess the impact of different dietary treatments on the diurnal patterns of proteolytic digestion capacity. Live feed (Rotifers and enriched *Artemia nauplii*) were fed three times a day at 07:15 a.m., 2:15 p.m. and 10:15 p.m., MDs were administrated every 15 min with an automatic feeding system. Diel variation of tryptic activity in the fed larvae showed a clear response on the administration of feed with increasing response in tryptic enzyme activity beyond feeding events. However, the activities in the morning and at noon revealed relatively high levels in comparison to the activity beyond the feeding event in the

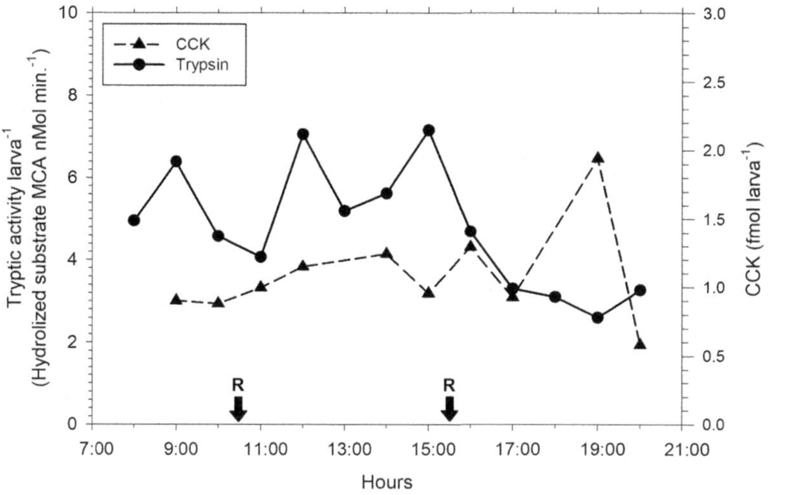

Fig. 4.5 Diurnal pattern of tryptic enzyme activity (nmol MCA min^{-1} larva^{-1}) and CCK (fmol larva^{-1}, excluding head) for Atlantic cod (*Gadus morhua*) larvae at 21 days post first-feeding. The arrows indicate a feeding event with rotifers (22 and 11 ml^{-1}, respectively). Data are presented as mean values of 5–10 larvae. Size distribution among individual samples was not significantly different (data not shown). Light was provided 24 h a day via indirect illumination of the rearing facility and a 100 lx light source above the tank. Redrawn from Tillner et al. (2013a)

evening. In contrast, tryptic enzyme activity remained significantly lower in larvae deprived of food compared to the fed groups throughout the day (Fig. 4.7a). Larvae in groups fed with *Artemia* nauplii and MD at 44 dph (Fig. 4.7b) showed a similar diurnal pattern in tryptic enzyme activity although the group MD was fed continuously. The results suggest that no matter what kind of diet is applied, sea bream larvae have apparently a quite variable digestive capacity with a decreasing trend towards the end of the day.

It is reasonable, that administration of feed needs to be considered in combination with the light regime (length of the photo phase and strength of illumination). Mata-Sotres et al. (2015) conducted a study to obtain a better understanding of the role of light cycle in modulating feeding daily rhythm during the larval development of sea bream. The potential rhythm was examined under normal light-dark (12:12, LD) cycle and permanent illumination (24:0, LL). A constant prey density was maintained during the 24 h-cycle to avoid the potential effect of temporary food availability on the larval feeding activity. Sea bream larvae exhibited a clear daily feeding rhythm under both LD and LL photoperiods. In LD treatment, the gut content started to increase when the light turned-on, peaked at the end of the photo phase and decreased quickly afterwards. Surprisingly, in LL treatment feeding activity exhibited the circadian pattern without the contribution of any photocycle or food availability cycle, and this was maintained during the whole larval stage (see Chap. 3 of this book). These results revealed that gilthead sea bream exhibits a clear circadian rhythm in feeding activity during larval

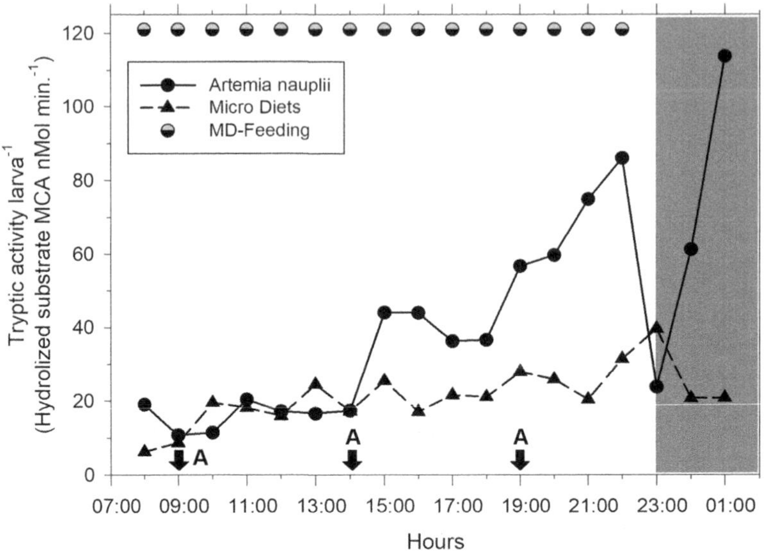

Fig. 4.6 Tryptic enzyme activity (nmol MCA min^{-1} $larva^{-1}$) in larval European sea bass (*Dicentrarchus labrax*) between 6 a.m. and 11 p.m. at 23 dph. Feeding treatments: *Artemia* nauplii and micro diets (MD; GEMMA Micro, Skretting). Data are represented as mean values of 5 individuals per sampling event. Black arrows indicate feeding events (*Artemia* nauplii). Micro diets were administrated hourly in excess between 8 a.m. and 10 p.m. with an automatic dispenser system (AMD-Feeder). Light was provided 16 h a day (7 a.m.–11 p.m.) and light intensity was 700 lx at the water surface during the experiment. The dark grey area indicates no light within the sampling period. Redrawn from Tillner et al. (2014)

development, and the employment of a continuous daylight photoperiod during larval rearing did not increase feeding activity in this species.

4.1.4 Conclusions and Recommendations

Under natural conditions, feeding activity in the earliest stages of fish has a strong circadian element and depends on natural light cycles. Food is not constantly available in the wild but may be restricted to particular times of the day in the corresponding season. Growth performance and survival rates depend on the availability of prey of proper quality and size and, to a less or more significant extent on climate and hydrographic conditions such as temperature, turbidity and currents. In addition, fish larvae must cope with the presence of predators, whose activities can be also restricted to certain times of the day. Therefore, fish have evolved time-keeping mechanisms to anticipate such cyclic, predictable events.

In contrast, under culture conditions it is expected that those most unfavourable conditions, which appear in natural environments causing high mortality, can be

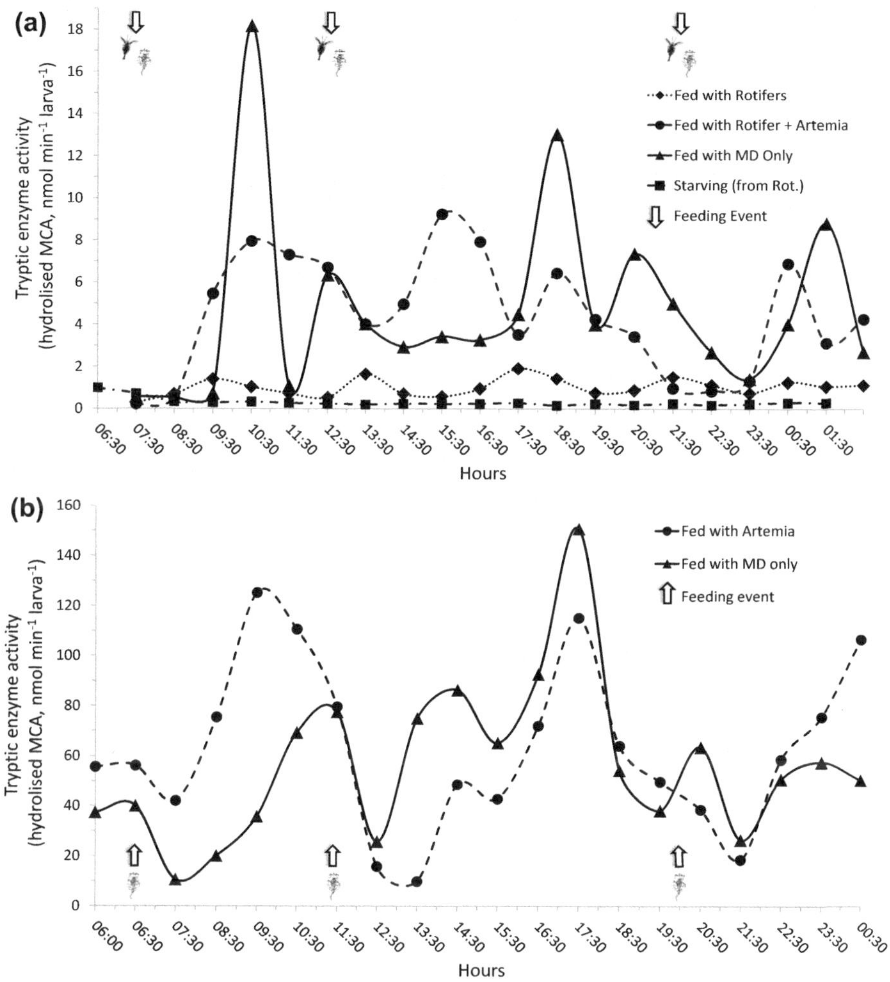

Fig. 4.7 Diel rhythm of tryptic enzyme activity (nmol MCA min^{-1} larva^{-1}) in **a** 26 dph and **b** 44 dph sea bream larvae. Comparison of different feeding regimes (**a**): rotifer, rotifer and *Artemia*, and micro diet fed groups compared to larvae deprived of food from the rotifer group. (**b**) Compares *Artemia* and micro diet fed groups. Sampling of each group took place between **a** 06:30–01:30 a.m. and **b** 00:30 a.m. respectively every full hour under 24 h of light and an intensity of 500 lx at the water surface. Data present mean values from individually measured larvae (n = 6 per sampling event). The arrows indicate the administration of rotifers and Artemia three times a day; the micro diet group was fed every 15 min until the end of the experiments (redrawn from Zeytin et al. 2016)

avoided in the larval stages. However, the survival of the larval stages in the most established marine aquaculture species, such as turbot, sea bream and sea bass, are still high and barely not more than 30% survival rate (from larvae to juveniles) has been achieved until present, including a high variability from batch to batch.

This is still surprising since rearing conditions are established to the best up-to-date knowledge; the fish larvae are swimming in a soup of food, there is no predation and temperature and illumination is supposed to be adjusted at optimal conditions. However, as discussed from López-Olmeda et al. (2011), under farming conditions, fish are challenged with "artificial" diets and feeding regimes, and inadequate feeding conditions can cause stress, alteration of normal behavioural patterns, poor performance and eventually diseases and death.

From the results presented above, produced under laboratory conditions, it is obvious that the daily digestive capacity is quite variable and limited, specifically in the younger stages of fish. The fact that larvae continue to feed despite having a full gut suggests that satiety factors are apparently absent in the first feeding stages and beyond (Kurokawa et al. 2000). Thus, it is questionable, if continuous feeding in combination with e.g., 24 h of light in the early stages is of advantage. Although plasticity of diel and circadian activity rhythms in juvenile and adult fish is well known (Reebs 2002), this does probably not apply with the same level to larval stages. Resting periods are common under natural conditions as demonstrated in the results above from field investigations and may be of advantage also under hatchery conditions.

It has been shown that fish larvae can accumulate trypsinogen in the pancreas in periods without feed availability preparing for the next meal (Ueberschär et al. 1992). This might be of advantage to react efficiently to the first meal of the day, and can explain the more pronounced reaction in proteolytic enzyme activity upon the first meal of the day; however, the storage capacity decreases under continuous feeding conditions and may be exhausted beyond a certain number of meals. According to the results presented above, it can be concluded, that the total daily digestive capacity is certainly dependant from the ontogenetic stage and further has a species-specific element. For example, larvae with a straight gut develop earlier CCK-expressing cells, and the reason may be that gut passage is faster compared to larvae with a coiled gut in order to improve efficiency in digestion (Rojas-García and Rønnestad 2002).

Feeding schedules in combination with light regimes which exceeds the digestion capacity may not only cause adverse effects on larval growth and performance but are questionable from the economical point of view. Only an optimal combination which is supposed to be species and age specific can maximize the growth rate and survival to meet economically viable conditions (Mata-Sotres et al. 2015). This is specifically true if formulated feed (MD) is being introduced into the feeding regimes. Microparticulate feed is getting increasingly into the focus of hatchery operators, since the performance of MD for early weaning or even as first feed is promising and is recently coming close to live feed, at least in advanced larval stages. MD needs to be administrated in short intervals since the food particles sink slowly to the bottom of the rearing tanks, becoming unavailable for the larvae. An economic feeding schedule is thus important with this type of feed. If it turns out that feeding events later the day in younger stages have no advantage concerning growth and performance, the feeding schedule should be adapted accordingly.

There are a number of studies dealing with the light regime and feeding events in larval rearing, which evaluates the gut fullness, and conclude a positive effect of the feeding schedule when the gut of the larvae is filled. However, because of the opportunistic behaviour of fish larvae to feed as long as light and food is available, the gut fullness is certainly an indicator of feeding activity but not necessarily a success indicator concerning the assimilation of the ingested feed. It is more appropriate, to evaluate the digestive enzyme capacity as a response to the number of meals in combination with various light regimes and the subsequent growth performance to reveal eventually the most advantageous feed and feeding strategy from the biological, as well as from the economical point of view. Recent studies emphasize that the potential variability of enzyme secretion along the whole day is an important factor to take into account for future studies (e.g. Mata-Sotres et al. 2016).

In this section, it was tried to highlight the impact of feeding rhythms, feeding time and type of food upon physiological indicators in the larval stages of fish, such as the digestive capacity, which obviously shows a diurnal/circadian rhythm in response to feeding events and beyond. Since the optimal feeding schedule changes with species, age and type of feed, it is a demanding task to evaluate the digestive capacity in various conditions and ontogenetic stages, but certainly worthwhile in order to overcome the "try and error" approach which is currently the common strategy in larval rearing under commercial scales.

4.2 Developmental Dynamics of Regulatory Mechanism and Food Cues Controlling Feeding Cycles

4.2.1 Introduction

Acquisition of the digestive function is one of the key factors of survival during the early life history of larval teleost. Digestive processes involve spatial and temporal coordination of a complex system of food signals, physiological responses and chemical modulators, orchestrating a complex system of neural, endocrine and local control mechanisms for a proper seek/ingestion, digestion and absorption of food. Despite the significant progress of the "omics era", the in-depth knowledge in larval fish stages on digestive processes and mechanisms involved in production, release and activity of digestive enzymes is still very poor.

Larval fish stages present ecological, morphological and physiological characteristics that differ from those of juvenile and adults. However, the apparent plasticity of fish larvae does not match with survival and this fact indicates a large influence of food availability. When controlled feeding protocols for artificial larval rearing are available but are not the appropriate ones, it can have important effects on growth, survival and individual robustness. Thus, a thorough understanding of digestive function and processing capacity from the onset of exogenous feeding can provide a better basis for the formulation of specific diets for larviculture. In adult fish, the digestive process starts with food ingestion, and includes the secretion of enzymes and fluids, mechanical and enzymatic digestion and absorption, and the regulation of the overall processes by several brain-gut axis (Rønnestad et al. 2013). This regulation, which is accomplished by neuronal and endocrine systems, is an interface between a primary stimulus and adjustments of gut functions, being food itself the most important primary stimulus (Holmgren and Olsson 2009).

The basic mechanisms of organ development are similar in all teleost, even though there are considerable differences regarding the relative timing during ontogeny. Ontogenetic changes determine the nutritional and physiological performances of a fish (Zambonino-Infante et al. 2008). A key question in larval ontogeny is when feeding and digestive regulatory loops become functional, but research in this area remains in its infancy. The scarce information available is mainly due to the lack of interest of frontier science for low impact zoo-technical fields and laborious physiological approaches. Working with fish larvae present many challenges, including their small size and difficulty in estimating the food intake, in combination with the diversity of peptide delivering routes, the nature of the peptides used, and the duration of the treatment, which render the results very difficult to compare (Hoskins and Volkoff 2012). This section will present the information available up to date and will also provide contemplations of applied research for artificial rearing.

4.2.2 Neuroendocrine Control of Appetite and Ingestion in Fish Larvae

Based on the digestive physiology of higher vertebrates and gastrointestinal human disease studies, it can be stated that the physiological regulation of satiation, appetite and ingestion involves a complex integration of peripheral and central signals. As in mammals, the brain centre that regulates ingestion in fish seems to be located in the hypothalamus, and works coordinated with the peripheral system, composed of tissues such as the GIT, pancreas and adipose tissue. Both systems are regulated by endogenous and external or environmental signals. Endogenous signals (metabolic and neuroendocrine) with either central or peripheral origin can be orexigenic (promoting appetite) or anorexigenic (inhibiting appetite). In artificial feeding, the external or environmental signals, as photoperiod or food availability, can be manipulated to ensure optimal rhythms and endocrine outputs (Gorissen et al. 2006; Hoskins and Volkoff 2012; Rønnestad et al. 2013; Tinoco 2015).

One of the key questions in larval stages is the start-up of the regulatory loops and the time required to become functional. Fish developmental profiles exhibit differences between species, but in general, some neural appetite regulatory factors are present at early developmental stages, although some factors develop in association with the metamorphosis (Table 4.1). In this context, it is not yet known whether altricial larvae possess an early system that regulates appetite and to what extent it depends on feedback from the GIT (for satiety) and/or from the body (nutrients/energy storage) (Rønnestad et al. 2013). Based on fish species studied to

Table 4.1 Time of detection of peptides involved in the digestive regulation in fish species

Peptide	Species	Age of detection (dph, days post-hatch)	References
Cholecystokinin (CCK)	Ayu (*Plecoglossus altivelis*)	0	Kamisaka et al. (2003)
	Japanese flounder (*Paralichthys olivaceus*)	2	Kurokawa et al. (2000)
	Atlantic halibut (*Hippoglossus hippoglossus*)	45–52	Kamisaka et al. (2001)
		7	Rojas-Garcia and Rønnestad (2002)
	Atlantic herring (*Clupea harengus*)	0	Kamisaka et al. (2005)
		0	Rojas-García et al. (2011)
	Spotted rose snapper (*Lutjanus guttatus*)	0	Moguel-Hernández et al. (2016)
	Bluefin tuna (*Thunnus thynnus*)	3	Kamisaka et al. (2002)
	Atlantic cod (*Gadus morhua*)	3	Kortner et al. (2011b)
Neuropeptide Y (NPY)	Red drum (*Sciaenops ocellatus*)	3	Webb et al. (2010)
Ghrelin	Spotted rose snapper (*Lutjanus guttatus*)	0	Moguel-Hernández et al. (2016)
	Atlantic halibut (*Hippoglossus hippoglossus*)	0	Manning et al. (2008)
	Atlantic cod (*Gadus morhua*)	3	Kortner et al. (2011b)
Orexin	Atlantic cod (*Gadus morhua*)	0	Xu and Volkoff (2009)
Preprosomatostatin 1 (PPSS 1)	Atlantic cod (*Gadus morhua*)	3	Kortner et al. (2011b)
CART	Atlantic cod (*Gadus morhua*)	3	Kortner et al. (2011b)

date, it appears that the molecular structure of appetite regulators is very well-conserved among vertebrates with regards to gene structure, amino acid composition, or 3D protein configuration, although it may exist group-specific differences (Copeland et al. 2011; Sundström et al. 2008).

Appetite-regulating hormones described in fish have been reviewed in several papers (Rønnestad et al. 2017; Volkoff et al. 2005, 2009a, b, 2010), nevertheless there are almost no studies of how neuropeptides and gut hormones affect appetite and ingestion in fish larvae. In Atlantic cod, it was demonstrated that genes involved in appetite regulation and digestion showed differential expression profiles after different feeding protocols, suggesting that may be tied directly to the type and quality of initial dietary constituents (protein, peptides, polyunsaturated fatty acids oils, polyamines, etc.) (Kortner et al. 2011a). In Atlantic halibut *Hippoglossus hippoglossus* the expression pattern of several appetite-controlling factors was analysed in response to feeding and also to a fast-refeed challenge (Gomes et al. 2015). Only peptide YY (PYY) expression was affected by feeding, being up-regulated 1 and 3 h after feeding, suggesting a role in satiety regulation in Atlantic halibut larvae. The expression patterns of the remaining factors analysed were intriguing, and therefore further exploration is needed to understand the role of each player in appetite control in this species. The effect of dietary lipids and feeding in putative anorexigenic and orexigenic genes has been studied in Senegalese sole larvae and post-larvae. A broad variety of expression patterns were observed, highlighting the complexity of appetite regulatory mechanisms in fish larval stages (Bonacic et al. 2016). Expression in pre-metamorphic larvae was generally less in accordance with the putative function of the appetite-controlling genes than in post-larvae, which suggest a yet underdeveloped regulatory system and a high somatic cell proliferation rate due to morphogenic changes which may mask the results.

To date, ghrelin is the only known orexigenic hormone in the fish gut (Murashita et al. 2009; Webb and Rønnestad 2011). However, species-specific actions on food intake have been described in adult fish. An orexigenic role for ghrelin was suggested in goldfish *Carassius auratus* and Atlantic salmon *Salmo salar* (Miura et al. 2009; Vikeså et al. 2015; Volkoff et al. 2005), while anorexigenic effects were described in rainbow trout *Oncorhynchus mykiss*. As for zebrafish *Danio rerio*, contradictory results have been observed and ghrelin has been shown to act both as an orexigenic and anorexigenic hormone (Amole and Unniappan 2009; Koven and Schulte 2012). Since ghrelin is produced primarily in the stomach of adult fishes, it is not clear what role it may have in developing fish larvae before the appearance of a fully functional stomach. In fish larvae active ghrelin has been found before the onset of exogenous feeding in several species (see Table 4.1), however its function is remains little explored. In Atlantic halibut larvae, ghrelin mRNA expression was detected at the time of hatching, even though the stomach in this species does not become functional until approximately 73 dph (Manning et al. 2008). In Atlantic cod, ghrelin mRNA was first detected at the cleavage stage of cod embryos (Xu and Volkoff 2009). These results suggest that ghrelin may be active in a developmental role even before the onset of exogenous feeding (Webb and Rønnestad 2011). In Senegalese sole post-larvae (23 dph), results suggested an orexigenic effect of ghrelin on food intake and, in accordance, a postprandial regulation of ghrelin effects when comparing fasted and fed post-larvae, suggesting that after the first feeding (in fed post-larvae) ghrelin down-regulation pathways were activated (Fig. 4.8) (Navarro-Guillén et al. 2017b). In addition, this was the first study to report response time to ghrelin, which was about 25 min after ghrelin administration (Fig. 4.9). Further studies using a physiological pulse dose of ghrelin in

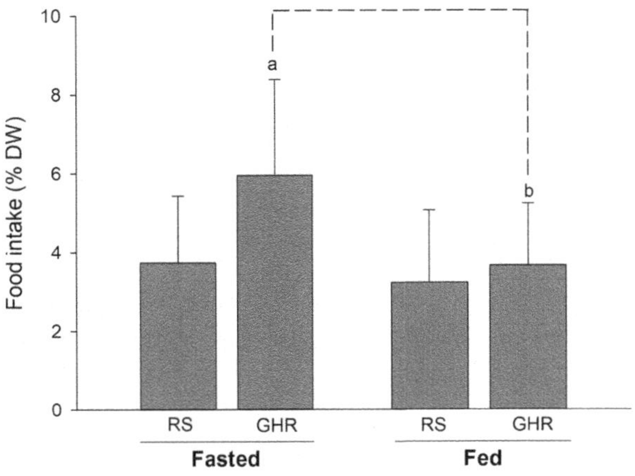

Fig. 4.8 Food intake ([(Rtotal/SRArtemia)/DWfish] × 100) of 27 dph Senegalese sole post-larvae tube-fed Ringer solution (RS, n = 8) or ghrelin (GHR, n = 15). Results are represented as mean ± SD. Different letters mean statistical differences ($P = 0.037$) between GHR-fasted and GHR-fed post-larvae (Navarro-Guillén et al. 2017b)

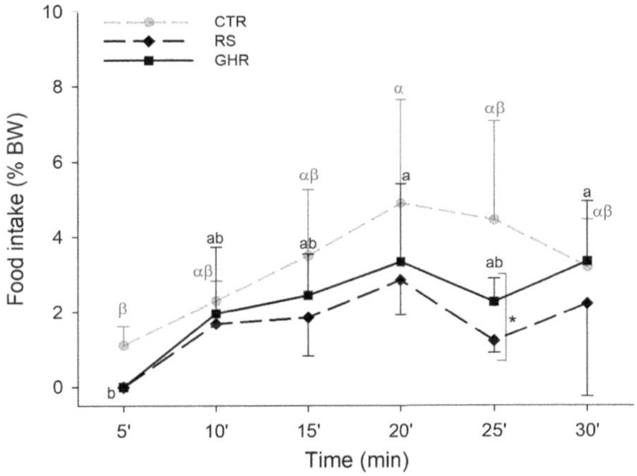

Fig. 4.9 Feed intake ([(Rfish/SRArtemia)/DWfish] × 100) of 23 dph Senegalese sole post-larvae along 30 min after been tube-fed. CTR (control, n = 8 per sampling point), RS (Ringer solution treatment, n = 8 per sampling point) and GHR (ghrelin treatment, n = 15 per sampling point). Results are represented as mean ± SD. Letters mean significant differences between each sampling point for each treatment. Asterisk indicates statistical differences between RS and GHR treatments at each sampling point (Navarro-Guillén et al. 2017b)

Senegalese sole juveniles or adults are needed to confirm its orexigenic function in this species.

Results above revealed that intra-stomach ghrelin administration appears to be not sufficient to determine accurately the hormonal control and regulation of feeding activity at early stages. In general, studies rely on the use of volunteer feeding protocols, which makes it difficult to control the bolus quality and quantity, and as consequence, postprandial endocrine response are variable. To completely control the quantity and quality of the food ingested techniques such as tube-feeding are necessary. Some fish larvae, such as Senegalese sole, exhibit very high growth rate during the first weeks after hatching; therefore high food availability is essential for a high food intake. Considerable research efforts have been made to elucidate what optimal rearing or nutritional conditions for these early life stages are that could improve juvenile quality. Despite the existence of several studies focused on larval nutrition, it is commonly found a high individual dispersion in food intake in fish larvae (Bonacic et al. 2016; Engrola et al. 2009, 2010; Mata-Sotres et al. 2015; Mai et al. 2009; Navarro-Guillén et al. 2017a; b, c; Rojas-Garcia and Rønnestad 2002; Tillner et al. 2013a, 2014). Hence, further studies to uncover the mechanisms regulating food consumption are necessary to decrease feeding variability and to improve larval feeding protocols.

4.2.3 Neuroendocrine Control of Digestive Function in Fish

After ingestion, food enters the digestive tract, where it is processed both mechanically and chemically (by acidic and/or alkaline enzymes). After being hydrolysed by these processes, nutrients are absorbed into the body, and the undigested matter is subsequently evacuated. This series of processes are tightly regulated and can be grouped into three phases: cephalic phase, gastric phase, and intestinal phase (described in Webb and Rønnestad 2011; Rønnestad et al. 2013).

In each phase, different endocrine and neural factors participate in regulating the mechanisms involved in the digestive process. Most commercially cultured fish species are altricial or agastric, suggesting that the intestinal phase is perhaps the most important phase in larval marine fish. In this phase, pancreas and gallbladder secretions are stimulated, which contain bicarbonate rich fluids and bile, respectively. These fluids neutralize the ingested food, emulsify lipids through the formation of micelles and contain pancreatic digestive enzymes (Rønnestad et al. 2013; Webb and Rønnestad 2011). Due to the importance of this phase in larval fish, it is supposed that the neuronal network and regulatory peptides that are necessary for its control become active before or at the onset of exogenous feeding. However, it has been suggested by Kurokawa et al. (2000) that teleost start exogenous feeding with a poor endocrine system regulating pancreatic enzyme secretion. The fact that larvae continue to feed despite having a full gut suggests that satiety factors may play a relatively minor role in the first feeding stages. Very limited information is available regarding the involvement of regulatory peptides during the larval stage (Drossou 2006; Kortner et al. 2011b; Navarro-Guillén et al. 2017a; Rønnestad et al. 2013; Tillner et al. 2013b, 2014). The most important gastrointestinal regulatory factors in larval fish and adults have been reviewed by Webb and Rønnestad (2011) and Holmgren and Olsson (2009). The appearance of few intestinal cells with endocrine functions in early larvae do not denote a regulatory capacity until the release of entero-hormones are sufficient to reach physiological levels under a minimum hormonal clearance, sustaining a premature endocrine framework to stimulate pancreatic maturation and gut elongation (e.g., CCK).

CCK is the most studied gut peptide hormone in fish larvae to date. CCK has a key regulatory effect on the intestinal phase of digestion; regulation of gallbladder contraction, pancreatic enzyme secretion, inhibition of gastric acid production, regulation of gastric emptying and gut motility (Liddle 2012; Rønnestad et al. 2013).

Analysis of CCK content and expression assessed by radioimmunoassay, in situ hybridization and quantitative RT-PCR has revealed important species-specific differences in the time of CCK-producing cells appearance in the larval GIT and CCK levels at the onset of first feeding (Rojas-García et al. 2011). In larvae that possess a coiled gut at onset of exogenous feeding, the presence of CCK-producing cells was identified just after the larvae started feeding (e.g., Atlantic halibut and Bluefin tuna *Thunnus thynnus*) (Rojas-Garcia and Rønnestad 2002). While in larvae with a straight gut, CCK-producing cells are present before start feeding and throughout the whole length of the gut (e.g., herring and ayu *Plecoglossus altivelis*) (Rønnestad et al. 2013, for more details see Table 4.1).

The available data suggest that CCK is involved in the pancreatic secretion in larval fish, since a regulatory loop between protease trypsin and CCK has been described in several marine species (Drossou 2006; Navarro-Guillén et al. 2017a; Tillner et al. 2013b, 2014). This regulatory loop seems to be regulated by the quantity of ingested food and the protein content, as well as by the peptide chain length of these proteins (Drossou 2006; Koven et al. 2002; Rojas-García et al. 2001; Rojas-Garcia and Rønnestad 2002; Tillner et al. 2014). In mammals, CCK is released into the systemic circulation which subsequently activates de CCK receptor A (CCK-RA) in the exocrine pancreas, stimulating the release of pancreatic enzymes into the digestive tract (Rojas-García et al. 2011). In zebrafish larvae, CCK-RA antagonist treatment reduced protease activity, but had no effect on phospholipase activity in 5 days post feeding (dpf) larvae. However, in 6 dpf zebrafish larvae, CCK-RA antagonist treatment caused significant reductions in both phospholipase and protease activities, demonstrating a marked ontogenetic difference in timing over the control of CCK-induced release of digestive enzymes (Hama et al. 2009). Nevertheless, information about the ontogeny and affinity of CCK receptors in fish larvae is scarce and much remains to be discovered with respect to CCK synthesis, release and actions. Koven et al. (2002) demonstrated that CCK stimulates pancreatic secretions in Atlantic herring larvae, and similar responses have also been seen in larvae of Atlantic halibut (Rojas-Garcia and Rønnestad 2002), European sea bass (Cahu et al. 2004), red drum *Sciaenops ocellatus* (Webb 2008), Atlantic cod (Tillner et al. 2013b) and Senegalese sole (Navarro-Guillén et al. 2017a). In this context, it is evident that the interaction CCK-trypsin must be highly dependent on the feeding behaviour, primarily on the amount of food ingested and the timing of the ingestion. For wild larval stages the food stimulus often is scarce, then, how does first-feeding larvae cope with long food deprivation? At least for CCK it has been demonstrate that most fish larvae possess a neural reserve to be used at first-feeding (Rojas-García et al. 2011). Several studies have demonstrated that fish larvae do not feed constantly under natural or laboratory conditions, exhibiting a circadian prandial pattern (Kotani and Fushimi 2011; Ma et al. 2006; Mata-Sotres et al. 2015; Navarro-Guillén et al. 2015; Østergaard et al. 2005). Moreover, recent studies reported that larvae digestive function follows a rhythm pattern modulated by the photoperiod and feeding time (Fujii et al. 2007; Mata-Sotres et al. 2016; Navarro-Guillén et al. 2017a, c; Rojas-García et al. 2011; Tillner et al. 2013b, 2014; Zeytin et al. 2016).

The analysis of how the regulatory loop of CCK and trypsin is affected by fasting, starvation and feeding regime shows some contradictory results. Rojas-García et al. (2011) described higher CCK levels in fed compared to fasted herring larvae, suggesting a release and recover/re-synthesis of CCK in the gut after feeding started and, a modulation of CCK levels by food intake. In accordance, Drossou (2006) described a feedback mechanism between CCK content and tryptic activity in Atlantic herring, Baltic cod and Atlantic halibut. In addition, the authors proposed that both parameters are short-term indicators for changes in the food supply, since starvation decreased CCK content and tryptic activity in the three species, to possibly minimize the energetic loss. It is well documented the feedback regulation of CCK secretion on pancreatic development (Rønnestad 2002). Thus, the occurrence of a transient endocrine framework during larval development might stimulate adipocytes growth and premature energy storage previous to the establishment of the regulatory loop for energy homeostasis (e.g., leptin). Navarro-Guillén et al. (2017a) demonstrated that Senegalese sole larvae were able to synchronize digestive functions to different feeding regimes. CCK levels were modulated by gut content and tryptic activity, and were lower when larvae were fed and higher when food was not available. This study also supported the existence of a regulatory loop between CCK and tryptic activity in pre- and post-metamorphic Senegalese sole, with simultaneous and opposite trends (Fig. 4.10). This reverse daily trend was clear at 6 and 20 dph and less evident in the older developmental stage analysed (32 dph), which suggests that other gut hormones as well as neuronal factors play a potential role in the regulation of digestion and that their roles may change during larval development.

In Atlantic cod larvae, however, CCK levels and tryptic enzyme activity revealed a reverse diurnal trend independently of the high and stable gut filling at 21 dpf for all feeding regimes, suggesting that CCK is synthesized and released as a stimulatory response when tryptic enzyme activity in the gut is low and vice versa (Tillner et al. 2013a). In sea bass larvae, Tillner et al. (2014) suggested that CCK is not dependent on food intake or digestive products in the gut, but follows a natural, endogenous rhythm. CCK levels fluctuated irrespective of high or low tryptic enzyme activity in both fed and starved groups, suggesting that there is not link between them. These results might be explained by an immature control of pancreatic enzyme secretion in sea bass larvae at 23 dph. This study suggests that the regulation of CCK on appetite-satiety is an innate condition (feature) to all teleost and necessary to trigger the premature regulatory loops at first-feeding.

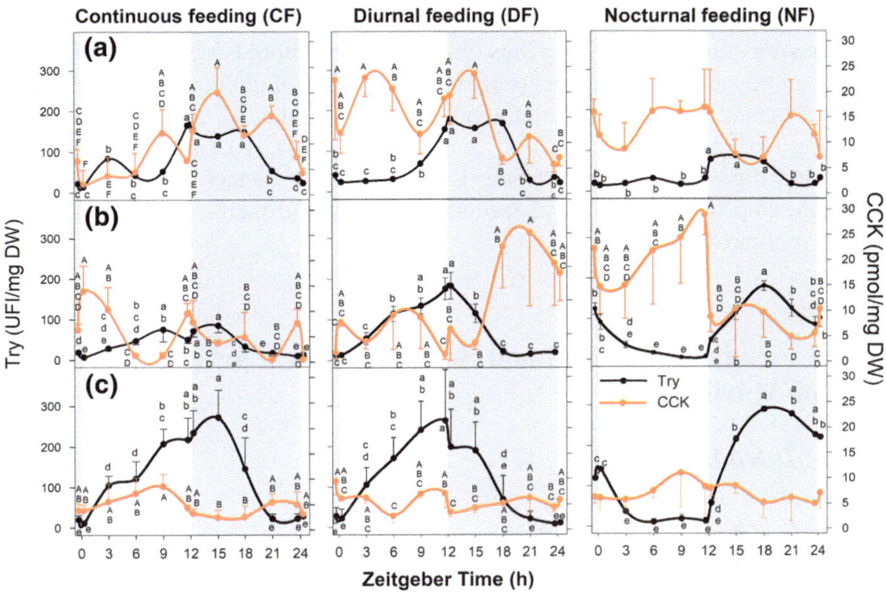

Fig. 4.10 Daily pattern of CCK and tryptic enzyme activity for Senegalese sole larvae at 6 (**a**), 20 (**b**) and 32 dph (**c**) from three different feeding regimes. Values are represented as means + SD. Different letters show significant differences between hours. Grey areas represent dark periods. (Navarro-Guillén et al. 2017a)

Subsequently, the frequency of food availability, feeding habits (carnivorous, herbivorous and filtering habits), length of food transit, food cues (e.g., viscosity, chemical complexity, free forms of polymers) and gut morphology (linear or coiled), contribute to set-up the transient or permanent regulatory-loops for digestive enzymes patterns (diurnal cycles).

4.2.4 Conclusions and Future Considerations

Fish larvae show a high dependency on the feeding and digestive processes to achieve high growth rates during the first weeks after hatching. Hence, it is important to understand the mechanisms involved in food intake and digestion regulation to improve larval rearing protocols. The studies cited in this section revealed that larval digestive function regulation is still poorly understood and many of the conclusions are contradictory between species and experiments. Understanding the basis that underlie the differences in digestive ontogeny between fish species and knowing which components are responsible for the release of appetite-controlling factors and digestion promoter factors in the larval stage will improve species-specific feeding protocols and MD development, to obtain high-quality juvenile. Ghrelin and CCK stimulation and release are most likely regulated by nutrient sensing. Thus, future studies should focus on identifying which and how dietary components activate the release of gut hormones, and which mechanisms are involved in the nutrient sensing and whether they are affected by ontogeny (for more details on food sensing see Sect. 4.3) Understand how food intake and digestive enzymes release is modulated by dietary composition is an important step to improve larval feeding efficiency and therefore, improve aquaculture profitability.

4.3 Considerations on Food Sensing in the Larval Stages of Marine Fish

4.3.1 Introduction

Sensing food is an important part of feeding, and it takes place both before and after ingestion. Food is recognized through a range of sensory systems involving chemical, visual and mechanical stimuli. Chemical sensing of nutrients is involved in detection, capture, the decision whether to swallow or reject the food, and in digestion and assimilation (Rønnestad et al. 2013). A wide range of literature in fish describes the morphological, electrophysiological and ligand-specific role of the chemical sensory systems related to olfaction and taste. The typical substances that stimulate the olfactory and gustatory sensory cells have a low molecular weight (<1000 Da) and are water-soluble, non-volatile, nitrogenous and amphoteric. They include amino acids (AA) and other similar substances, nucleotides and betaines (Hara 1993; Morais 2017). Since fish sense chemicals in water, it may appear to be difficult to determine the specific role of each taste/smell system in a particular behaviour (Marui and Caprio 1992). However, in anatomical terms these senses are clearly separated, since olfaction signals are transmitted directly to the CNS by neurons of the first cranial nerve (*n. olifactorius*), while gustatory information is detected by specialized epithelial cells (i.e., taste cells) and transmitted to the CNS by neurons of the cranial nerves VII (*n. facialis*), IX (*n. glossopharyngeal*) or X (*n. vagus*) (Marui and Caprio 1992). In practical terms, olfaction detects long-range stimuli and is involved in search behaviour, while gustation detects the close chemical stimuli (Pavlov and Kasumyan 2002) and is related to ingestion and swallowing. Furthermore, recent advances mainly in mammals, have demonstrated that chemosensors linked to olfaction and taste are also present in the GIT. These chemosensors evaluate food composition during transit through the GIT, with a major impact on digestion, appetite and coordination of post-absorptive metabolism (Gribble 2012). In this section, we introduce nutrient sensing, discuss the physiological implications of these systems, and provide holistic recommendations for future research on nutrient sensing in fish larvae.

4.3.2 Olfaction/Smell

Chemical sensory cells related to food are located in the olfactory epithelium (Boglione et al. 2003; Hara 1993, 2005), solitary cells and taste buds (Døving and Kasumyan 2009). The olfactory sensory cells start to differentiate at hatching, and play an active role in food detection in most fish larvae from first feeding (Knutsen 1992; Kolkovski et al. 1997). Olfaction provides directional stimuli which, in combination with the detection of water currents by the lateral line system, enable the fish to locate the odour source. The small water-soluble molecules trigger feed search behaviour and attract the fish to swim closer until food is located (Valentinčič 2005). This is an important consideration for MD design, since some leaching of water-soluble compounds such as AA may serve as attractants and can improve the ingestion and acceptability of formulated feeds for fish larvae (Kolkovski et al. 2000). Valentinčič et al. (1999) showed that solutions of L-alanine and L-proline provoked reflexive biting/snapping behaviour of rainbow trout alevins sixth days prior the first exogenous feeding. Although excessive nutrient leakage should be avoided, food particles may therefore not be sealed tightly.

AA that stimulate olfaction in fish, even in mixtures, are species-specific and may vary greatly both with respect to specificity and concentration, and without any apparent pattern. Species tested are typically very sensitive to chemical stimuli, and the threshold of detection for some AA is below 10^{-9} M (Døving and Kasumyan 2009). Fish have been grouped according to their olfactory cells sensitivity to different AA into wide or limited response-range species (Hara 1993; Marui and Caprio 1992). Therefore, the diet administered to fish larvae should take into consideration each species' sensitivity and how well the chemical substances serve as attractants. However, very few of these data-sets are available for larval stages. The following data depict some recent results in this research area.

Behavioural studies have been conducted using early and late larval stages of European sea bass to test the attraction for various free AA and natural extracts on their swimming activity (Sommerfeld 2014). In a controlled environment, different age groups of larvae (37, 51, 58 and 65 dph) were exposed to dissolved mixtures of various AA (glycine–betaine–alanine–arginine, glycin–betain and alanine–glycine–histidine–prolin, 10^{-1}, 10^{-3} and 10^{-7} M; all AA applied as L-racemate), natural extracts from shrimp, squid and *Calanus*, and artificial sea water (control), and their behaviour was monitored for 30 min (Sommerfeld 2014). The fish larvae's swimming behaviour was recorded using two video cameras, and every five minutes a photo was taken to count the number of larvae in the area were the attractants were introduced (observation window). The swimming speed of the larvae was evaluated using the video recordings and a multi-tracking program (based on MatLab by Hedrick 2008).

The more advanced larval stages showed a clear trend to approach the attractants (10^{-7} M mix of glycine–betaine–alanine–arginine), while the younger larvae (37 dph) did not show a clear reaction to the attractants (Fig. 4.11). This may

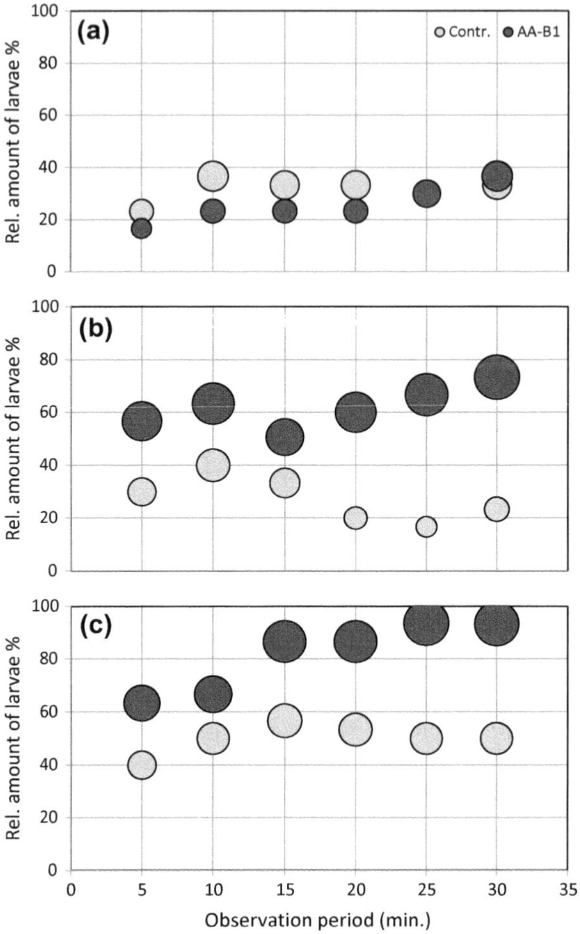

Fig. 4.11 Relative abundance of seabass larvae in the observation window of the aquarium within 30 min in % of the total number of larvae (n = 10). Size and position of the bubbles indicate the abundance of larvae at **a** 37 dph, **b** 58 dph and **c** 65 dph. Dark bubbles represent Glycine-Betaine-Alanine-Arginine treatment (concentration 10^{-7} M), and light grey bubbles the saline water control. Counting was performed every 5 min. Redrawn from Sommerfeld (2014)

suggest that the ability to smell is undeveloped in early stages, which is supported by Boglione et al. (2003), stating that olfaction becomes rather relevant in older larvae.

The squid extract (Fig. 4.12) yielded similar results as for the Glycine-Betaine-Alanine-Arginine mix (Fig. 4.11). The youngest stage (37 dph) did not react differently to the control group, while the older larvae (51 and 58 dph) moved towards and swim around in the attractant release area.

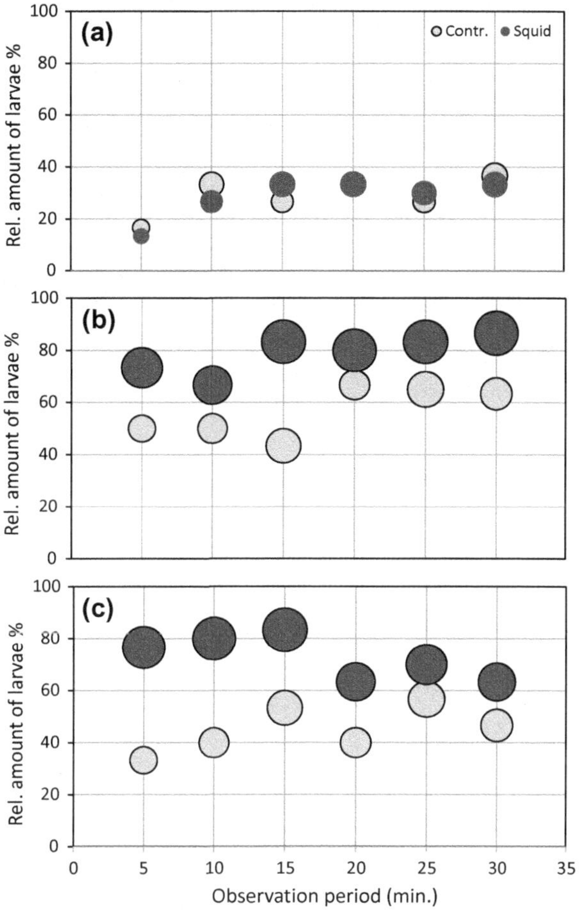

Fig. 4.12 Relative abundance of seabass larvae in the observation window of the aquarium within 30 min in % of the total number of 30 larvae. Size and position of the bubbles indicate the abundance of **a** 37 dph, **b** 51 dph and **c** 58 dph larvae. Black bubbles represent the treatment with Squid extract and light grey bubbles represent the saline water control. Counting was performed every 5 min. Redrawn from Sommerfeld (2014)

In summary, Sommerfeld (2014) showed that positive attraction of some AA mixtures and natural hydrous extract of squid increases with the development of seabass larvae. However, some other AA mixes and concentrations tested were less successful; e.g., Glycin-Betaine did not yield a trend towards increased abundance in the observation window for all age groups (Sommerfeld 2014).

The effect of free AA and natural extracts on the tryptic enzyme activity has been evaluated in sea bass larvae. Hanken (2015) designed a classical non-conditioned response study, similar to the experiments of Pavlov (1927). The main goal was to identify substances that have the potential to stimulate trypsinogen secretion and consequently facilitate the tryptic enzyme activity in the larval gut. Four different solutions were tested: two AA mixtures (betaine–alanine–arginine–glycine (BAAG) and glycine–betaine (GB); all AA were applied as L-racemate) and two natural extracts (*Artemia* rearing water (ARW) and *Mytilus edulis* (MY) extract). Sea bass larvae were exposed to these solutions for a total period of 30 min and samples collected after 10, 20 and 30 min at the age of 14–28 dph.

The trend to higher tryptic enzyme activity was observed when the larvae were exposed to BAAG (Fig. 4.13c) and ARW (Fig. 4.14c) at 28 dph. However, no clear trend was observed for larvae exposed to GB or MY. In addition, no obvious response was observed for early ages (10 and 20 dph). These observations support once more that early larval stages may have a limited ability to smell and/or other additional triggers, such as mechanical stimuli inside of the gut, are required to induce the secretion of trypsinogen (Pedersen and Andersen 1992). These results show the importance of exploring the role of different attractants to improve the digestibility, palatability and attractiveness of MD. Such studies are therefore crucial to characterize which substances and concentrations are most effective in triggering physiological responses in fish larvae.

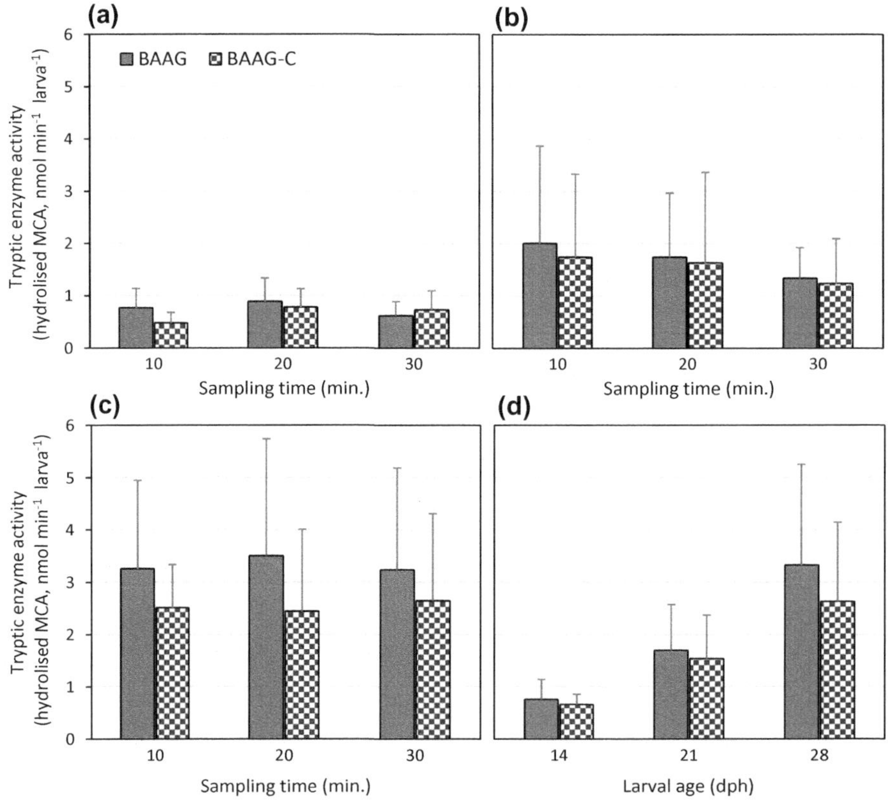

Fig. 4.13 Tryptic enzyme activity of sea bass larvae with BAAG (betaine–alanine–arginine–glycine, 10^{-4} M) as trigger solution at **a** 14 dph, **b** 21 dph, **c** 28 dph and **d** the comparison of the pooled tryptic enzyme activities versus the control group over the total experimental period (14, 21 and 28 dph). Larvae, exposed to BAAG and control (no trigger), were sampled 10, 20 and 30 min after exposure. Grey bars represent the results with BAAG as a trigger, and the spotted bars the control (BAAG-C). Values are calculated as mean and standard deviations of 18 larvae from 3 replicate tanks. Redrawn from Hanken (2015)

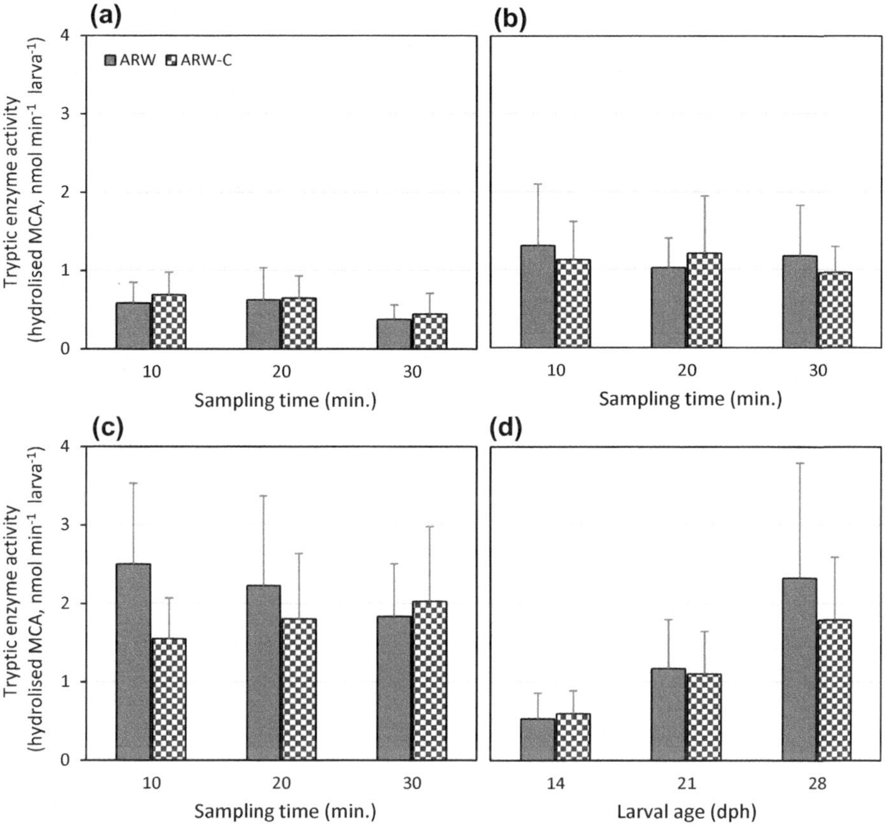

Fig. 4.14 Tryptic enzyme activity of sea bass larvae exposed to ARW (*Artemia* rearing water) as trigger solution at **a** 14 dph, **b** 21 dph, **c** 28 dph and **d** the comparison of the pooled tryptic enzyme activities versus the control group (ARW-C) over the total experimental period (14, 21 and 28 dph). Sampling of larvae, which were exposed to ARW, and control larvae (without any trigger) was performed after 10, 20 and 30 min. Grey bars represents the results from the trigger (ARW), the spotted bars the control (ARW-C). Values are calculated as mean and standard deviations of 18 larvae from 3 replicate tanks. Redrawn from Hanken (2015)

4.3.3 Gustation/Taste

The gustatory or taste system is sensitive to fewer number and sometimes different chemical stimuli than olfaction. These sensors may also require higher concentrations of the substance for activation. This may reflect the need for higher sensitivity during food search behaviour than during the ingestion and swallowing phases of feeding (Morais 2017). The external (extra-oral) taste buds play a role in the preliminary determination of suitable food, and activate reflex responses such as turning and biting or snapping (Valentinčič 2005). Taste receptors in the outer mouth area detect a wider range of substances than the receptors in the inner (oral–pharyngeal) buds (Kasumyan 1993). The latter are involved in the final decision on whether to swallow or reject the ingested food item (Pavlov and Kasumyan 2002). In some species, taste buds are present prior to the mouth opening, and the number of cells usually starts to increase days or weeks after the onset of exogenous feeding (Boglione et al. 2003; Sánchez-Amaya et al. 2007). For this reason, it has been suggested that the involvement of taste in food acceptance is more relevant towards the post-larval phase (Boglione et al. 2003).

4.3.4 Nutrient Sensing in the Gut

Recent advances in mammalian research have shown that chemosensors that were only linked to olfaction and taste are also present in the GIT (Reimann et al. 2012). The chemical sensory systems in the GIT are localized within cells in the mucosal layer, and are mainly found on the apical side of enteroendocrine cells (EEC) but also in brush cells, enterocytes and nerves. Recent studies have led to a rapidly improved understanding of the underlying mechanisms involved in GIT sensing, handling of luminal contents and the signalling pathways involved in the release of gut hormones and downstream processes. They have also contributed to a better understanding of how the gut–brain axis controls food intake, energy and glucose metabolism (Rønnestad et al. 2014). The nutrient sensing system has been linked to mucosal protection (Akiba and Kaunitz 2011) as well as physiological and health effects of intestinal microbiota (Schéle et al. 2016). These links have therefore provided novel information on how the intestine responds to small luminal molecules, substrate specificity and sensitivity of chemosensors in the duodenal mucosa and the related afferent neuronal and hormonal signalling pathways underlying food-induced gut hormone release. While GIT nutrient sensing is a "hot topic" in human nutrition very little is known in fish, and in the larval stages it still remains to be explored. However, given the importance of the nutrient sensing system for feed acceptance, digestibility, immunology, gut health, appetite and growth, it is important to advance this research field in fish larvae. A challenge when working with teleost (more than 30,000 species) is their extreme diversity in ecology, habitats, food requirements, and digestive system morphology. A possible starting point is to explore these systems in model species and species that are commercially important for the aquaculture industry. In the sections below, the nutrient sensing systems that have been explored in the GIT of mammals and that warrant studies of fish larvae are described.

Fish have often been used as a model for taste research in many classical electrophysiological studies, due to their high sensitivity to substances. Some species have highly specialized taste organs (e.g., barbels in catfishes) that are very suitable for studies of how taste nerves respond to nutrients (Yasuoka et al. 2004). Subsequently, large data-sets on the response of fish taste nerves to a wide range of compounds are already available. However, almost no information exists on fish-specific taste receptors and transduction pathways (see review by Morais 2017).

As mentioned above, the signalling pathways originating from nutrients in the teleost GIT are unknown, both in the adult and larval stages.

4.3.5 The Nutrient Sensing Mechanisms

Most of the chemical sensing in the GIT described in mammals relies on G-protein-coupled receptors (GPCRs), but the Solute Carriers (SLC; transporters) act to some extent as sensors too (Rønnestad et al. 2014; Tolhurst et al. 2012). The GPCRs involved in sensing nutrients belong to classes A and C (Reimann et al. 2012), and are cell surface receptors that can be activated by a wide range of components present in the luminal chyme. Typically, nutrients bind to the extracellular face of the receptor, often located at the apical side, and activate intracellular G proteins, generating a cascade-like downstream signalling pathway. This initiates the secretion of GI peptides or aromatic amines into the submucosal space. SLCs transfers solutes (nutrients) coupled to a transmembrane electrochemical gradient of ions, and sensing is often linked to electrogenic activity that alters the transmembrane potential, which subsequently enhances voltage-gated Ca^{2+} influx and subsequent Ca^{2+}-induced stimulation of peptide hormone secretion (see review by Reimann et al. 2012). The release of GI peptides induces local mucosal autocrine and paracrine effects. Most of these signals are transferred through receptors in the vagal, splanchnic and intrinsic afferent nerves.

4.3.6 Sensing of Protein, Peptides and Amino Acids

The mechanisms underlying protein sensing are still debated. It seems improbable that intact proteins are sensed, thereafter digestion is required to sense small peptides and AA (Gabriel and Uneyama 2013; Gribble 2012), although oligopeptides also activate sensing pathways (Choi et al. 2007a, b; Diakogiannaki et al. 2013).

There are several known AA sensors in mammals. The taste receptor type 1 heterodimer, T1R1-T1R3 (known in humans as the umami, "savory" taste receptor) is activated by natural isoforms of several aliphatic L-AA and glutamate in the monosodium form (MSG) (Nelson et al. 2002). In humans, T1R1-T1R3 specifically responds to MSG and Asp, whereas in rats and mice T1R1-T1R3 senses Phe, Leu, and Glu but not Trp (Nelson et al. 2002). Sensing by the T1R1-T1R3 receptor is enhanced by the nucleotides inosine monophosphate (IMP) or guanosine monophosphate (GMP). Studies indicate that this sensor activates CCK release from EEC (Daly et al. 2013). Heterologous expression of medaka *Oryzias latipes* T1R1-T1R3 revealed that the receptor responds to some L-amino acids such as L-Arg and L-Ser but not to D-Arg. However, the addition of IMP to L-Arg did not significantly influence activation (Oike et al. 2007), possibly indicating a lower importance of nucleic acids to luminal sensing in teleosts. This can also be explain by the low T1R1/T1R3 sequence similarity between teleosts and mammals, for e.g., zebrafish and Atlantic salmon T1R1T1R3 share only around 33–40% of their identity with the mammalian homologue gene (Gomes et al. 2016). The information available on substrate specificity of this receptor in fish has recently been reviewed by Morais (2017). To our best knowledge, no studies have demonstrated the presence of T1R1–T1R3 in the GIT of fishes. However, the downstream signalling effectors of T1R1–T1R3, α-gustducin and α-transducin, have been detected in the GIT of European sea bass (Latorre et al. 2013) using immunohistochemistry, nevertheless no studies have been successful on identifying α-gustducin in the teleost genome. Moreover, for T1R1–T1R3, as well as for the other sensory systems discussed below, ligand specificity needs to be established together with their functional characterization.

The CaSR (calcium sensing receptor; CaR) detects luminal aromatic AA such as Phe and Trp and some aliphatic and polar AA in mammals. While Ca^{2+} is the primary ligand for this receptor, CaSR can also bind di-, tri- and oligopeptides (Diakogiannaki et al. 2013). There are indications that peptides (peptone; a mix of different-sized peptides) can trigger GLP-1 secretion via the CaSR receptor (Hara 2005). CaSR is expressed in duodenal CCK-secreting EEC but not in absorptive epithelial cells. Administration of L-Phe increased CCK plasma concentrations, decreasing food intake in humans, which is probably mediated by CaSR (Liou et al. 2011) and in combination with T1R1–T1R3 (Daly et al. 2013). The LPAR5 (lysophospholipid 5 receptor; also named GPR92 or GPR93), seems to act as a receptor for protein hydrolysates and peptides (Choi et al. 2007a, b), although the mechanism involved is somewhat unclear (Tolhurst et al. 2012). The GPRC6A receptor is a broadly tuned receptor activated by several substances, including basic L-α-amino acids (Haid et al. 2011) and has been shown to promote GLP-1 secretion (Bystrova et al. 2010). Several metabotropic glutamate receptors, mGluRs (belonging to GPCR class C) are also present in the GIT. These are activated by luminal glutamate and are involved in gastric and duodenal defence and barrier protection (Kaji et al. 2013), in addition to stimulating gastric acid and pepsinogen secretion (Uneyama et al. 2006).

Sensing of luminal AA is also linked to SLCs. The H^+ coupled oligopeptide transporter PepT1 have been suggested to be an intestinal peptide chemosensor (termed as “transceptor”), stimulating both CCK (Liou et al. 2011) and GLP-1 secretion (Diakogiannaki et al. 2013) in response to peptides. The spatio-temporal expression of PepT1 mRNA has been analysed in zebrafish (Verri et al. 1993), Atlantic cod (Amberg et al. 2008) and grass carp (Liu et al. 2013) during ontogeny. In Atlantic cod, PepT1 mRNA expression was not affected by prey type, with the exception for larvae larger than 0.15 mg dry mass, with an increased PepT1 expression in zooplankton-fed compared to rotifer-fed larvae. However, no other studies have focused on this topic and the role of PepT1 in nutrient sensing during early stages of fish remains unknown.

Efforts to characterise these AA sensory systems in the GIT of teleosts have only just started. Preliminary studies in Atlantic salmon have identified several homologue genes for many of the vertebrate receptors, including several taste receptors

(T1Rs), CasR, LPAR5, GPRC6A and mGluRs, and efforts to describe their spatial expression along the GIT are also underway (Gomes et al. 2016; Rønnestad et al. 2016). Expression of tas1r2-like and tas1r3-like receptors has been reported in the gut of rainbow trout (Polakof and Soengas 2013). To the best of our knowledge, no studies in fish larvae exist.

4.3.7 Carbohydrate and Monosaccharide Sensing

Carbohydrate sensing in mammals requires the presence of di- or monomers. In carnivorous fish, carbohydrates are considered of less importance than they are for omnivorous and herbivorous fish, at least in quantitative metabolic terms. However, in the early larval stages, ingestion of algae and zooplankton with a chitinous exo-skeleton (a long-chain N-containing polymer of glucose) indicates that such carbohydrate sensing may also be of relevance in the GIT of fish.

In mammals, sensory systems include the sweet taste receptor heterodimer T1R2–T1R3 that is broadly sensitive to naturally-occurring sweet substances, including sweet-tasting D-AA. However, it fails to trigger incretin (GLP-1) release from EEC in vivo, casting doubt on the contribution of T1R2–T1R3 to direct intestinal glucose sensing (Young et al. 2013). The sodium-dependent glucose transporter 1 (SGLT-1), a key glucose transporter, is also targeted as a sweet sensor (Roeder et al. 2014). Other glucose transporters such as SGLT3 may play a minor chemosensory role (Parker et al. 2012). There is conflict evidence regarding glucose sensing, incretin secretion, and epithelial absorption of glucose by the GLUT2 transporter (Stumpel et al. 2001). There are several studies in fish related to glucose sensing, but few of them focus on the intestine (summarized by Conde-Sieira and Soengas 2016). Components of both SGLT-1 and glucokinase are found in enterocytes and EEC of rainbow trout (Polakof et al. 2010) and GLUT2, tas1r2-like, tas1r3-like receptors have also been identified (Polakof and Soengas 2013). These authors demonstrated that the gut of rainbow trout can detect simple sugars including glucose, galactose and mannose, and respond to the stimuli by changing the expression levels of glucose-sensing proteins; although the responses were different from those of other vertebrates (Polakof and Soengas 2013).

Several challenges related to the activation of digestion in fish larva culture still remain unsolved. These are particularly related to the onset of first feeding and weaning. When food items have been ingested and swallowed, the digestive process is activated, with release of digestive enzymes and other secreta. While much work on characterizing the digestive system in larval stages has targeted proteins, amino acids and lipids, very few studies have focused on carbohydrates. Future studies on the GIT nutrient sensing systems should also focus on the potential activation by substances, such as chitin, that are normally encountered when larvae feed in nature and are therefore present in large amounts in the GIT.

4.3.8 Lipid and Fatty Acid Sensing

Fat is a strong stimulus for the release of several GI peptides in mammals, including CCK, GIP and GLP-1 (Tolhurst et al. 2012). The sensory mechanisms involved, at least for triacylglycerol (TAG), appear to depend on prior hydrolysis (Foster et al. 2014), with little information regarding signalling pathways for other lipid components such as phospholipids and cholesterol. Digestion of TAG from most marine food sources mainly releases long-chain fatty acids (LCFA) which, at least in mammals, stimulate CCK secretion, with the secretory response correlating with non-esterified FA (NEFA) chain length between C12–22 and with the degree of saturation (Harden et al. 2012). Short-chain fatty acids (SCFA), including acetate, propionate, and butyrate, are present at high concentrations in the hindgut of mammals (Topping and Clifton 2001) due to production of the microbiota.

The GPCR fat-sensitive receptors in mammals are all highly expressed in the EEC, and are categorized according to their sensitivity to NEFA chain length. Free fatty acid receptors, FFA1 and FFA4, activated by medium-chain FA and LCFA, mediate some of the hormonal responses to luminal FA (Edfalk et al. 2008). FFA2 and FFA3 are NEFA receptors that selectively respond to SCFA with FFA2 activated by C2–C3 FFAs, whereas FFA3 responds to C3–C5 (Akiba et al. 2015; Reimann and Gribble 2013). GPR119 is not activated by FFA, but rather by lipid derivatives like oleoylethanolamide (OEA) and lysophosphatidyl choline (Tolhurst et al. 2012). GPR119, expressed in L and K cells, also correlates with elevated plasma concentrations of the incretins GIP and GLP-1 after administration of these lipid substrates in rats, further supported by impairment of incretin release in GPR119 knockout mice (Harden et al. 2012). CD36 (Cluster of differentiation 36), a LCFA transporter suggested to act as a fat-sensor, is also expressed in the GIT and appears to facilitate secretion of CCK in EEC (Sundaresan et al. 2013). The research on fatty acid sensing in fish has recently been summarized by Conde-Sieira and Soengas (2016). Similar to the studies of glucose sensors most of this work focused on cellular sensory systems and in key tissues like the hypothalamus and brain, but to a less extent in the GIT. In Senegalese sole post-larvae it has been shown that orally administered fatty acids enhance anorectic potential but do not activate central fatty acid sensing (Velasco et al. 2017). These studies can serve as a basis for research into specific GIT fatty acid sensory systems in fish larvae.

4.3.9 Other Sensory Systems

The classical oral taste sensors include sweet, sour, bitter, salt and umami. As discussed above the taste of "sweet" and "umami" involve several nutrient sensory systems. The sensing of sour and salt is coupled to ion-sensitive channels and may play important roles also in the GIT of fish, but these systems are beyond the scope of this overview. Bitter taste is mediated by T2Rs receptors (related to GPCRs class

A, rhodopsin-like), that comprise a group of 20–35 members in mammalian species. These receptors are activated by ligands that may be harmful or toxic to the organism and with a broad or narrow ligand specificity (Meyerhof et al. 2010). T2R gene family in teleosts is more diverse than those in tetrapods. Dong et al. (2009) proposed that the T2R gene repertoire is closely related to the dietary habits of individual species and is also associated with adaptations to dietary changes during the species' life history. It has also been suggested that vertebrates have a conserved gustatory mechanism whereby T2Rs respond to aversive tastants (Oike et al. 2007). How this relates to sensory systems and the functional response in the GIT systems in fish, including their larval stages, remains to be shown.

4.3.10 Conclusions and Recommendations

This section has focused on providing a brief status and recommendations for future research on the nutrient sensing systems in the GIT. The information available emphasises that nutrients are not only substrates in the metabolic pathways, but are active regulators of physiological processes themselves. Together with mechanical sensors, the GIT sensory systems comprise the first part of the signalling pathways that underlie neuronal signals and release of gut hormones and thus mediate the digestive process. Therefore, future research work on nutrient sensing, endocrine control, as well as studies related to digestive enzymes (see Sects. 4.1 and 4.2) in fish larvae can provide more insight into the mechanisms involved and how they are linked. This approach can provide a better overall understanding of digestion and may help to solve challenges related to first feeding and efficient weaning of fish larvae. It is also necessary to understand the role of these sensory systems and the interactions between the larval gut and the luminal bacteria; the microbiome. Future studies should also focus on understanding how different nutrients affect the gut-brain axis and feed intake. The latter topic also calls for a holistic approach to study how food sensing is controlled, which should also include a broader set of systems involved in these processes, since it involves a range of cellular systems in various tissues. Such studies ought to include, for instance, current investigations of fatty acids and glucose sensors in the hypothalamus (Conde-Sieira and Soengas 2016). Progress in this field should also enable to better understand how specific nutrients and feed ingredients affect feed intake, appetite, digestion, gut health and growth in fish.

Acknowledgements B.U. was supported by funds of the "Deutsche Bundesstiftung Umwelt" (DBU), Grant No. AZ 30586. C.N.G was supported by funds of MINECO (Spain) + FEDER, Grant BES-2012-051956 and Projects AGL2011-23722 and AGL2014-52888-R. I.R and A.G were supported by funds from the Regional Research Fund West, Norway (RFFV; No. 247978 SalmoFeedPlus) and Cargill Inc.

References

Akiba Y, Kaunitz JD (2011) Luminal chemosensing in the duodenal mucosa. Acta Physiol 201:77–84

Akiba Y, Kaji I, Kaunitz JD (2015) Short-chain fatty acid chemosensing and transport in the duodenal epithelium. Digest Dis Sci 60:2552–2553

Ali MA (ed) (1992) Rhythms in fishes. Plenum Press, New York

Amberg JJ, Myr C, Kamisaka Y et al (2008) Expression of the oligopeptide transporter, PepT1, in larval Atlantic cod (*Gadus morhua*). Comp Biochem Physiol B 150:177–182

Amole N, Unniappan S (2009) Fasting induces preproghrelin mRNA expression in the brain and gut of zebrafish, *Danio rerio*. Gen Comp Endocrinol 161:133–137

Bass J (2012) Circadian topology of metabolism. Nature 491(7424):348–356

Barahona-Fernandes MH (1979) Some effects of light intensity and photoperiod on the sea bass larvae (*Dicentrarchus labrax* (L.) reared at the Centre Oceanologique de Bretagne. Aquaculture 17:311–321

Boglione C, Giganti M, Selmo C et al (2003) Morphoecology in larval fin-fish: a new candidate species for aquaculture, *Diplodus puntazzo* (Sparidae). Aquacult Int 11:17–41

Bonacic K, Campoverde C, Gomez-Arbones J et al (2016) Dietary fatty acid composition affects food intake and gut-brain satiety signaling in Senegalese sole (*Solea senegalensis*, Kaup 1858) larvae and post-larvae. Gen Comp Endocrinol 228:79–94

Boujard T, Leatherland JF (1992) Circadian-rhythms and feeding time in fishes. Environ Biol Fishes 35:109–131

Büke E (2002) Sea bass (*Dicentrarchus labrax* L., 1781) seed Production. Turk J Fish Aquat Sc 2.1:61–69

Bystrova MF, Romanov RA, Rogachevskaja OA et al (2010) Functional expression of the extracellular-Ca^{2+}-sensing receptor in mouse taste cells. J Cell Sci 123:972–982

Cahu C, Rønnestad I, Grangier V et al (2004) Expression and activities of pancreatic enzymes in developing sea bass larvae (*Dicentrarchus labrax*) in relation to intact and hydrolyzed dietary protein; involvement of cholecystokinin. Aquaculture 238:295–308

Choi S, Lee M, Shiu AL et al (2007a) Identification of a protein hydrolysate responsive G protein-coupled receptor in enterocytes. Am J Physiol Gastrointest Liver Physiol 292:G98–G112

Choi S, Lee M, Shiu AL et al (2007b) GPR93 activation by protein hydrolysate induces CCK transcription and secretion in STC-1 cells. Am J Physiol Gastrointest Liver Physiol 292: G1366–G1375

Conde-Sieira M, Soengas JL (2016) Nutrient sensing systems in fish: impact on food intake regulation and energy homeostasis. Front Neurosci 10:603

Copeland DL, Duff RJ, Liu Q et al (2011) Leptin in teleost fishes: an argument for comparative study. Front Physiol 2:26

Da Silva RF, Kitagawa A, Sánchez Vázquez FJ (2016) Dietary self-selection in fish: a new approach to studying fish nutrition and feeding behaviour. Rev Fish Biol Fisheries 26(1):39–51

Daly K, Al-Rammahi M, Moran A et al (2013) Sensing of amino acids by the gut-expressed taste receptor T1R1-T1R3 stimulates CCK secretion. Am J Physiol Gastrointest Liver Physiol 304: G271–G282

Diakogiannaki E, Pais R, Tolhurst G et al (2013) Oligopeptides stimulate glucagon-like peptide-1 secretion in mice through proton-coupled uptake and the calcium-sensing receptor. Diabetologia 56:2688–2696

Dong D, Jones G, Zhang SY (2009) Dynamic evolution of bitter taste receptor genes in vertebrates. BMC Evol Biol 9:12

Dou S, Seikai T, Tsukamoto K (2000) Feeding behavior of Japanese flounder larvae under laboratory conditions. J Fish Biol 56:654–666

Døving KB, Kasumyan AO (2009) Chemoreception. In: Finn RN, Kapoor BG (eds) Fish larval physiology. Science Publishers, Enfield, NH, USA, pp 331–394

Drossou A (2006) CCK concentration and tryptic activity in early stages of finfish larvae. PhD thesis, Leibniz Institut für Meereswissenschaften. IFM-GEOMAR, Christian Albrechts Universität Kiel

Drossou A, Ueberschär B, Rosenthal H et al (2006) Ontogenetic development of the proteolytic digestion activities in larvae of *Oreochromis niloticus* fed with different diets. Aquaculture 256:479–488

Edfalk S, Steneberg P, Edlund H (2008) Gpr40 is expressed in enteroendocrine cells and mediates free fatty acid stimulation of incretin secretion. Diabetes 57:2280–2287

Edgar RS, Green EW, Zhao Y et al (2012) Peroxiredoxins are conserved markers of circadian rhythms. Nature 485(7399):459–464

Engrola S, Mai M, Dinis MT et al (2009) Co-feeding of inert diet from mouth opening does not impair protein utilization by Senegalese sole (*Solea senegalensis*) larvae. Aquaculture 287:185–190

Engrola S, Dinis MT, Conceição LEC (2010) Senegalese sole larvae growth and protein utilization is depressed when co-fed high levels of inert diet and *Artemia* since first feeding. Aquacult Nutr 16:457–465

Ferron A, Leggett WC (1994) An appraisal of condition measures for marine fish larvae. Adv Mar Biol 30:217–303

Forward RB, Tankersley RA, Burke JS (1996) Endogenous swimming rhythms of larval Atlantic menhaden, *Brevoortia tyrannus* Latrobe: implications for vertical migration. Mar Eco Prog Ser 204:195–207

Foster SR, Roura E, Thomas WG (2014) Extrasensory perception: odorant and taste receptors beyond the nose and mouth. Pharmacol Ther 142(1):41–61

Fujii A, Kurokawa Y, Kawai SI et al (2007) Diurnal variation of tryptic activity in larval stage and development of proteolytic enzyme activities of malabar grouper (*Epinephelus malabaricus*) after hatching. Aquaculture 270:68–76

Gabriel AS, Uneyama H (2013) Amino acid sensing in the gastrointestinal tract. Amino Acids 45:451–461

Gisbert E, Fernández I, Villamizar N et al (2014) European sea bass larval culture. In: Sánchez-Vázquez FJ (ed) Biology of European Sea Bass. Wiley, London, pp 162–202

Gomes AS, Jordal AEO, Olsen K et al (2015) Neuroendocrine control of appetite in Atlantic halibut (*Hippoglossus hippoglossus*): changes during metamorphosis and effects of feeding. Comp Biochem Physiol A 183:116–125

Gomes AS, Jordal A, Aksnes A et al (2016) Do fish possess a sensing system for amino acids in the gastrointestinal tract? In: 28th Conference of European comparative endocrinologists, Leuven, Belgium

Gorissen M, Flik G, Huising M (2006) Peptides and proteins regulating food intake: a comparative view. Anim Biol 56:447–473

Govoni JJ, Boehlert GW, Watanabe Y (1986) The physiology of digestion in fish larvae. Environ Biol Fishes 16(1–3):59–77

Gribble FM (2012) The gut endocrine system as a coordinator of postprandial nutrient homoeostasis. Proc Nutr Soc 71:456–462

Haid D, Widmayer P, Breer H (2011) Nutrient sensing receptors in gastric endocrine cells. J Mol Histol 42:355–364

Hama K, Provost E, Baranowski TC et al (2009) In vivo imaging of zebrafish digestive organ function using multiple quenched fluorescent reporters. Am J Physiol Gastrointest Liver Physiol 296:G445–G453

Hanken I (2015) Effect of free amino acids and natural extracts on the tryptic enzyme activity of European sea bass larvae (*Dicentrarchus labrax* L.). Master thesis, University of Hamburg, p 49

Hara TJ (1993) Chemoreception. CRC Press, Boca Raton, Ann Arbor, USA

Hara TJ (2005) Olfactory responses to amino acids in rainbow trout: revisited. Science Publishers Inc, Enfield, NH, USA

Harden CJ, Jones AN, Maya-Jimenez T et al (2012) Effect of different long-chain fatty acids on cholecystokinin release in vitro and energy intake in free-living healthy males. Br J Nutr 108:755–758

Hedrick TL (2008) Software techniques for two- and three-dimensional kinematic measurements of biological and biomimetic systems, Bioinspirat Biomimet 3(3):034001. https://doi.org/10.1088/1748-3182/3/3/034001 (Epub 1 Jul 2008)

Holmgren S, Olsson C (2009) The neuronal and endocrine regulation of gut function. In: Bernier NJ, Kraak GVD, Farrell A, Brauner C (eds) Fish physiology. Academic Press, Burlington, U.S.A.

Hoskins LJ, Volkoff H (2012) The comparative endocrinology of feeding in fish: insights and challenges. Gen Comp Endocrinol 176:327–335

Hoss DE, Checkley DM Jr, Settle LR (1989) Diurnal buoyancy changes in larval Atlantic menhaden (*Brevoortia tyrannus*). Rapp P-V Reun Cons Int Explor Mer 191:105–111

Houde ED (1987) Fish early life dynamics and recruitment variability. Am Fish Soc Symp 2:17–29

Kaji I, Akiba Y, Kaunitz JD (2013) Digestive physiology of the pig symposium: involvement of gut chemosensing in the regulation of mucosal barrier function and defense mechanisms. J Anim Sci 91:1957–1962

Kamisaka Y, Totland GK, Tagawa M et al (2001) Ontogeny of cholecystokinin-immunoreactive cells in the digestive tract of Atlantic halibut, *Hippoglossus hippoglossus*, larvae. Gen Comp Endocrinol 123:31–37

Kamisaka Y, Kaji T, Masuma S et al (2002) Ontogeny of cholecystokinin-immunoreactive cells in the digestive tract of bluefin tuna, *Thunnus thynnus*, larvae. Sarsia 87:258–262

Kamisaka Y, Fujii Y, Yamamoto S et al (2003) Distribution of cholecystokinin-immunoreactive cells in the digestive tract of the larval teleost, ayu, *Plecoglossus altivelis*. Gen Comp Endocrinol 134:116–121

Kamisaka Y, Drivenes Ø, Kurokawa T et al (2005) Cholecystokinin mRNA in Atlantic herring, *Clupea harengus*—molecular cloning, characterization, and distribution in the digestive tract during the early life stages. Peptides 26:385–393

Kasumyan AO (1993) Behavioral responses of early juveniles of Siberian sturgeon, *Acipenser baeri*, and stellate sturgeon, *A. stellatus* (Acipenseridae), to gustatory stimulating substances. Jpn J lchthyol 33:85–97

Knutsen JA (1992) Feeding-behavior of north-sea turbot (*Scophthalmus maximus*) and dover sole (*Solea solea*) larvae elicited by chemical stimuli. Mar Biol 113:543–548

Kolkovski S, Arieli A, Tandler A (1997) Visual and chemical cues stimulate microdiet ingestion in sea bream larvae. Aquacult Int 5:527–536

Kolkovski S, Czesny S, Dabrowski K (2000) Use of krill hydrolysate as a feed attractant for fish larvae and juveniles. J World Aquacult Soc 31:81–88

Kortner TM, Overrein I, Øie G et al (2011a) The influence of dietary constituents on the molecular ontogeny of digestive capability and effects on growth and appetite in Atlantic cod larvae (*Gadus morhua*). Aquaculture 315:114–120

Kortner TM, Overrein I, Øie G et al (2011b) Molecular ontogenesis of digestive capability and associated endocrine control in Atlantic cod (*Gadus morhua*) larvae. Comp Biochem Physiol A 160:190–199

Kotani T, Fushimi H (2011) Determination of appropriate feeding schedules from diel feeding rhythms in finfish larviculture. Aquaculture 315:104–113

Koven W, Rojas-García C, Finn R et al (2002) Stimulatory effect of ingested protein and/or free amino acids on the secretion of the gastro-endocrine hormone cholecystokinin and on tryptic activity, in early-feeding herring larvae, *Clupea harengus*. Mar Biol 140:1241–1247

Koven W, Schulte P (2012) The effect of fasting and refeeding on mRNA expression of PepT1 and gastrointestinal hormones regulating digestion and food intake in zebrafish (*Danio rerio*). Fish Physiol Biochem 38:1565–1575

Kurokawa T, Suzuki T, Andoh T (2000) Development of cholecystokinin and pancreatic polypeptide endocrine systems during the larval stage of Japanese flounder, *Paralichthys olivaceus*. Gen Comp Endocrinol 120:8–16

Kulczykowska E, Sánchez-Vázquez FJ (2010) Neurohormonal regulation of feed intake and response to nutrients in fish: aspects of feeding rhythm and stress. Aquac Res 41:654–667

Latorre R, Mazzoni M, De Giorgio R et al (2013) Enteroendocrine profile of alpha-transducin immunoreactive cells in the gastrointestinal tract of the European sea bass (*Dicentrarchus labrax*). Fish Physiol Biochem 39:1555–1565

Lauff M, Hofer R (1984) Proteolytic enzymes in fish development and the importance of dietary enzymes. Aquaculture 37(4):335–346

Liddle RA (2012) Regulation of pancreatic secretion. In: Johnson LR (ed) Physiology of the gastrointestinal tract, 15 ed. Academic Press, Burlington

Liou AP, Chavez DI, Espero E et al (2011) Protein hydrolysate-induced cholecystokinin secretion from enteroendocrine cells is indirectly mediated by the intestinal oligopeptide transporter PepT1. Am J Physiol Gastrointest Liver Physiol 300:G895–G902

Liu Z, Zhou Y, Feng J et al (2013) Characterization of oligopeptide transporter (PepT1) in grass carp (*Ctenopharyngodon idella*). Comp Biochem Physiol B 164:194–200

López-Olmeda JF, Noble C, Sánchez-Vázquez FJ (2011) Does feeding time affect fish welfare? Fish Physiol Biochem 38:143–152

Ma AJ, Liu XZ, Xu YJ et al (2006) Feeding rhythm and growth of the tongue sole, *Cynoglossus semilaevis* Günther, during its early life stages. Aquac Res 37:586–593

Mai MG, Engrola S, Morais S et al (2009) Co-feeding of live feed and inert diet from first-feeding affects *Artemia* lipid digestibility and retention in Senegalese sole (*Solea senegalensis*) larvae. Aquaculture 296:284–291

Manning AJ, Murray HM, Gallant JW et al (2008) Ontogenetic and tissue-specific expression of preproghrelin in the Atlantic halibut, *Hippoglossus hippoglossus* L. J Endocrinol 196:181–192

Marui T, Caprio J (1992) Teleost gustation. Fish and fisheries series. Springer, Dordrecht, The Netherlands, pp 171–198

Mata-Sotres JA, Martínez-Rodríguez G, Pérez-Sánchez J et al (2015) Daily rhythms of clock gene expression and feeding behavior during the larval development in gilthead *seabream Sparus aurata*. Chronobiol Int 32(8):1061–1074

Mata-Sotres JA, Moyano FJ, Martínez-Rodríguez G et al (2016) Daily rhythms of digestive enzyme activity and gene expression in gilthead seabream (*Sparus aurata*) during ontogeny. Comp Biochem Physiol A 197:43–51

Meyerhof W, Batram C, Kuhn C et al (2010) The molecular receptive ranges of human TAS2R bitter taste receptors. Chem Senses 35:157–170

Miura T, Maruyama K, Kaiya H et al (2009) Purification and properties of ghrelin from the intestine of the goldfish, *Carassius auratus*. Peptides 30:758–765

Moguel-Hernández I, Peña R, Andree KB et al (2016) Ontogeny changes and weaning effects in gene expression patterns of digestive enzymes and regulatory digestive factors in spotted rose snapper (*Lutjanus guttatus*) larvae. Fish Physiol Biochem 42(5):1319–1334

Morais S (2017) The physiology of taste in fish: potential implications for feeding stimulation and gut chemical sensing. Rev Fish Sci Aquac 25:133–149

Murashita K, Kurokawa T, Nilsen TO et al (2009) Ghrelin, cholecystokinin, and peptide YY in Atlantic salmon (*Salmo salar*): molecular cloning and tissue expression. Gen Comp Endocrinol 160:223–235

Navarro-Guillén C, Moyano FJ, Yúfera M (2015) Diel food intake and digestive enzyme production patterns in *Solea senegalensis* larvae. Aquaculture 435:33–42

Navarro-Guillén C, Rønnestad I, Jordal AEO et al (2017a) Involvement of cholecystokinin (CCK) in the daily pattern of gastrointestinal regulation of Senegalese sole (*Solea senegalensis*) larvae reared under different feeding regimes. Comp Biochem Physiol A 203:126–132

Navarro-Guillén C, Yúfera M, Engrola S (2017b) Ghrelin in Senegalese sole (*Solea senegalensis*) post-larvae: Paracrine effects on food intake. Comp Biochem Physiol A 204:85–92

Navarro-Guillén C, Yúfera M, Engrola S (2017c) Daily feeding and protein metabolism rhythms in Senegalese sole post-larvae. Biol Open 6:77–82
Nelson G, Chandrashekar J, Hoon MA et al (2002) An amino-acid taste receptor. Nature 416:199–202
Oike H, Nagai T, Furuyama A et al (2007) Characterization of ligands for fish taste receptors. J Neurosci 27:5584–5592
Okauchi M, Oshiro T, Kitamura S et al (1980) Number of rotifer, *Brachionus plicatilis*, consumed daily by a larva and juvenile of porgy, *Acanthopagrus schlegeli*. Bull Natl Res Inst Aquac 1:39–45 (in Japanese with English abstract)
Østergaard P, Munk P, Janekarn V (2005) Contrasting feeding patterns among species of fish larvae from the tropical Andaman Sea. Mar Biol 146:595–606
Parker HE, Adriaenssens A, Rogers G et al (2012) Predominant role of active versus facilitative glucose transport for glucagon-like peptide-1 secretion. Diabetologia 55:2445–2455
Pavlov IP (1927) Conditioned reflexes; an investigation of the physiological activity of the cerebral cortex. Oxford University Press, Humphrey Milford, London
Pavlov DS, Kasumyan AO (2002) Feeding diversity in fishes: trophic classification of fish. Jpn J lchthyol 42:S137–S159
Pedersen BH, Andersen KP (1992) Induction of trypsinogen secretion in herring larvae (*Clupea harengus*). Mar Biol 112:559–565
Pedersen BH, Ugelstad I, Hjelmeland K (1990) Effects of a transitory, low food supply in the early life of larval herring (Clupea harengus) on mortality, growth and digestive capacity. Mar Biol 107(1):61–66
Picapedra PHS, Lansac-Tôha FA, Bialetzki A (2015) Diel vertical migration and spatial overlap between fish larvae and zooplankton in two tropical lakes, Brazil. Braz J Biol 75(2):352–361
Pittendrigh CS (1993) Temporal organisation: reflections of a Darwinian clock-watcher. Annu Rev Physiol 55:16–54
Polakof S, Alvarez R, Soengas JL (2010) Gut glucose metabolism in rainbow trout: implications in glucose homeostasis and glucosensing capacity. Am J Physiol Regul Integr Comp Physiol 299: R19–R32
Polakof S, Soengas JL (2013) Evidence of sugar-sensitive genes in the gut of a carnivorous fish species. Comp Biochem Physiol B 166:58–64
Reebs SG (1991) Do fish sleep at night? FAMA 14(4):59–60
Reebs SG (2002) Plasticity of diel and circadian activity rhythms in fishes. Rev Fish Biol Fisher 12:349–371
Reebs SG (2011) Circadian rhythms in fish. In: Farrell AP (ed) Encyclopedia of fish physiology: from genome to environment, vol 1. Academic Press, San Diego, pp 736–743
Reimann F, Tolhurst G, Gribbl FM (2012) G-protein-coupled receptors in intestinal chemosensation. Cell Metab 15:421–431
Reimann F, Gribble FM (2013) A tag to track short chain fatty acid sensors. Endocrinology 154:3492–3494
Roeder PV, Geillinger KE, Zietek TS et al (2014) The role of SGLT1 and GLUT2 in intestinal glucose transport and sensing. PLoS ONE 9(2):e89977
Rojas-García CR, Rønnestad I, Ueberschär B (2001) Combined sensitive analytical methods for cholecystokinin levels and tryptic activity in individual fish larvae. J Exp Mar Biol Ecol 265:01–115
Rojas-García CR, Rønnestad I (2002) Cholecystokinin and tryptic activity in the gut and body of developing Atlantic halibut larvae: evidence for participation in the regulation of protein digestion. J Fish Biol 61:973–986
Rojas-García CR, Morais S, Rønnestad I (2011) Cholecystokinin (CCK) in Atlantic herring (*Clupea harengus* L.)—Ontogeny and effects of feeding and diurnal rhythms. Comp Biochem Physiol A 158:455–460
Rønnestad I (2002) Control and efficiency of digestive function of marine fish larvae. Avances en Nutrición Acuícola, Cancún, Quintana Roo, México

Rønnestad I, Yúfera M, Ueberschär B et al (2013) Feeding behaviour and digestive physiology in larval fish: current knowledge, and gaps and bottlenecks in research. Rev Aquacult 5(Suppl. 1): S59–S98

Rønnestad I, Akiba Y, Kaji I et al (2014) Duodenal luminal nutrient sensing. Curr Opin Pharmacol 19:67–75

Rønnestad I, Jordal AEO, Gomes AS (2016) Nutrient sensing in Atlantic Salmon—Exploring the sensory systems for amino acids in the gastrointestinal tract. FASEB J 30:1020–1022

Rønnestad I, Gomes AS, Murashita K et al (2017) Appetite-controlling endocrine systems in teleosts. Front Endocrinol 8:73

Sánchez-Amaya ML, Ortiz-Delgado JB, García-López A et al (2007) Larval ontogeny of redbanded seabream *Pagrus auriga* Valenciennes, 1843, with special reference to the digestive system. A histological and histochemical approach. Aquaculture 263:259–279

Sánchez-Vázquez FJ, Madrid JA, Zamora S (1995a) Circadian rhythms of feeding activity in sea bass, *Dicentrarchus labrax* L: dual phasing capacity of diel demand-feeding pattern. J Biol Rhythms 10:256–266

Schéle E, Grahnemo L, Anesten F et al (2016) Regulation of body fat mass by the gut microbiota: possible mediation by the brain. Peptides 77:54–59

Sclafani M, Taggart CT, Thompson KR (1993) Condition, buoyancy and the distribution of larval fish: implications for vertical migration and retention. J Plankton Res 15(4):413–435

Shapiro CM, Hepburn HR (1976) Sleep in a schooling fish, *Tilapia mossambica*. Physiol Behav 16:613–615

Shoji J, Maehara T, Tanaka M (1999) Diel vertical movement and feeding rhythm of Japanese mackerel larvae in the central Seto inland sea. Fish Sci 65:726–730

Sommerfeld D (2014) Dissolved amino acids and natural extracts as attractant for sea bass larvae (*Dicentrarchus labrax* Linnaeus, 1758). Master thesis, University of Rostock, 57 pp

Stumpel F, Burcelin R, Jungermann K et al (2001) Normal kinetics of intestinal glucose absorption in the absence of GLUT2: evidence for a transport pathway requiring glucose phosphorylation and transfer into the endoplasmic reticulum. Proc Natl Acad Sci USA 98:11330–11335

Sundaresan S, Shahid R, Riehl TE et al (2013) CD36-dependent signaling mediates fatty acid-induced gut release of secretin and cholecystokinin. FASEB J 27:1191–1202

Sundström G, Larsson TA, Brenner S et al (2008) Evolution of the neuropeptide Y family: new genes by chromosome duplications in early vertebrates and in teleost fishes. Gen Comp Endocrinol 155:705–716

Thorpe JE (ed) (1978) Rhythmic activity of fishes. Academic Press, London, UK, pp 69–90

Tillner R, Rønnestad I, Harboe T et al (2013a) Hormonal control of tryptic enzyme activity in Atlantic cod larvae (*Gadus morhua*): involvement of cholecystokinin during ontogeny and diurnal rhythm. Aquaculture 402:133–140

Tillner R, Rønnestad I, Harboe T et al (2013b) Evidence for a regulatory loop between cholecystokinin (CCK) and tryptic enzyme activity in Atlantic cod larvae (*Gadus morhua*). Comp Biochem Physiol A 166:490–495

Tillner R, Rønnestad I, Dhert P et al (2014) The regulatory loop between gut cholecystokinin and tryptic enzyme activity in sea bass (*Dicentrarchus labrax*) larvae is influenced by different feeding regimes and trigger substances. Aquaculture 420–421:139–146

Tinoco AB (2015) Implicación de los péptidos leptina, colecistoquinina y ghrelina, y del derivado lipídico oleiletanolamida en la regulación del comportamiento alimentario en los teleósteos. PhD thesis, Universidad Complutense de Madrid, Spain

Tolhurst G, Reimann F, Gribble FM (2012) Intestinal sensing of nutrients. Handb Exp Pharmacol 209:309–335

Topping DL, Clifton PM (2001) Short-chain fatty acids and human colonic function: roles of resistant starch and nonstarch polysaccharides. Physiol Rev 81:1031–1064

Ueberschär B (1988) Determination of the nutritional condition of individual marine fish larvae by analyzing their proteolytic enzyme activities with a highly sensitive fluorescence technique. Meeresforsch Rep mar Res 32:144–154 (Ber. dt. wiss. Kommn Meeresforsch.)

Ueberschär B, Pedersen BH, Hjelmeland K (1992) Quantification of trypsin with radioimmunoassay in herring larvae (*Clupea harengus*) compared with a highly sensitive fluorescence technique to determine tryptic enzyme activity. Mar Biol 113(3):469–473

Ueberschär B (1993) Measurement of proteolytic enzyme activity: significance and application in larval fish research. In: Walther BT, Fyhn HJ (Eds) Physiological and Biochemical Aspects of Fish Development. University of Bergen Norway, pp. 233–239

Ueberschär B (1995) The use of tryptic enzyme activity measurement as a nutritional condition index: laboratory calibration data and field application. ICES mar Sci Symp 201:119–129

Uneyama H, Niijima A, San Gabriel A et al (2006) Luminal amino acid sensing in the rat gastric mucosa. Am J Physiol Gastrointest Liver Physiol 291:G1163–G1170

Valentinčič T, Lamb CF, Caprio J (1999) Expression of a reflex biting/snapping response to amino acids prior to first exogenous feeding in salmonid alevins. Physiol Behav 67(4):128–139

Valentinčič T (2005) Olfactory discrimination in fishes. Science Publishers Inc, Enfield, NH, USA

Velasco C, Bonacic K, Soengas JL et al (2017) Orally administered fatty acids enhance anorectic potential but do not activate central fatty acid sensing in Senegalese sole post-larvae. J Exp Biol 220:677–685

Verri T, Kottra G, Romano A et al (1993) Molecular and functional characterisation of the zebrafish (*Danio rerio*) PEPT1-type peptide transporter. FEBS Lett 549:115–122

Vikeså V, Nankervis L, Remø SC et al (2015) Pre- and postprandial regulation of ghrelin, amino acids and IGF1 in Atlantic salmon (*Salmo salar* L.) at optimal and elevated seawater temperatures. Aquaculture 438:159–169

Vitaterna MS, Takahashi JS, Turek FW (2001) Overview of circadian rhythms. Alcohol Res Health 25(2):85–93

Volkoff H, Canosa LF, Unniappan S et al (2005) Neuropeptides and the control of food intake in fish. Gen Comp Endocrinol 142:3–19

Volkoff H, Unniappan S, Kelly SP (2009a) The endocrine regulation of food intake. In: Bernier NJ, Kraak GVD, Farrell A, Brauner C (eds) Fish physiology. Academic Press, Burlington, U.S.A.

Volkoff H, Xu M, Macdonald E (2009b) Aspects of the hormonal regulation of appetite in fish with emphasis on goldfish, Atlantic cod and winter flounder: notes on actions and responses to nutritional, environmental and reproductive changes. Comp Biochem Physiol A 153:8–12

Volkoff H, Hoskins LJ, Tuziak SM (2010) Influence of intrinsic signals and environmental cues on the endocrine control of feeding in fish: potential application in aquaculture. Gen Comp Endocrinol 167:352–359

Webb KA (2008) Cholecystokinin and the ontogeny of digestion in the red drum (*Sciaenops ocellatus*). PhD thesis, The University of Texas at Austin

Webb KA Jr, Khan IA, Nunez BS et al (2010) Cholecystokinin: molecular cloning and immunohistochemical localization in the gastrointestinal tract of larval red drum, *Sciaenops ocellatus* (L.). Gen Comp Endocrinol 166:152–159

Webb K, Rønnestad I (2011) Regulation of digestive processes. In: Holt GJ (ed) Larval fish nutrition. Wiley, NJ

Werner RG, Blaxter, JHS (1981) The effect of prey density on mortality, growth and food consumption in larval herting (*Clupea harengus* L). Rapp P-V Réun Cons int Explor Mer 178:405–408

Wurtsbaugh WA, Neverman D (1988) Post-feeding thermotaxis and daily vertical migration in a larval fish. Nature 333:846–848

Xu M, Volkoff H (2009) Molecular characterization of ghrelin and gastrin-releasing peptide in Atlantic cod (*Gadus morhua*): cloning, localization, developmental profile and role in food intake regulation. Gen Comp Endocrinol 160:250–258

Yamamoto S (1996) Diurnal rhythm of feeding activity and estimation of daily ration of the larval red spotted grouper *Epinephelus akaara*. Nippon Suisan Gakkaishi 62:399–405 (in Japanese with English abstract)

Yamamoto S, Fujii Y, Murata M (2003) Developmental changes in diurnal feeding activity and daily food consumption in ayu, *Plecoglossus altivelis*, larvae. Suisanzoshoku 51:73–80 (in Japanese with English abstract)

Yamamoto S, Sugino H, Chuuriki K et al (2005) Developmental changes in diel feeding activity and estimation of daily food consumption in hatchery-reared larval and juvenile marbled sole *Pleuronectes yokohamae*. Aquacult Sci 53:157–165 (in Japanese with English abstract)

Yasuoka A, Aihara Y, Matsumoto I et al (2004) Phospholipase C-beta 2 as a mammalian taste signaling marker is expressed in the multiple gustatory tissues of medaka fish, *Oryzias latipes*. Mech Dev 121:985–989

Young RL, Chia B, Isaacs NJ et al (2013) Disordered control of intestinal sweet taste receptor expression and glucose absorption in type 2 diabetes. Diabetes 62:3532–3541

Zambonino-Infante JL, Gisbert E, Sarasquete C et al (2008) Ontogeny and physiology of the digestive system of marine fish larvae. In: Cyrino JEO, Bureau D, Kapoor BG (eds) Feeding and digestive functions of fishes. Science Publishers Inc, Enfield, USA

Zeytin S, Schulz C, Ueberschär B (2016) Diurnal patterns of tryptic enzyme activity under different feeding regimes in gilthead sea bream (*Sparus aurata*) larvae. Aquaculture 457:85–90

Chapter 5
Phospholipids in Marine Larval Rearing

Keshuai Li, Rolf Erik Olsen, Yang Jin and Yngvar Olsen

Abstract Recent studies have shown that fish larvae require not only a certain quantitative amount of dietary phospholipid (PL) in their feed, but they also depend on the quality of the dietary PL, and their docosahexaenoic acid (DHA) content is particularly important for normal growth and functional development. The most commonly used live feeds in aquaculture, rotifers (*Brachionus* sp.) and brine shrimp (*Artemia* sp.), do not contain adequate amounts of DHA in their PL. Therefore, there is an emerging need to learn more on how PL of live feed organisms can be efficiently enriched. In this chapter, we discussed the factors that could affect the enrichment of DHA in PL of live feed and suggested some strategies that could increase the DHA levels in PL of rotifers. The mechanism behind the PL requirement of fish larvae is not well understood and the overall objective of our studies has been to obtain more knowledge of ontogenesis of PL synthesis capability of early stages of Atlantic cod *Gadus morhua*. Transcriptome analysis of larvae in different stages was carried out using microarray, to evaluate the effect of development on the expression of key genes of PL biosynthesis. Moreover, labeled lipid precursors were tube fed to cod larvae to evaluate their capacity of PL synthesis. The larvae showed relatively high biosynthesis ability of PL compared to neutral lipids. Our overall data suggested that besides the possible limited de novo PL synthesis ability in the intestine, other metabolic constraints should also be considered.

Keywords Live feed · DHA · Enrichment · Phospholipid biosynthesis
Larvae

K. Li (✉) · R. E. Olsen · Y. Jin · Y. Olsen
Norwegian University of Science and Technology, NTNU, Trondheim, Norway
e-mail: keshuai.li@biomar.no

Present Address:
K. Li
BioMar Norway, Havnegata 9, Pirsenteret I, 7010 Trondheim, Norway

M. Yúfera (ed.), *Emerging Issues in Fish Larvae Research*,
https://doi.org/10.1007/978-3-319-73244-2_5

5.1 Phospholipid of Live Feed

Live feeds are important food supply for many species of interest in aquaculture during their early life stages. Rotifers (e.g., *Brachionus* Nevada and *Brachionus* Cayman) and brine shrimp (*Artemia* sp.) are the most commonly used live feed organisms in aquaculture due to their relatively cost-efficient protocols for stable mass production (Conceiçao et al. 2010). However, the content of n-3 highly unsaturated fatty acids (HUFA), particularly docosahexaenoic acid (DHA, 22:6n3) and eicosapentanoic acid (EPA, 20:5n3), in both rotifers and *Artemia* are insufficient for many marine larval fish species. Techniques to produce and enrich live feed with lipids that contain high levels of n-3 HUFA have therefore been established over the last decades (Coutteau and Sorgeloos 1997; Dhert et al. 2001), but there are still questions raised on the lipid quality of live feeds used for larval fish species that are attractive for cultivation in northern countries, such as Atlantic cod *Gadus morhua* L. (Garcia et al. 2008; O'Brien-MacDonald et al. 2006).

Juveniles produced using marine copepod species as live feed generally grow better and copepods could be used as a reference for the nutritional requirements of marine fish larvae (Evjemo et al. 2003; Hamre 2006). Copepod nauplii and copepodites usually have high levels of n-3 HUFA, and specifically high contents of DHA in their glycerolphospholipid (PL) (Bell et al. 2003; Li et al. 2015a; Tocher et al. 2008), which may contain up to 40% DHA of their total fatty acids (Overrein 2010). The n-3 HUFAs seem to be more beneficial for larval growth and development when they are incorporated in PL than in neutral lipids, as shown for European sea bass *Dicentrarchus labrax* (Gisbert et al. 2005), gilthead seabream *Sparus aurata* (Izquierdo et al. 2001) and for cod (Kjørsvik et al. 2009; Wold et al. 2007). The PL associated DHA and EPA has also showed higher bioavailability for rodents and humans (Cansell et al. 2003; Rossmeisl et al. 2012; Schuchardt et al. 2011).

Rotifers, *Artemia* and copepods, contain PL in similar quantitative amounts (around 40–60 mg g^{-1} dry weight) (Bergvik et al. 2012; Harel et al. 1999; Olsen 2004; Rainuzzo et al. 1994a). It appears difficult to modify the quantitative contents of PL (e.g., mg PL g^{-1} dry weight) in live feed organisms since they are mainly structural lipids associated to membranes (Coutteau and Sorgeloos 1997; McEvoy et al. 1996; Rainuzzo et al. 1994b). It is mainly the contents of triacylglycerol (TAG) that are increased during lipid enrichment of rotifers and *Artemia*. On the other hand, it appears possible, within limits, to manipulate the fatty acid composition of PL, including the percentage of DHA (McEvoy et al. 1996; Rainuzzo et al. 1994a). Among the numerous studies of n-3 HUFA enrichment in the live feed, very few have analyzed enrichment of PL of live feed (Fernandez-Reiriz and Labarta 1996; Frolov et al. 1991; Guinot et al. 2013b; Olsen et al. 2014; Sargent et al. 1999). Fish larvae may have specific requirements to their dietary PL composition, and there is an emerging need to learn more on how PL of live feed organisms can be efficiently enriched by n-3 HUFA. Phosphatidylcholine (PC) and phosphatidylethanolamine (PE) are the most abundant PLs of eukaryotic membranes, and for rotifers, PC and PE contributes up to 99% of the total PL (Rainuzzo et al. 1994b). Our enrichment studies are therefore focused on DHA content in PC and PE.

The main focus of this chapter is on two main challenges related to nutrition of cod and other marine fish larvae with the following objectives: to increase the content of DHA in PC and PE in cultivated rotifers and *Artemia*; to improve the understanding of the specific requirement of dietary PL in the early stages of fish larvae.

5.2 Manipulation of Fatty Acid in PC and PE of Live Feed

5.2.1 Rotifer Enrichment

Marine PL versus TAG (triacylglycerol) Most of the enrichment procedures are developed using TAG or fatty acid ethyl esters as dietary lipids and few studies have used PL-based diets for lipid and fatty acid enrichment of rotifers, and little is therefore known about the roles of dietary PC and PE on rotifer lipid accumulation (Rainuzzo et al. 1994b). In early experiments, the *Brachionus* Nevada lineage exhibited very high rate of TAG accumulation during short-term enrichment by an oil emulsion extracted from roe of Atlantic halibut, and dietary PL was effectively digested and incorporated as TAG in the rotifer body (Rainuzzo et al. 1994b). A later experiment was designed to test whether it was more efficient to manipulate the DHA levels in PC and PE of rotifers by using enrichment diets based on marine PL rather than that based on marine TAG. Two commonly used rotifer lineage were included, because there may be differences in lipid metabolism between rotifer lineage (Li et al. 2015a).

Long-term enrichment techniques were used (Rainuzzo et al. 1994b), where efficient rotifer enrichment of n-3 HUFA can be achieved by feeding a complete diet over time, a diet with relatively low lipid content and a high percentage n-3 HUFA of total fatty acids, which supported fast growth and reproduction of the rotifer cultures. A steady state of growth and a constant biochemical composition can be reached in 4–5 days of semi-continuously cultivation at high growth rate (~ 0.4 day^{-1}). More than 90% of the rotifer individuals will then be produced in the period.

Rotifer cultures of *Brachionus* Nevada and Cayman were grown semi-continuously, fed live *Rhodomonas baltica* algae, and cultures were diluted daily (20% day^{-1}). The enrichment diets Marol E (TAG-rich) or cod roe emulsion (PL-rich) were added to both cultures (Nevada 40 ng ind^{-1} day^{-1} and Cayman 25 ng ind^{-1} day^{-1}, respectively). After 4 days of enrichment at a dilution rate of 20% day^{-1}, the percentage DHA of the total fatty acids in rotifers enriched with Marol E were generally higher than that enriched with Cod roe emulsion (Fig. 5.1a, b). However, the DHA levels in copepods were still much higher than in enriched

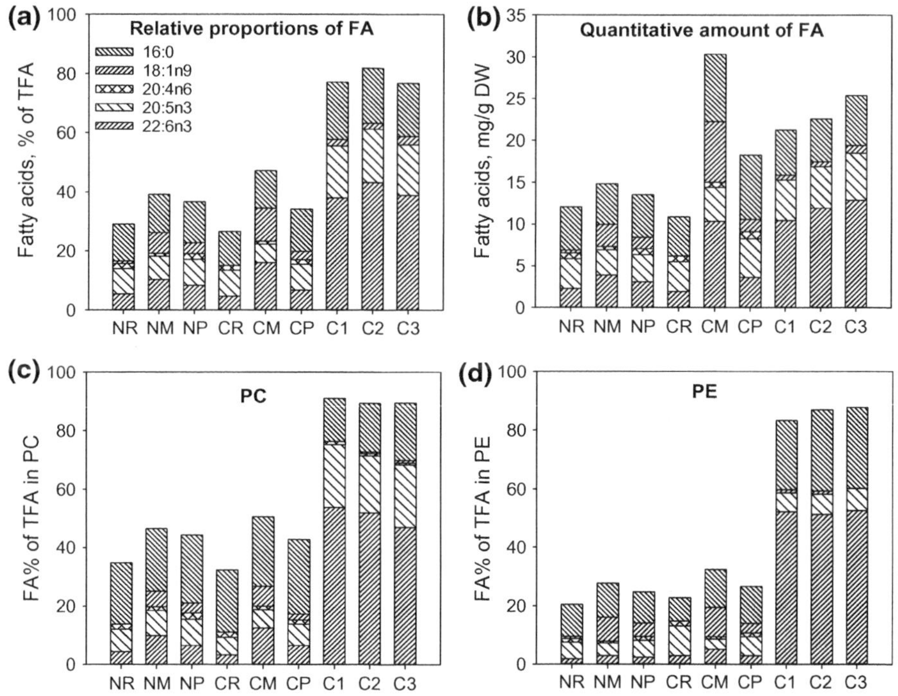

Fig. 5.1 Selected fatty acid (FA) composition of total fatty acid (TFA) in total lipid (**a** and **b**), PC (**c**) and PE (**d**) of enriched rotifers compared to copepods. *NR* Nevada cultivated with *Rhodomonas baltica*; *NM* Nevada enriched by Marol E; *NP* Nevada enriched by PL emulsion; *CR* Cayman cultivated with *R. baltica*; *CM* Cayman enriched with Marol E; *CP* Cayman enriched with PL emulsion. *C1* copepod size between 300 and 400 μm; *C2* copepod size between 400 and 500 μm; *C3* copepod size between 700 and 1000 μm. Data from Li et al. (2015a)

rotifers ($P < 0.05$). Nevertheless, Cayman rotifers fed Marol E (CM) showed comparable ($P > 0.05$) quantitative amounts of DHA to natural harvested copepods (Fig. 5.1b). The Cayman lineage appeared to be slightly more efficient in incorporating HUFAs than the Nevada lineage.

The PC-DHA (DHA in PC) and PE-DHA (DHA in PE) levels of copepods were up to 50%, far higher than the enriched rotifers (Fig. 5.1c, d). For rotifers, the highest PC-DHA (12.5%) and PE-DHA (5.1%) levels were found in Cayman rotifers enriched with Marol E. The increase of DHA in PL of rotifers was mainly due to an increase of DHA in PC, whereas the PE-DHA was only weakly affected by dietary lipids and the enrichment procedure. The percent PC-DHA levels in rotifers were positively related to the total dietary DHA levels ($P < 0.0001$), independent of the PC-DHA content in the diets. Diets based on marine PL were not more efficient than diets based on marine TAG for the DHA manipulation of PC and PE of rotifers.

Multigain *versus* DHA selco (DSelco) Among the selected fatty acids shown in Fig. 5.1b, the content of 18:1n9 was apparently far higher in CM rotifers than in copepods. It might be possible to further increase the DHA percentage in total fatty acid and in PC and PE by reducing competing fatty acids, such as 18:1n9 in the enrichment diets. Assuming that 18:1n9 may compete with DHA in incorporating into PL, the DHA enrichment efficiency was compared by two commercial diets; Multigain containing low 18:1n9 and high 16:0 and DSelco containing high 18:1n9 and low 16:0 (Li and Olsen 2015). Total lipid and the fatty acid composition in total lipid, PC and PE of rotifers were measured throughout the enrichment period of about 24 h (Fig. 5.2). At 0 h, a similar DHA ration (20 ng ind^{-1}) from DSelco or Multigain was provided for enrichment. For the multigrain treatment, another portion was added at 12 h to achieve similar total lipid ration (130–150 ng ind^{-1}). The highest DHA levels in PC (23.0%) and PE (16.5%) in *Brachionus* Cayman were found after 24 h enrichment with Multigain, and levels were higher than obtained by DSelco (Fig. 5.2 b, c), and also higher than that obtained by Marol E or Cod roe enrichment (Li et al. 2015a; Olsen et al. 2014).

It could be suggested that the higher percentage fraction of DHA provided using Multigain rather than DSelco explained why PC and PE enrichments were most efficient for incorporating DHA in PC and PE of the rotifer, but the DHA ration given up to 12 h of enrichment was the same for both diets. A second alternative explanation might be that a lipid emulsion, which constitutes mainly lipids, is more poorly assimilated and therefore metabolically less efficient for transferring DHA to rotifer PL than a slightly more complete enrichment feed, with some proteins added (13%, from algae and yeast) and with high DHA level. The enhanced efficiency might be a combination of these effects, and a Multigain-type diet would then in all events be most efficient for PL enrichment of DHA and HUFA.

The Marol E enrichment studies put, however, some doubts about this simple explanation. The enrichment emulsion (Marol E) contained more than 40% DHA in total lipid, and the DHA% in total lipid of enriched rotifers could reach 22.3–26.2%. However, the maximum DHA% in PL was still only about 12% (Li et al. 2015a). This suggested that provision of diets with high DHA content might not alone be enough to obtain high DHA levels in PL of rotifers.

The results accordingly suggested that MUFA was replaced by PUFA in PL of rotifers during the enrichment process. High contents of 18:1n9 in the Marol E and short chain n-3 PUFAs like 18:3n3 and 18:4n3 in *R. baltica* may have resulted in a relatively lower level of DHA incorporated into PL. Beside this, the high 16:0 levels in Multigain may have a stimulating function for the biosynthesis of PL of growing and reproducing rotifers. It has been reported that 16:0 may increase phospholipid synthesis in cultured Caco-2 cells (Van Greevenbroek et al. 1995) and reduce the intracellular lipid droplet formation in Arctic char (Olsen et al. 2000). In addition, there are indications that 16:0 may facilitate the incorporation of dietary DHA into PC of human plasma (Subbaiah et al. 1993). We therefore suggest that an efficient enrichment diet should have high content of DHA, a low content of MUFA and C18 PUFAs, and most likely a high content of 16:0.

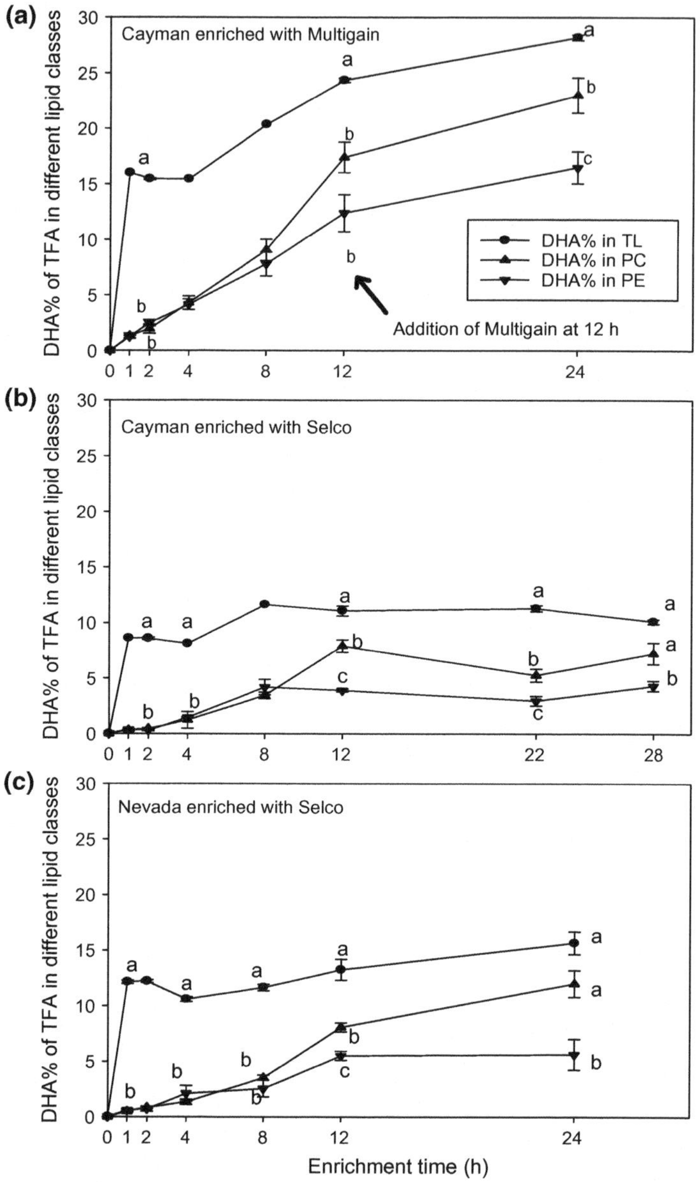

Fig. 5.2 Changes in DHA% in total lipid (circles), PC (triangles) and PE (inverted triangles) as a function of time in *Brachionus* Cayman enriched with Multigain (**a**) and DSelco (**b**) and in *Brachionus* Nevada enriched with Dselco (**c**). Error bars express standard errors of the mean. Different letter inserted for data points for specific time indicate significantly different values ($P < 0.05$). Figure modified from Li and Olsen (2015)

Stability Post-enrichment The enriched percentage DHA in PC and PE was stable at 10 °C for at least 24 h post enrichment under starving conditions ($P > 0.05$), whereas a significant ($P < 0.05$) decrease of DHA was observed during starvation at 20 °C. However, even after starvation for 24 h at 20 °C, the DHA levels in total lipid, PC and PE were 21.4, 14.3 and 10.5%, respectively, which can still be suitable for many warm water species cultured above 20 °C (Li and Olsen 2015).

Enrichment Time The enrichment time is another decisive factor for the rate of fatty acid replacement in phospholipids of the rotifers and *Artemia*. Previous studies have shown that prolonged enrichment time was more effective than increased oil rations in boosting n-3 HUFA content in TL of rotifers (Rodriguez et al. 1996). Rotifers are commonly enriched only for a short period (for example, 2 h) in some hatcheries and laboratory studies (Maehre et al. 2013; Rehberg-Haas et al. 2015). Our results revealed that total lipid content and DHA% in total fatty acids increased most efficiently during the first hour of enrichment, whereas the DHA in PC and PE showed a much slower rate of increase during the same time period (Fig. 5.2). Because of this relatively slow increase of DHA in PC and PE compared to that in TL, a prolonged enrichment time (24 h) is suggested to obtain high DHA or HUFA content of rotifer PL. Contrary, very short enrichment time, for example 2 h, will not result in very much DHA or HUFA enrichment in the PL of rotifers, which can be important for experimental purposes (Olsen et al. 2014).

Low Feeding Ration to Keep Rotifers Lean Short-term enrichment (within 24 h) strategies tend to produce rotifers with high lipid content (Rainuzzo et al. 1994b), which has not been recommended for cold water fish larvae, because high rotifer mortality may occur when fat rotifers are transferred to cold water (Olsen et al. 1993). Fat rotifers contain a higher TAG fraction than more lean rotifers, because the PL content per dry matter remains fairly constant independent of the total lipid contents. A high TAG content of the rotifers may also not be optimal for early fish larvae. DHA is normally believed to be esterified in the *sn*-2 position of TAG (Sargent et al. 1999, 2002). During digestion processes in fish larvae, the hydrolyzed fatty acids from *sn*-1 and *sn*-3 positions of dietary TAG could then dilute the DHA% in the enterocyte, resulting in lower DHA levels in the resynthesized PL through lyso-PL (Lands cycle) pathway. It is noteworthy that younger stages of *Acartia* and other copepods have very low levels of TAG (Fig. 5.3), and this may affect the digestion pattern of fish larvae. To produce copepod-like live feed (Fig. 5.3), it is therefore an issue to both minimize dietary lipid rations and rotifer lipid level besides having high DHA enrichment of PL, which may seem to be a contradiction, or at least a challenge. It will, however, not be possible to reduce rotifer TAG to the levels of copepods. The importance of this difference in lipid class composition for rotifers and copepods are further outlined below.

Enrichment Method to Produce Rotifers with High DHA in PL Our study has shown that it is, nevertheless, possible to enrich the DHA levels in both PC and PE of rotifers to relatively high levels, and that these levels can be kept stable for at least 24 h at 10 °C. Based on our overall results, we suggest the enrichment time

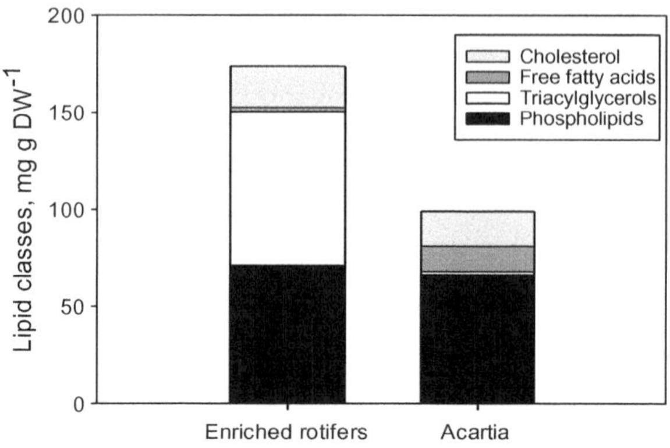

Fig. 5.3 Main lipid classes of short-term enriched rotifers and cultivated copepods (*Acartia tonsa*). Data from Olsen et al. (2014)

Table 5.1 Schematic description of methods used to produce live feed diets containing similar total lipid, total fatty acids and DHA, but different percentage DHA levels in phospholipids

Experimental rotifer based larval diet	Production procedure
Low percentage DHA in PL Short-term n-3 HUFA enrichment of rotifers cultured on low-DHA diet just before application	Cultivation feed: 1.2 μg ind^{-1} day^{-1} Baker's yeast *Saccharomyces cerevisiae* and 0.5 mL 10^6 ind^{-1} day^{-1} *Nannochloropsis* sp. paste (Reed Mariculture). Feed for short-term enrichment: 0.16 μg ind^{-1} lipid emulsion Marol E, 2 h of incubation
High percentage DHA in PL Long-term n-3 HUFA enrichment during cultivation	Cultivation feed: 1.2 μg ind^{-1} day^{-1} Baker's yeast *Saccharomyces cerevisiae*, 0.5 mL 10^6 ind^{-1} day^{-1} *Nannochloropsis* sp. paste (Reed Mariculture), and 0.09 μg ind^{-1} day^{-1} lipid emulsion Marol E

Table modified from Olsen et al. (2014)

should be at least 24 h, in order to maximize the efficiency of the PL enrichment. Beside this, because of the effect of high TAG mentioned above, the feeding rations of lipids should be as low as possible to reduce the total lipid contents of the rotifers. The enrichment diet should have a high percentage content of DHA. The possible stimulating effect of 16:0 and inhibiting effect of 18:1n9, 18:2n-3 and 18:4 n-3 can be important, but needs further investigation for final confirmation.

Method to Produce Live Feed with Variable DHA in PL The main challenge of using live feed diets to study PL nutrition of fish larvae is to design diets where one principal component is varied whereas the remaining components are kept relatively constant over some time. We have described a method to produce rotifers containing similar amount of total lipid, total fatty acids, DHA and n-3 HUFA, and PL content per dry weight ((Olsen et al. 2014), see Table 5.1). With these components constant, it was possible to obtain rotifers with variable DHA% in total PL fatty acids using this procedure. This can be achieved by providing the same feed components for all rotifer treatments, but different feeding strategies involving a combination of long-term and short-term enrichment techniques as described above. The basis of the method for producing rotifers with low DHA% in PL is the "lag phase" in DHA accumulation in PL of rotifers during enrichment (Fig. 5.2). The effects of a variable DHA% in PL or TAG in live feed could therefore be conveniently investigated by feeding fish larvae with live feed organisms enriched using this principal method, which must be adapted to the specific experimental situation.

5.2.2 *Artemia* Enrichment

Artemia shows very much the same patterns of fatty acid and HUFA enrichment in PC and PE as shown for rotifers (Fig. 5.4). Most studies of DHA enrichment of PL in *Artemia* have shown that DHA ends up in TAG and that very little DHA are incorporated into PL (<3% of total fatty acid in PL) (Coutteau and Mourente 1997; Guinot et al. 2013a; Jin et al. 2014; Sargent et al. 1999).

Changes of HUFAs in PC and PE of *Artemia franciscana* nauplii and juveniles were studied following enrichment and subsequent starvation. Multigain was provided in a concentration of 200 mg L^{-1} at the start of enrichment of the *Artemia* culture with a density of 120 ind mL^{-1}. An extreme lag phase of 1 h in the accumulation of DHA, EPA and ARA (arachidonic acid, 20:4 n6) in PC and PE after initiation of enrichment was found for *Artemia* (Fig. 5.4), suggesting slower HUFA incorporation rate in PL compared to rotifers which showed a similar delay in the HUFA incorporation. The highest DHA levels were found 8 h post enrichment in total lipid (8.00%), PC (1.96%) and PE (2.01%) of *Artemia*.

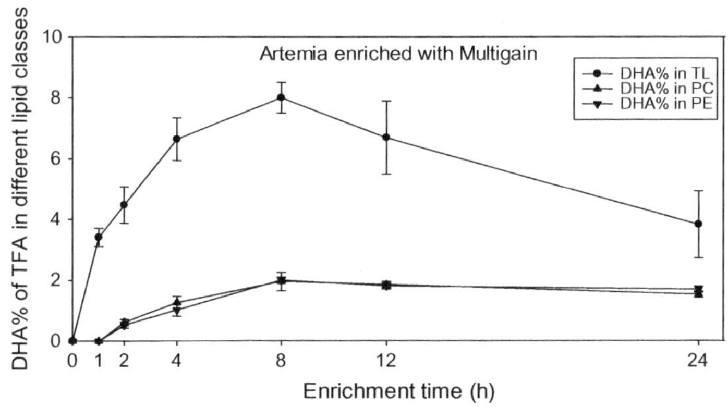

Fig. 5.4 Changes in DHA% in total lipid (circles), PC (triangles) and PE (inverted triangles) as a function of time in *Artemia franciscana* enriched with Multigain. Error bars express standard errors of the mean, data from Jin et al. (2014)

Also the enrichment time affected the relative HUFA content and the overall fatty acid composition in total lipid relative to that of PL in *Artemia*, as shown for rotifers. A very short enrichment time resulted in a relatively low HUFA enrichment of PL, while total lipid became enriched very rapidly. To obtain an optimum HUFA enrichment of PL in *Artemia*, we suggest that the enrichment time of *Artemia* nauplii in commercial hatcheries should be at least 8 h. The lag phase in HUFA enrichment opens also possibilities to produce *Artemia* nauplli with variable relative HUFA enrichments in PL and TAG, as achieved for rotifers described above.

There is an apparent constrain of DHA enrichment in PL of *A. franciscana*, which is most likely related to strain or its genetic characteristics. One obstacle is their ability to retroconvert DHA into EPA both during enrichment and subsequent starvation (Navarro et al. 1999), making it extremely difficult to achieve high DHA content in PL of *Artemia*. On the other hand, the DHA content in total lipid decreased much faster than that in PL of *Artemia* upon starvation post-enrichment. The use of *Artemia* juveniles, which could contain higher levels of HUFA in PL in relative terms, may therefore be a good alternative for some marine fish species. Their relatively low levels of DHA in total lipid can additionally be manipulated by short-term enrichment with an emulsified lipid diet just before use. However, the big size of *Artemia* juvenile and extra cost for cultivation will likely limit its application in commercial hatcheries.

5.3 Essentiality of Phospholipid for Fish Larvae

The importance of dietary PL for fish larvae was demonstrated in the early 1980s (Kanazawa et al. 1981, 1983). The beneficial effects found were mainly associated with growth, survival rates, digestive functions, occurrence of deformities, and stress resistance in larval and juvenile stages of various species of fish (Cahu et al. 2009; Coutteau et al. 1997; Tocher et al. 2008). Fish larvae seem dependent on a dietary supply of PL between 2 and 12% of the diet (Tocher et al. 2008) for normal growth and functional development. However, The PL requirements appear to decline as the fish grow bigger, and no requirements have generally been observed in fish larger than 5 g (Cahu et al. 2009; Tocher et al. 2008). Intestinal steatosis (lipid droplet accumulation) were found in much higher ratio (13 out of 16 vs.1 out of 18) in the fry stages (20 days post swim-up stage) compared to juvenile stages (146 days post swim-up stage) of rainbow trout *Oncorhynchus mykiss* when a PL-deficient diet was fed to the fish (Dapra et al. 2011). Dietary PL was also found to enhance fatty acid transport, up take and antioxidant capacity of large yellow croaker *Larimichthys crocea* larvae at transcriptional level (Cai et al. 2016). However, the beneficial effects on survival and growth performance of large yellow croaker were only restricted to larval stages (Feng et al. 2017). The mechanism of essentiality of PL for larval and early juvenile fish is not clear, however, the requirement of dietary PL was suggested to be due to the limited ability to biosynthesize PL de novo, which is necessary for lipoprotein synthesis and transport from the enterocytes when dietary PL is insufficient (Cahu et al. 2009; Tocher et al. 2008). However, little is known about the PL biosynthesis capability in marine fish larvae.

PC and PE are the most abundant PLs of eukaryotic membranes, accounting for more than 50% of the total PL species (Kent 2005). PC is also an important component of lipoproteins, which is required for chylomicrons assembly in the enterocytes and normal very low-density lipoprotein (VLDL) secretion from the hepatocytes, playing an important role in lipid absorption and transport. In mammals, the PC used in chylomicrons can originate from either dietary, biliary or de novo synthesis (Mansbach 1977). The liver of healthy Caucasians produces 7–22 g PC every day that is excreted into the intestinal lumen through the bile (Northfield and Hofmann 1975), and a 20 g mouse can secret 23 mg PC daily (Kuipers et al. 1997). However, early life stages of many fish species have incomplete developed livers during the period of first feeding and it has been suggested that these as mentioned also have limited capacity for de novo synthesis of PL (Cahu et al. 2009; Tocher et al. 2008). Deficiency in endogenous PC could therefore lead to low lipoprotein synthesis and lipid accumulation in the enterocytes. As a result, supplementation of dietary PL could improve the intestinal lipid absorption for many species of fish larvae (Dapra et al. 2011; Fontagne et al. 1998; Olsen et al. 2003).

The biochemically limiting steps for PL synthesis in fish larvae are unknown. The metabolic pathways appear, however, to be the same as for mammals (Fig. 5.5) (Sargent et al. 2002). In mammals, the major biosynthesis pathway of PC is the CDP-choline (cytidine diphosphate-choline) pathway with choline kinase (CK) producing phosphocholine, followed by CTP:phosphocholine cytidylyltransferase (CT) producing CDP-choline, and accomplished by CDP-choline: *sn*-1, 2-diacylglycerol cholinephosphotransferase (CPT) producing PC. The second step catalyzed by CT is considered to be the rate-limiting step in the pathway under normal physiological conditions. However, the last step catalyzed by CPT can become rate-limiting if the supply of di-acylglycerol (DAG) is restricted (Gibellini and Smith 2010). Alternatively, PC can be synthesized by methylation of phosphatidylethanolamine (PE), catalyzed by phosphatidylethanolamine N-methyltransferase (PEMT), which mainly operates in the liver and contributes to 30–40% of PC in hepatocytes (DeLong et al. 1999; Reo et al. 2002; Sundler and Akesson 1975). PE is synthesized through the CDP-ethanolamine pathway, which is similar as the CDP-choline pathway. In mitochondria, PE is synthesized by decarboxylation of phosphatidylserine (Borkenhagen et al. 1961).

Diacylglycerol (DAG) is an important intermediate, common for the synthesis of both TAG and PC (PE). It can be generated by acylation of glycerol-3-phosphate (G-3-P) to phosphatidic acid (PA) which is subsequently dephosphorylated by phosphatidic acid phosphohydrolase (PAP or lipin), termed the G-3-P pathway. Alternatively, DAG can be formed by reacylation of *sn*-2-monoacylglycerol (2-MAG) by monoacylglycerol acyltransferase (MGAT), termed the 2-MAG pathway.

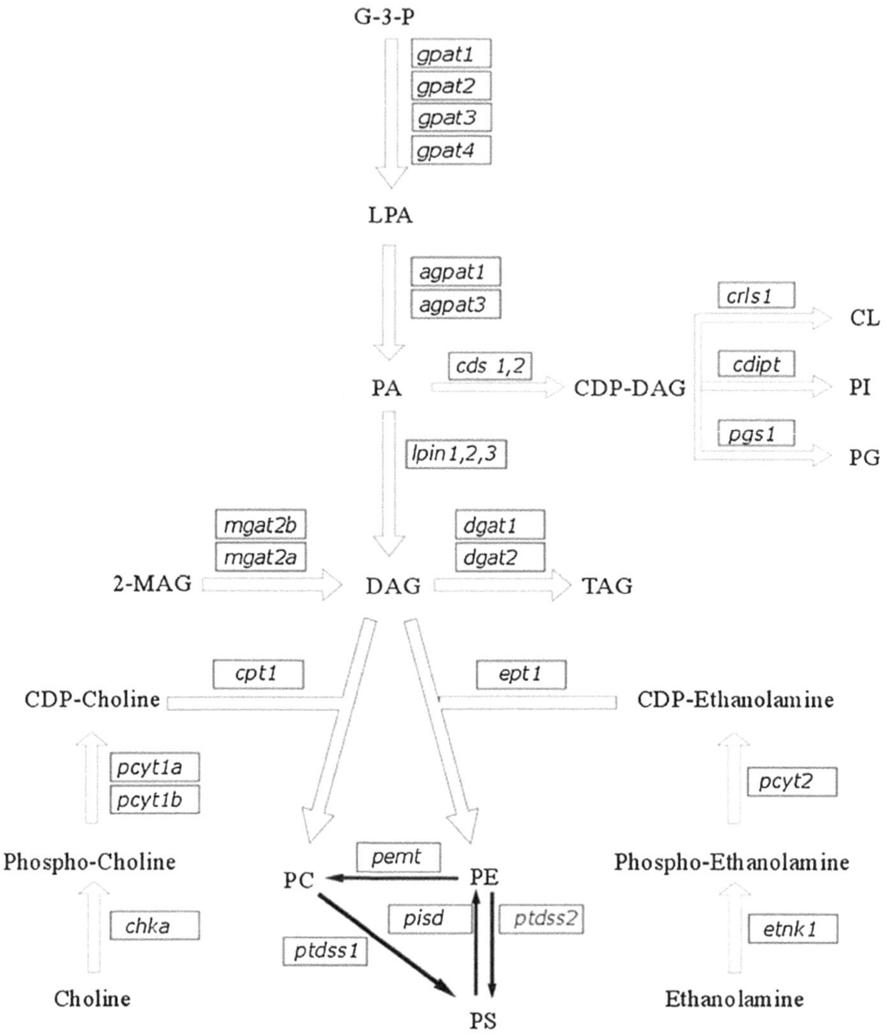

Fig. 5.5 Pathways of phospholipid de novo biosynthesis, modified from Li et al. (2015b). *PC* phosphatidylcholine; *PE* phosphatidylethanolamine; *PS* phosphatidylserine; *PI* phosphatidylinositol; *PG* phosphatidylglycerol; *CL* cardiolipin; *G-3-P* glycerol-3-phosphate; *LPA* 2-acyl-glycerol-3-phosphate; *PA* phosphatide acid; *2-MAG* 2-monoacylglycerol; *DAG* diacylglycerol; *TAG* triacylglycerol. CHKA choline kinase α; *PCYT1A/B* phosphate cytidylyltransferase 1 α/β; *chpt1* choline phosphotransferase 1; *ETNK1* ethanolamine kinase; *PCYT2* phosphate cytidylyltransferase 2; *EPT1* ethanolamine phosphotransferase 1; *PEMT* phosphatidylethanolamine N-methyltransferase; *PISD* phosphatidylserine decarboxylase; *Ptdss1* phosphatidylserine synthase 1; *Ptdss2* phosphatidylserine synthase 2; *CRLS1* cardiolipin synthase 1; *Cdipt* CDP-diacylglycerol-inositol 3-phosphatidyltransferase; *PGS1* phosphatidylglycerophosphate synthase 1; *GPAT1,2,3,4* sn-1-glycerol-3-phosphate acyltransferase; *AGPAT1,2,3 acyl-CoA* 1-acylglycerol-3-phosphate acyltransferase; *Lipin 1,2,3* PA phosphatase; *MGAT* MAG acyltransferase; *DGAT* DAG acyltransferase

Considering the metabolic mechanisms that can be involved, the fatty acid composition in PL can be modified through PL synthesis and modification through the pathway of de novo synthesis and the land cycle (PL to lyso-PL to PL) pathway, respectively. Many acylases and transacylases involved in these pathways do not have absolute specificities, which mean that the levels, or fractions, of fatty acids available from the diet might have a significant influence on the fatty acid composition in PL of the enriched rotifers (Sargent et al. 1999). However, some degree of selectivity on fatty acids or fatty acid associated structures in phospholipid synthesis and modification has been reported, especially for the key enzymes involved in respective pathways, the CPT and lysophosphatidycholine acyltransferase (LPCAT).

Studies using rat liver microsomes have shown that CPT preferred 1-16:0-2-DHA-DAG for synthesis of PC over its *sn*-1-stearoyl counterparts, and 18:1n9 was potentially a strong competitor for the *sn*-2 position in this DAG structure (Holub 1978). Human LPCAT3 enzyme showed preference towards 18:2n6-CoA and 20:4n6-CoA as acyl donors and lyso-PC with saturated fatty acid at the *sn*-1 position as substrates (Kazachkov et al. 2008). Unfortunately, the selectivity of these enzymes in rotifers has not been described. If a similar selectivity exists in live feed organisms like rotifers and *Artemia*, the non-HUFA (non-highly unsaturated fatty acid) composition of the enrichment diets might affect the efficiency of DHA enrichment in PLs as discussed in Sect. 5.2.1.

5.4 Phospholipid Metabolism in Cod Larvae

5.4.1 Implication of DHA% in Live Feed Diet and Cod Larvae

First feeding experiments of cod larvae were carried out by feeding with short-term (2.1% DHA in PL) or long-term (9.4% DHA in PL) enriched rotifers (see method in Table 5.1), and cultivated nauplii of *Acartia tonsa* (30% DHA in PL) (Olsen et al. 2014). An increasing gradient of DHA% in PL of live feeds was achieved as described above (Sect. 5.2.1). Meanwhile, the respective DHA% in total lipid was similar in all three diets (26.2, 22.3, and 28.4–39.4% of total fatty acids). The dominant fatty acids in PL of 17 days post-hatching (dph) larvae were highly correlated to the fatty acid composition in the dietary PL (Fig. 5.6), but the regressions with DHA in total lipids were poorer.

A similar study, using the same enrichment strategies, but emulsions containing less DHA, resulted in lower DHA% in total lipid of rotifers (6.7–11.6%). However, after 18 days of feeding, the DHA levels in cod larvae were not significantly different ($P > 0.05$, 14.4–20.9% vs. 15.3–17.1%). The only dietary origin of DHA was via rotifers in both studies, which suggested that DHA was selectively incorporated and retained in PL of cod larvae, especially in the PE fractions, which contained close to 50% DHA before first feeding and 30% at 18 dph. A threshold DHA level to maintain membrane fluidity and other functions may exist in the PL of cod larvae.

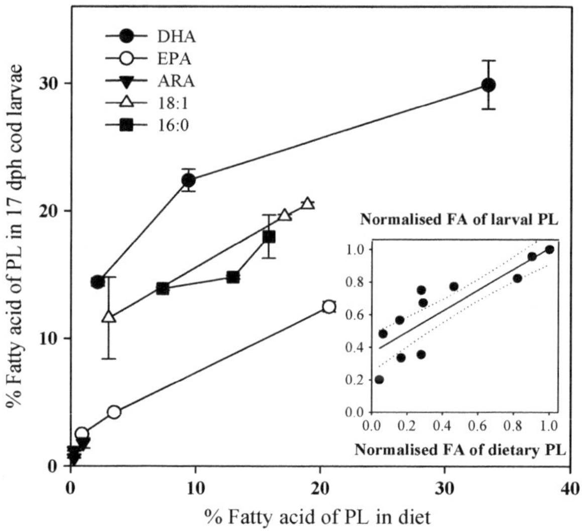

Fig. 5.6 Composition of dominant fatty acids in PL (% of PL fatty acids) in 17 dph Atlantic cod *Gadus morhua* larvae as a function of the respective fatty acid composition in the dietary PL. Inset shows the similar relative composition after normalizing contents of the individual fatty acid to the respective maximal value, with regression line and 95% CI included, data from Olsen et al. (2014)

Our analytical data allow some further speculations on how the fatty acid composition and position of dietary PL and TAG may affect PL synthesis in fish larvae. The main pathways of de novo synthesis of PL based on dietary TAG and through re-acylation of lyso-PL are summarized in Fig. 5.7. The anabolic synthesis of PL and re-acylation of lyso-PL take place in the enterocytes, from which TAG and PL are transported to different tissues as lipoproteins.

If we assume that the n-3 HUFA of a dietary TAG is mainly esterified at the *sn*-2 position (Sargent et al. 1999, 2002), the n-3 HUFA will be retained in the MAG after digestion in the intestine and in DAG after re-esterification in the enterocytes. Most of this DAG will be re-esterified further to TAG and a small proportion can be synthesized to PC via CDP-choline pathway as shown in mammals and in Atlantic salmon, *Salmo salar* (Lehner and Kuksis 1996; Oxley et al. 2007). Other exchange mechanisms in between the PL molecules may take place as well, but it follows that both de novo synthesized PL and TAG will retain a majority of their original n-3 HUFA at the sn-2 position because it will not be removed from the glycerol backbone. If PL of tissues are de novo synthesized based on dietary marine TAG, we should expect a similar n-3 HUFA composition in the dominant PL molecules of fish tissues as in the diet and a positive correlation between percentage n-3 HUFA content in dietary TAG or total lipids and tissue PL of fish. However, no such correlation was found for DHA of PL.

It appears that the principal precursor of PL synthesis in larval enterocyte is dietary PL, forming lyso-PL and a FFA after digestion. These components are taken up in enterocytes and lyso-PL undergoes reacylation to form PL. The fatty acid composition and position of dietary PL are very important for the fatty acid profile

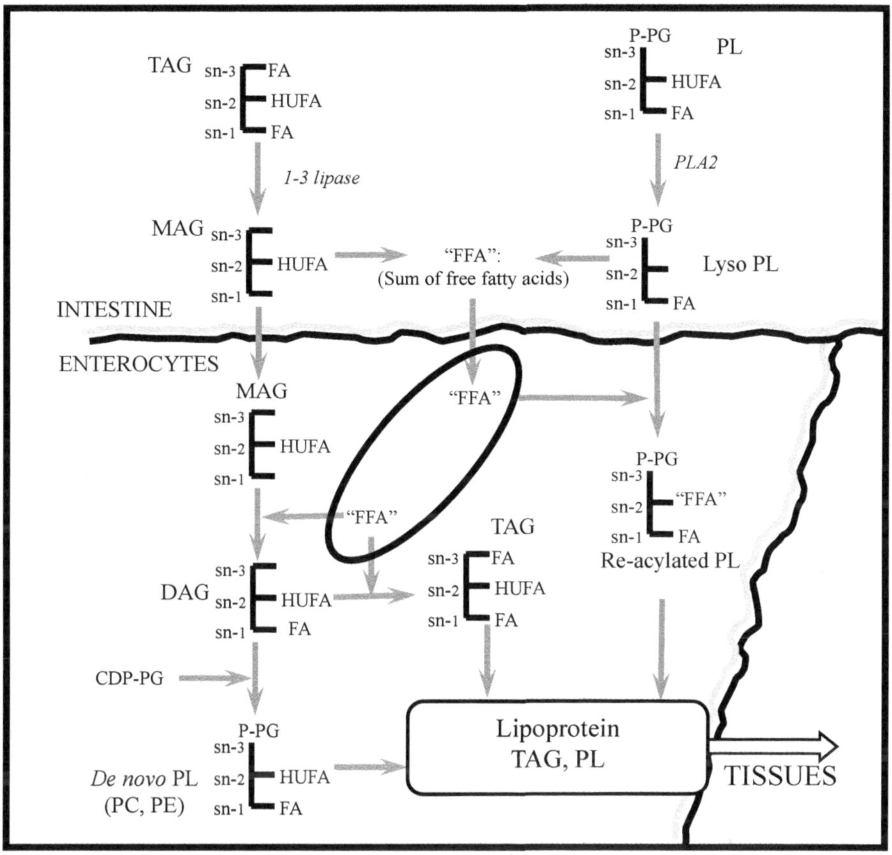

Fig. 5.7 Schematic view of the digestion and metabolism of TGA and PL in the intestine and enterocytes. *TAG* triacylglycerides (with HUFA at the sn-2 position); *MAG* monoacylglycerides; *DAG* diacylglycerides; *PL* phospholipid (with HUFA at the sn-2 position); *PG* polar group (choline, ethanolamine, inositol, and serine); *CDP* cytidine diphosphate; *CDP-PG* activated choline cytosine with polar group (here choline or ethanolamine); *PC* phosphatidylcholine; *PE* phosphatidylethanolamine; de novo *PL* phospholipids synthesized based on dietary TAG; re-acylated *PL* phospholipids synthesized based on dietary PL; *ΣFFA* pool of free fatty acids, saturated, monounsaturated, and polyunsaturated, including HUFA; 1–3 lipase and PLA2 are digestive enzymes. Figure modified from Olsen et al. (2014)

of larval PL, and accordingly also for larval membrane functionality, cellular metabolism, and ultimately for growth and performance of larvae and juveniles. Phospholipase A2 (PLA2) is generally believed to be the active digestive enzyme for PL, which removes the sn-2 fatty acid (usually an n-3 HUFA) from the PL molecule. In the re-acylation process in the enterocytes, the composition of FFA available ($\sum$FFA), the pool of FFA, includes saturated, monounsaturated, and polyunsaturated fatty acids) and the specificity of the enzymes catalyzing the reacylation process affect the n-3 HUFA content of PL. The $\sum$FFA pool is related to the total lipid content of the diet or its TAG content. If total lipids or TAG is low, like in copepod nauplii (Fig. 5.3), the sn-2 fatty acid of the dietary PL will dominate $\sum$FFA and the pool will be smaller. If total lipid or TAG is high, the shorter chain fatty acids from sn-1 and sn-3 position of TAG will compete with the reacylation process, resulting in lower n-3 HUFA content.

However, the digestion, absorption and resynthesis of TAG and PL in fish larvae are very complicated metabolic processes (Li 2015). First of all, the DHA in TAG of fish-oil-enriched rotifers was found to be retained in *sn*-3 positions rather than *sn*-2 position (Ando et al. 2004). If most of the DHA was esterified in *sn*-3 of the rotifers provided, no information is available on the fatty acids that remained in the *sn*-2 position after digestion. Secondly, the main neutral lipase in cod is bile-activated lipase (BAL) and its specificity remains to be determined. It was found that the purified BAL of cod possessed 1, 3-specificity toward TAG (Gjellesvik 1991; Gjellesvik et al. 1992), whereas Lie and Lambertsen (1985) suggested that lipolytic activity in cod intestinal juice hydrolyzed HUFAs in the *sn*-2 position of TAG. Besides, the activity of PLA2 in fish larvae is still uncertain, and Sæle et al. (2011) showed that expression of the PLA2 enzyme species secreted from the pancreas remained low in cod until day 62. At last, but not least, the enzyme selectivity for DAG species to either TAG or PC is not considered, which will be discussed later.

The lipid profile of live feed and cod larvae provided evidence for that DHA incorporated in PL, rather than TAG, was more beneficial to cod larvae for their new synthesis of tissue PL as most DHA originated from dietary PL. However, the conclusion on the ability of de novo synthesis of PL needs further investigation.

5.4.2 Expression of the Related Genes Involved in the Pathway

PL is found to be an essential dietary component for fish larvae (Cahu et al. 2009; Tocher et al. 2008). Transcriptome analysis of different stages of cod larvae was therefore carried out using a recently developed cod microarray with the objective to obtain more knowledge on ontogenesis of lipid metabolism, especially the pathway of de novo PL synthesis during early stages of cod larvae (Li et al. 2015b).

There are more than 100 genes involved in metabolism of PL. Most of the genes encoding de novo PL biosynthesis pathway, the Lands cycle pathway (PL to lyso-PL to PL) and the PL turnover pathway were not significantly up-regulated ($P > 0.05$) up to 60 dph (Fig. 5.8). One study (Dapra et al. 2011) showed that the intestinal *cpt* (choline phosphotransferase, CDP-choline pathway), *lpcat* (lysophosphatidylcholine acyltransferase, Lands cycle pathway) and *dgat* (diacyl-glycerol O-acyltransferase, DAG to TAG) mRNA levels were not affected by PL deficiency in fry and juveniles of rainbow trout whereas only *apoB* (apolipoprotein B) was up-regulated in the PL depleted treatment during fry stage. In our study,

Gene	Description
De novo PL synthesis	
GPAT3	1-acyl-sn-glycerol-3-phosphate acyltransferase theta-B
AGPAT4	1-acyl-sn-glycerol-3-phosphate acyltransferase delta-like
PPAP2C	lipid phosphate phosphohydrolase 2-like
LPIN2	phosphatidate phosphatase LPIN2-like
LPIN1	phosphatidate phosphatase LPIN1-like
mogat2-a	2-acylglycerol O-acyltransferase 2-A-like
Dgat1	diacylglycerol O-acyltransferase 1
Dgat1	diacylglycerol O-acyltransferase 1-like
dgat2	diacylglycerol O-acyltransferase 2-like
PCYT1B	choline-phosphate cytidylyltransferase B-like
ETNK1	ethanolamine kinase 1-like
Lipase	
CEL	Bile salt-activated lipase-like protein
PLA2G1B	Phospholipase A2
PLA2G1B	phospholipase A2-like
Pla2g1b	phospholipase A2-like
PLB1	Phospholipase B1, membrane-associated-like
PLB1	Phospholipase B1, membrane-associated-like
abhd12	monoacylglycerol lipase ABHD12-like
acsf2	acyl-CoA synthetase family member 2, mitochondrial-like
ACSS3	acyl-CoA synthetase short-chain family member 3, mitochondrial-like
SLC27A2	Very long-chain acyl-CoA synthetase
pld3	phospholipase D3-like
PLA2G3	group 3 secretory phospholipase A2 precursor
PLA2G4C	cytosolic phospholipase A2 zeta-like
PLA2G4C	cytosolic phospholipase A2 zeta-like
Plcg2	1-phosphatidylinositol-4,5-bisphosphate phosphodiesterase gamma-2
PLA2G10	group 10 secretory phospholipase A2-like
PNPLA8	calcium-independent phospholipase A2-gamma-like
Transport	
MTTP	microsomal triglyceride transfer protein large subunit-like
apoB-100	apolipoprotein B-100-like
apoB	Apolipoprotein B-100
apoA-IV	Apolipoprotein A-IV
apoA-IV	apolipoprotein A-IV-like
apoA-I	Apolipoprotein A-I-like

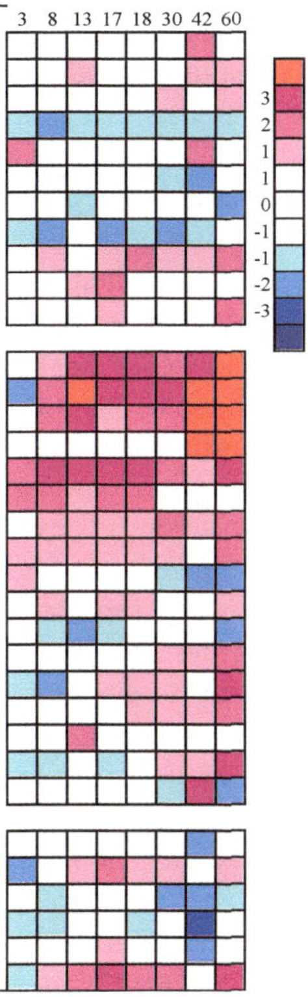

Fig. 5.8 Summary of the significantly regulated genes ($P < 0.01$, at least in one of the sampling points) that were related to phospholipid metabolism in Atlantic cod *Gadus morhua* larvae. From left to right, the squares indicate expression levels at 3, 8, 17, 18, 32, 40 and 60 dph, and the data are normalized to 1 dph. The different style indicates log2-transformed gene expression ratios. Data from Li et al. (2015b)

both *apoB* and *apoAI* (apolipoprotein AI) showed significant up-regulation at 17 dph, and *apoAI* was also significantly up-regulated at 13 and 60 dph, suggesting that an elevated lipoprotein assembly might have occurred at these stages (Fig. 5.8). The expression of *cpt* in cod larvae did not show any regulation up to 60 dph. Other studies have indicated that the enzyme activity of CPT in cells was in excess (Gibellini and Smith 2010; Vance 2002), which likely means that transcriptional regulation of *cpt* is not necessary for the production of PC.

Of the PC synthesis genes, only *pcyt1b* (phosphate cytidylyltransferase 1β) showed a significant up-regulation at 17 dph in our study. *Pcyt1b* encodes for CT (cytidylyltransferase) β2 and CTβ3 in mice and CTβ1 and CTβ2 in humans. CTβ is expressed at the same level as CTα in the brain, whereas CTβ expression in other tissues is only about 10% of CT (Gibellini and Smith 2010). Recent studies with *Pcyt1b*$^{-/-}$ mice showed no obvious neurological problems, but reduced PC synthesis in distal axons and less neurite branching was observed in neurons cultured from CTβ2-deficient mice (Strakova et al. 2011; Vance and Vance 2009).

The up-regulation of *pcyt1b* in the present study may be related to the need for biosynthesis of PL in the brain of cod larvae due to low DHA levels of the PL in rotifers. There are evidences that lyso-PC may be a preferred form to pass through the blood-brain barrier and carry DHA to the brain (Lagarde et al. 2001). The expression of *pcyt1b* of cod larvae was found to be down-regulated during the live feed period when they were fed copepods, their natural prey, which contains much higher DHA levels in PL compared to rotifers (Elin Kjørsvik communication). To our knowledge, no information on CTβ has been reported in any fish species. Even in mammals, there is very little information about CTβ compared to the dominant CTα, which is encoded by *pcyt1a* (phosphate cytidylyltransferase 1α). As they have similar membrane-binding domain, CTβ is probably also regulated by a similar translocation process as CTα (Sugimoto et al. 2008).

An increase in PL content per individual larvae is expected due to cell division and growth, especially during the fast growth stages in the early life. However, very few genes involved in the de novo synthesis pathways of PL were significantly regulated throughout the live feed period and up to 60 dph. One reason for this may be because the larvae had sufficient dietary PL from live feed and commercial diets, with no need for extensive de novo PL synthesis. It has been suggested that exogenous PC can inhibit de novo PC synthesis (Mansbach 1977).

Secondly, non-transcriptional regulation of key genes involved in the pathway might be another explanation. For example, the activity of CTα involved in the rate-limiting step of PC biosynthesis can be regulated efficiently by translocation on and off membranes. The active form of CTα is membrane bound and de-phosphorylation can promote activation of CTα (Sugimoto et al. 2008). Therefore, cells could increase PC biosynthesis by translocation of CTα to membranes without changing the expression level of *pyct1a* (Golfman et al. 2001).

Thirdly, it is important to note that the gene expression levels of the current study were normalized to 1 dph. No up-regulation does not imply low gene expression level, meaning that the gene expression level of cod larvae may be high already at 1 dph or before hatching. A recent study reported gene expression levels of cod eggs in six stages (Kleppe et al. 2014), and by analyzing their raw data, we found that the expression of *pyct1a* was generally higher in the egg stages than in the larvae stages. Finally, due to the small size of cod larvae, RNA was extracted from the whole body, which could overlook tissue specific gene expression patterns. Another study with small size marine fish larvae, large yellow croaker, showed no differences in the transcription levels of *pcyt1a*, *cpt1* and *dgat2* (DAG acyltransferase 2), when whole larvae were analyzed after feeding trial with different content of dietary PL (Feng et al. 2017). Studies with bigger fish larvae, such as salmon fry (2.5 g), makes the separation of intestine and liver applicable. The intestinal transcriptome analysis by microarray showed that the 10 g parr had higher expression of *cpt1* compared to that of fry (2.5 g) (De Santis et al. 2015). Quantitative real-time PCR (qPCR) showed that many genes involved in the biosynthesis of PC, PE and PS were lower expressed in intestine in the fry stage compared to the parr stage, including *chka* (choline kinase α), *cpt1* for PC synthesis; *etnk1* (ethanolamine kinase), *pcyt2* (phosphate cytidylyltransferase 2), and *ept1* (ethanolamine phosphotransferase 1) for PE synthesis; CDP-DAG-serine O-phosphatidyltransferase for PS synthesis (Carmona-Antonanzas et al. 2015). The fold change of these genes was generally within 1 and the less sensitive genome-wide techniques, such as microarray, may have overlooked such changes. However, the expression of the rate-limiting gene for PC synthesis, *pcyt1*, analyzed by qPCR showed no significant differences between the parr stage and the fry stage (Carmona-Antonanzas et al. 2015), which was probably due to the post-transcriptional regulation as mentioned above.

To summarize, current studies suggested that there is no clear genetic constrains of undertaking de novo PL synthesis in cod larvae, and larvae appeared to have relatively high capability of PL biosynthesis already at 1 dph. This capability may regrettably be related to cell division and growth, rather than to lipoprotein assembly. Expression of many genes involved in PL synthesis in the intestine of salmon parr was higher compared to that of the small fry, suggesting higher capability of PL synthesis in the later developmental stage (parr), consistent with the hypothesis that the requirement of PL for fish larvae is associated with the insufficient ability to biosynthesize PL for chylomicron formation and lipid transport.

5.4.3 Biochemical Study with Radiolabeled Tracers

As mentioned in the previous section, many of the regulatory mechanisms of PC biosynthesis work at non-transcriptional levels (Sugimoto et al. 2008), and biochemical or enzymological studies are therefore needed to elucidate if marine fish larvae are capable to synthesis PL de novo. In vitro studies using intestinal segments were not applicable due to the small size of marine fish larvae. A tube feeding method, as described by Rønnestad et al. (2001), was used to deliver 2-Oleoyl-[1,2,3-^{3}H]glycerol and [^{14}C(U)] glycerol together with bovine serum albumin (BSA) bound 16:0 and DHA, with or without choline chloride, to the foregut of anesthetized cod larvae (Li et al. 2016b). Choline was chosen because it is an essential nutrient for animals, and its main fate is the biosynthesis of PC via the CDP-choline pathway (Li and Vance 2008). Inadequate choline intake can lead to fatty liver or muscle damage in humans (Fischer et al. 2007) and liver dysfunction, reduced growth and poor feed efficiency in many fish species (Millikin 1982). Supplementation of 0.2% choline of dry weight in the diet can improve growth performance, intestinal enzymes activities and feed utilization of blunt snout bream fed high-lipid diet (Li et al. 2016a). One of the obvious functions of choline was the stimulating effect of PC biosynthesis as shown in isolated rat hepatocytes where supplementation of 0.5–2 mM choline could stimulate PC synthesis 2–3 folds via the CDP-choline pathway (Sundler and Akesson 1975). The metabolism of these components in the larvae were thereafter monitored through short time

(0–4 h) following injection. Anesthesia can reduce digestive functions as has previously been reported in zebrafish *Danio rerio* (Hama et al. 2009), but the injected mixtures would be absorbed directly without being further digested in the larvae.

After 0.5 h of incubation, the percentage of total [^{3}H] label in 2-MAG was already very low in both control and choline treatments (Table 5.2). The highest proportion of [^{3}H] radioactivity was found in TAG, followed by PC at 0.5 h post incubation. There was an increasing trend in [^{3}H] PC and a decreasing trend in [^{3}H] TAG, and the PC:TAG ratio increased with increasing incubation time. The choline supplementation did not increase the percentage of [^{3}H] PC significantly, but the percentage of [^{3}H] TAG was reduced significantly due to the increased radioactivity of other phospholipids, including PE, PS and PI. The PC:TAG ratio was therefore significantly (P = 0.038) higher in the choline treatment.

The predominant lipid classes derived from [^{14}C] glycerol were TAG and PC, followed by PE (Table 5.2). A significant ($P < 0.05$) increase in [^{14}C] PC and a decreasing percentages of [^{14}C] TAG over time were also observed (Table 5.2). There was generally a higher recovery of [^{14}C] in PC, PE, and PS & PI in the choline treated fish compared to controls at all the sampling times, but these effects were not significant. However, significant reductions in [^{14}C] TAG for the choline treated fish resulted in an increase in the PC:TAG ratio (P = 0.012). T-tests were performed to compare the distribution of label in lipid classes between the 2-MAG pathway and the G-3-P pathway. The percentages incorporation of radioactivity in all the phospholipid classes; PC, PE and PS & PI for the G-3-P pathway were significantly higher than that derived from 2-MAG pathway. Accordingly, significantly higher PC:TAG ratio and significantly lower radioactivity in TAG was found in G-3-P pathway compared to 2-MAG pathway.

The very low percentage of [^{3}H] label in 2-MAG after 0.5 h of incubation (Table 5.2), suggested fast absorption and utilization of the 2-MAG precursor by cod larvae. In vitro studies using intestinal segments of rainbow trout have shown that only 0.5–1% of the initial radioactivity of fatty acids added to the luminal side passed through the intestinal epithelium after about 2 h (Geurden et al. 2009). It is likely that the biosynthesis of PL and TAG in this study occurred mainly in the enterocytes of cod larvae within the first hour post injection. Within the first hour, the molecular ratio of PC:TAG obtained from the 2-MAG and the G-3-P pathways were 0.44–0.74 and 1.02–2.06, respectively (Table 5.2). It was relatively high compared to that found in other studies with mammals and salmon (0.04–0.22 derived from 2-MAG pathway, 0.27–1.43 derived from G-3-P pathway) (Lehner and Kuksis 1992; Oxley et al. 2007). DAG is a common precursor for both the 2-MAG and the G-3-P pathways. Its fatty acid composition is important for the selectivity of CPT channeling it further into PC production (Mantel et al. 1993).

The fatty acids 16:0 and 22:6n3 were chosen as the combination believed to favor PC synthesis. They are also the most abundant fatty acids in the natural prey for cod larvae (Li et al. 2015a). The CPT of rat liver has showed highest selectivity for DAG species containing 16:0 in the *sn*-1 position and PUFA in the *sn*-2 position (Holub 1978; Mantel et al. 1993; Morimoto and Kanoh 1978), and salmon intestine showed higher PC:TAG ratio when 16:0 was available as substrate

Table 5.2 Distribution of radioactivity of lipid classes, shown as percentage values of total lipid ± standard error (n = 6) in Atlantic cod *Gadus morhua* larvae

[^{3}H] 2-MAG	0.5 h		1 h		2 h		4 h	
	Control	Choline	Control	Choline	Control	Choline	Control	Choline
PC	21.5 ± 2.83^{a}	18.7 ± 3.87^{a}	24.4 ± 1.85^{a}	28.2 ± 3.45^{ab}	30.1 ± 5.82^{ab}	30.2 ± 3.10^{ab}	30.1 ± 2.41^{ab}	38.9 ± 1.37^{b}
PE	5.19 ± 0.67^{a}	5.98 ± 1.33^{a}	4.69 ± 0.43^{a}	6.90 ± 0.34^{ab}	7.32 ± 0.99^{ab}	7.45 ± 0.97^{ab}	6.69 ± 1.11^{a}	11.6 ± 2.04^{b}
PS&PI	1.77 ± 0.26^{a}	1.76 ± 0.54^{a}	1.37 ± 0.29^{a}	2.79 ± 0.27^{ab}	3.53 ± 1.17^{ab}	4.33 ± 1.11^{ab}	2.23 ± 0.59^{ab}	5.40 ± 1.35^{b}
MAG	5.41 ± 1.16	7.42 ± 2.42	3.81 ± 0.61	5.36 ± 1.11	2.47 ± 0.80	3.99 ± 0.50	2.68 ± 0.37	2.65 ± 0.22
DAG	7.63 ± 1.26^{ab}	11.5 ± 1.50^{b}	9.38 ± 0.97^{ab}	9.43 ± 0.54^{ab}	7.53 ± 0.62^{ab}	7.95 ± 0.93^{ab}	7.66 ± 0.71^{ab}	6.69 ± 0.50^{a}
TAG	51.1 ± 3.65^{b}	46.4 ± 8.22^{ab}	51.8 ± 1.81^{b}	40.6 ± 3.29^{ab}	43.9 ± 7.55^{ab}	38.8 ± 5.77^{ab}	44.4 ± 4.16^{ab}	28.1 ± 5.11^{a}
PC:TAG	0.44 ± 0.09^{a}	0.50 ± 0.15^{a}	0.48 ± 0.05^{a}	0.74 ± 0.14^{a}	0.83 ± 0.30^{ab}	0.90 ± 0.22^{ab}	0.74 ± 0.13^{a}	1.77 ± 0.45^{b}
[^{14}C] Glycerol	0.5 h		1 h		2 h		4 h	
	Control	Choline	Control	Choline	Control	Choline	Control	Choline
PC	26.0 ± 1.60^{a}	28.4 ± 1.79^{ab}	32.5 ± 0.99^{abc}	35.6 ± 2.46^{bc}	36.3 ± 3.76^{bc}	37.1 ± 1.82^{bc}	36.1 ± 1.64^{bc}	38.4 ± 2.45^{c}
PE	18.1 ± 2.38	20.2 ± 2.03	17.5 ± 1.37	18.6 ± 1.65	15.2 ± 1.69	16.4 ± 1.05	16.9 ± 1.89	19.6 ± 1.24
PS&PI	6.78 ± 0.61	4.65 ± 0.50	4.90 ± 0.76	6.51 ± 0.32	7.34 ± 1.77	7.43 ± 0.70	6.04 ± 1.60	8.77 ± 2.03
MAG	3.94 ± 0.60	4.89 ± 1.29	3.53 ± 0.49	3.68 ± 0.68	2.08 ± 0.66	3.92 ± 0.63	3.08 ± 0.39	3.48 ± 0.88
DAG	6.99 ± 0.79	8.43 ± 0.59	7.65 ± 0.37	8.19 ± 0.44	6.15 ± 0.47	6.76 ± 0.80	7.25 ± 0.55	6.54 ± 1.13
TAG	29.4 ± 3.99^{b}	23.4 ± 3.73^{ab}	26.3 ± 0.74^{ab}	18.4 ± 1.57^{ab}	24.2 ± 5.22^{ab}	17.6 ± 1.72^{ab}	20.2 ± 2.60^{ab}	14.0 ± 1.75^{a}
PC:TAG	1.02 ± 0.21^{a}	1.43 ± 0.34^{a}	1.24 ± 0.05^{a}	2.06 ± 0.30^{ab}	1.88 ± 0.62^{ab}	2.19 ± 0.25^{ab}	1.97 ± 0.30^{ab}	3.02 ± 0.49^{b}

Values in the same row not sharing a superscript are significantly different ($P < 0.05$). Data from Li et al. (2016b)

PC phosphatidylcholine; *PE* phosphatidylethanolamine; *PS&PI* phosphatidylserine and phosphatidylinositol; *MAG* monoacylglycerides; *DAG* diacylglycerides; *TAG* triacylglycerides; *PC TAG* ratio between PC and TAG

(Oxley et al. 2007). Moreover, addition of 16:0 to diets containing high linseed oil significantly reduced intestinal steatosis in Arctic char *Salvelinus alpinus* (Olsen et al. 2000). The high PC:TAG ratio found after 2 and 4 h of incubation in the present study was probably related to TAG being used as an energy source whereas PC was used for membrane or structural purposes.

The results clearly showed that cod larvae intestines have a high metabolic rate, and are fully capable to synthesize PL through both the 2-MAG and G-3-P pathways when dietary components that could be directly assimilated were injected. Furthermore, supplementation of choline chloride significantly increased the PC synthesis over TAG (PC:TAG ratio) ($P < 0.05$). Cod larvae were able to synthesize PC to similar levels as TAG, agreeing with the relatively fast reacylation process of MAG to TAG in the enterocytes found for salmon juvenile (Oxley et al. 2005, 2007). We therefore suggest that cod larvae might have relatively high ability to synthesize PC from 2-MAG and glycerol precursors under the conditions of the tube-feeding study where these components were injected and might be assimilated directly without further need for digestion.

5.4.4 Final Evaluation of the Requirement of Dietary PL of Cod Larvae

As mentioned above, PL is regarded to be an essential dietary component for marine fish larvae during their early life stages and the specific requirement of PL of fish larvae has been suggested to originate in an inefficient ability to biosynthesis PL de novo (Cahu et al. 2009; Tocher et al. 2008). This hypothesis is supported by the higher expression level of genes involved in the PL biosynthesis in the intestine of salmon parr compared to that of salmon fry. However, the tube-feeding study showed that cod larvae were able to biosynthesis PL de novo efficiently from 2-MAG and glycerol precursors when these components were injected and might be assimilated directly without further need for digestion.

Early studies with Atlantic cod larvae suggested that these larvae might have low ability to digest neutral lipid classes due to insufficient suitable lipases, bile acids or both (Olsen et al. 1991). Bile activated lipase (BAL) was found to be the main functional lipase for digestion of neutral lipids, and its enzyme activity and gene expression was found to be low in the early stages of cod larvae, and "adult type" digestion of neutral lipid was suggested to occur when the pyloric caeca started to develop (from 20 mm SL) (Kortner et al. 2011; Sæle et al. 2010). The incorporation of radiolabeled isotopes in the tube-feeding experiment was made independent of digestion, and then the PL synthesis pathway could be studied without influence of the first enzymatic steps during digestion. It is possible that with normal feed the substrates for de novo PL synthesis in the enterocytes becomes limited and dietary PL becomes important. The requirements of dietary PL can also be a particular result of the high growth rate of fish larvae, the need of PL may not be fulfilled even with a relatively high capacity of de novo synthesis.

Bile PC produced by the liver has been suggested to be a preferred source of chylomicron PC as compared to exogenous PC (Mansbach 1977). A recent study showed that the requirement of intact PC is not only due to the availability of PC for chylomicron surface formation, but it activates protein kinase C zeta in the form of lyso-phosphatidylcholine in the intestine, which plays an important role in chylomicron assembly and lipid transport (Siddiqi and Mansbach 2015). This, to some extent, suggests that intact PC, either in the form of dietary or biliary, is more important than intestinal de novo synthesized PC for lipoprotein assembly. It is possible that the requirement of intact PC in the early stage of fish larvae is not only due to the immature function of enterocytes for de novo synthesizing PC, but also a low input of bile PC from the immature liver. In the later stages, sufficient biliary PC is produced by the liver and dietary PC is not required for the fish. This could be partly supported by the low gene expression of *pemt* from hatching up to 60 dph of cod larvae, suggesting low PC production in the liver via PEMT pathway in the early stages (Li et al. 2015b). However, further studies are needed to quantify the enzyme activity in the PEMT pathway, CDP-choline pathway and factors affecting PC biosynthesis ability in fish larvae. Addition of choline increased the PC:TAG ratio, however, its function either in simulating PC synthesis or TAG catabolism or both needs further investigation.

In summary, Atlantic cod larvae at 30 dph (9.90 mm SL) showed relatively high PC:TAG ratio from both 2-MAG and glycerol precursors within the first hour of tube feeding, suggesting they might have comparable biosynthesis ability of PC and TAG under the conditions of the tube-feeding study. Supplementation of choline significantly increased the PC:TAG ratio, which could be helpful for the formation of lipoproteins and prevent steatosis. The dietary requirement of intact PC for fish larvae may be a combination effect of immature intestine and immature liver, resulting in low PC inputs from on-site de novo synthesis and from biliary PC. Understanding the mechanisms behind will contribute to the development of better feed and welfare for marine fish larvae. The protocol for DHA enrichment in PL described earlier will provide live feed with more suitable lipid nutrition and contribute to the development of efficient juvenile production of marine species in general.

References

Ando Y, Kobayashi S, Sugimoto T et al (2004) Positional distribution of n-3 highly unsaturated fatty acids in triacyl-sn-glycerols (TAG) of rotifers (*Brachionus plicatilis*) enriched with fish and seal oils TAG. Aquaculture 229:275–288

Bell JG, McEvoy LA, Estevez A et al (2003) Optimising lipid nutrition in first-feeding flatfish larvae. Aquaculture 227:211–220

Bergvik M, Overrein I, Bantle M et al (2012) Properties of Calanus finmarchicus biomass during frozen storage after heat inactivation of autolytic enzymes. Food Chem 132:209–215

Borkenhagen L, Kennedy EP, Fielding L (1961) Enzymatic formation and decarboxylation of phosphatidylserine. J Biol Chem 236:PC28–PC30
Cahu CL, Gisbert E, Villeneuve LAN et al (2009) Influence of dietary phospholipids on early ontogenesis of fish. Aquac Res 40:989–999
Cai Z, Feng S, Xiang X et al (2016) Effects of dietary phospholipid on lipase activity, antioxidant capacity and lipid metabolism-related gene expression in large yellow croaker larvae (*Larimichthys crocea*). Comp Biochem Physiol B 201:46–52
Cansell M, Nacka F, Combe N (2003) Marine lipid-based liposomes increase in vivo FA bioavailability. Lipids 38:551–559
Carmona-Antonanzas G, Taylor JF, Martinez-Rubio L et al (2015) Molecular mechanism of dietary phospholipid requirement of Atlantic salmon, Salmo salar, fry. Biochim Biophys Acta 1851:1428–1441
Conceiçao LEC, Yúfera M, Makridis P et al (2010) Live feeds for early stages of fish rearing. Aquac Res 41:613–640
Coutteau P, Mourente G (1997) Lipid classes and their content of n-3 highly unsaturated fatty acids (HUFA) in Artemia franciscana after hatching, HUFA-enrichment and subsequent starvation. Mar Biol 130:81–91
Coutteau P, Sorgeloos P (1997) Manipulation of dietary lipids, fatty acids and vitamins in zooplankton cultures. Fresh Biol 38:501–512
Coutteau P, Geurden I, Camara MR et al (1997) Review on the dietary effects of phospholipids in fish and crustacean larviculture. Aquaculture 155:149–164
Dapra F, Geurden I, Corraze G et al (2011) Physiological and molecular responses to dietary phospholipids vary between fry and early juvenile stages of rainbow trout (*Oncorhynchus mykiss*). Aquaculture 319:377–384
De Santis C, Taylor JF, Martinez-Rubio L et al (2015) Influence of development and dietary phospholipid content and composition on intestinal transcriptome of Atlantic Salmon (*Salmo salar*). PLoS One 10:e0140964
DeLong CJ, Shen YJ, Thomas MJ et al (1999) Molecular distinction of phosphatidylcholine synthesis between the CDP-choline pathway and phosphatidylethanolamine methylation pathway. J Biol Chem 274:29683–29688
Dhert P, Rombaut G, Suantika G et al (2001) Advancement of rotifer culture and manipulation techniques in Europe. Aquaculture 200:129–146
Evjemo J, Reitan K, Olsen Y (2003) Copepods as live food organisms in the larval rearing of halibut larvae (*Hippoglossus hippoglossus* L.) with special emphasis on the nutritional value. Aquaculture 227:191–210
Feng S, Cai Z, Zuo R et al (2017) Effects of dietary phospholipids on growth performance and expression of key genes involved in phosphatidylcholine metabolism in larval and juvenile large yellow croaker, *Larimichthys crocea*. Aquaculture 469:59–66
Fernandez-Reiriz MJ, Labarta U (1996) Lipid classes and fatty acid composition of rotifers (*Brachionus plicatilis*) fed two algal diets. Hydrobiologia 330:73–79
Fischer LM, Dacosta KA, Kwock L et al (2007) Sex and menopausal status influence human dietary requirements for the nutrient choline. Am J Clin Nutr 85:1275–1285
Fontagne S, Geurden I, Escaffre AM et al (1998) Histological changes induced by dietary phospholipids in intestine and liver of common carp (*Cyprinus carpio* L.) larvae. Aquaculture 161:213–223
Frolov AV, Pankov SL, Geradze KN et al (1991) Influence of the biochemical-composition of food on the biochemical-composition of the rotifer *Brachionus plicatilis*. Aquaculture 97:181–202
Garcia AS, Parrish CC, Brown JA (2008) Use of enriched rotifers and *Artemia* during larviculture of Atlantic cod (*Gadus morhua* Linnaeus, 1758): effects on early growth, survival and lipid composition. Aquac Res 39:406–419
Geurden I, Jutfelt F, Olsen RE et al (2009) A vegetable oil feeding history affects digestibility and intestinal fatty acid uptake in juvenile rainbow trout *Oncorhynchus mykiss*. Comp Biochem Physiol A 152:552–559

Gibellini F, Smith TK (2010) The Kennedy pathway-*de novo* synthesis of phosphatidylethanolamine and phosphatidylcholine. IUBMB Life 62:414–428

Gisbert E, Villeneuve L, Zambonino-Infante J et al (2005) Dietary phospholipids are more efficient than neutral lipids for long-chain polyunsaturated fatty acid supply in European sea bass *Dicentrarchus labrax* larval development. Lipids 40:609–618

Gjellesvik DR (1991) Fatty acid specificity of bile salt-dependent lipase enzyme recognition and super-substrate effects. Biochim Biophys Acta 1086:167–172

Gjellesvik DR, Lombardo D, Walther BT (1992) Pancreatic bile salt dependent lipase from cod (*Gadus morhua*) purification and properties. Biochim Biophys Acta 1124:123–134

Golfman LS, Bakovic M, Vance DE (2001) Transcription of the CTP: phosphocholine cytidylyltransferase alpha gene is enhanced during the S phase of the cell cycle. J Biol Chem 276:43688–43692

Guinot D, Monroig O, Hontoria F et al (2013a) Enriched on-grown Artemia metanauplii actively metabolise highly unsaturated fatty acid-rich phospholipids. Aquaculture 412:173–178

Guinot D, Monroig O, Navarro JC et al (2013b) Enrichment of Artemia metanauplii in phospholipids and essential fatty acids as a diet for common octopus (*Octopus vulgaris*) paralarvae. Aquacult Nutr 19:837–844

Hama K, Provost E, Baranowski TC et al (2009) In vivo imaging of zebrafish digestive organ function using multiple quenched fluorescent reporters. Am J Physiol Gastrointest Liver Physiol 296:G445–G453

Hamre K (2006) Nutrition in cod (*Gadus morhua*) larvae and juveniles. ICES J Mar Sci 63:267–274

Harel M, Ozkizilcik S, Lund E et al (1999) Enhanced absorption of docosahexaenoic acid (DHA, 22: 6n-3) in Artemia nauplii using a dietary combination of DHA-rich phospholipids and DHA-sodium salts. Comp Biochem Physiol B 124:169–176

Holub BJ (1978) Differential utilization of 1-palmitoyl and 1-stearoyl homologs of various unsaturated 1, 2-diacyl-sn-glycerols for phosphatidylcholine and phosphatidylethanolamine synthesis in rat-liver microsomes. J Biol Chem 253:691–696

Izquierdo MS, Tandler A, Salhi M et al (2001) Influence of dietary polar lipids' quantity and quality on ingestion and assimilation of labelled fatty acids by larval gilthead seabream. Aquacult Nutr 7:153–160

Jin Y, Li K, Olsen Y (2014) Time kinetics of fatty acid changes in phospholipids following enrichment and starvation of *Artemia franciscana* with a main focus on docosahexaenoic acid. Aquacult Nutr

Kanazawa A, Teshima SI, Inamori S et al (1981) Effects of phospholipids on growth, survival rate, and incidence of malformation in the larval Ayu. Mem Fac Fish Kagoshima Univ 30:301–310

Kanazawa A, Teshima SI, Inamori S et al (1983) Effects of dietary phospholipids on growth of the larval Red Sea Bream and Knife Jaw. Mem Fac Fish Kagoshima Univ 32:109–114

Kazachkov M, Chen QL, Wang LP et al (2008) Substrate preferences of a lysophosphatidylcholine acyltransferase highlight its role in phospholipid remodeling. Lipids 43:895–902

Kent C (2005) Regulatory enzymes of phosphatidylcholine biosynthesis: a personal perspective. BBA-Mol Cell Biol Lipids 1733:53–66

Kjørsvik E, Olsen C, Wold PA et al (2009) Comparison of dietary phospholipids and neutral lipids on skeletal development and fatty acid composition in Atlantic cod (*Gadus morhua*). Aquaculture 294:246–255

Kleppe L, Edvardsen RB, Furmanek T et al (2014) Global transcriptome analysis identifies regulated transcripts and pathways activated during oogenesis and early embryogenesis in Atlantic cod. Mol Reprod Dev 81:619–635

Kortner TM, Overrein I, Oie G et al (2011) Molecular ontogenesis of digestive capability and associated endocrine control in Atlantic cod (*Gadus morhua*) larvae. Comp Biochem Physiol A 160:190–199

Kuipers F, Oude Elferink RP, Verkade HJ et al (1997) Mechanisms and (patho) physiological significance of biliary cholesterol secretion. Subcell Biochem 28:295–318

Lagarde M, Bernoud N, Brossard N et al (2001) Lysophosphatidylcholine as a preferred carrier form of docosahexaenoic acid to the brain. J Mol Neurosci 16:201–204
Lehner R, Kuksis A (1992) Utilization of 2-monoacylglycerols for phosphatidylcholine biosynthesis in the intestine. Biochim Biophys Acta 1125:171–179
Lehner R, Kuksis A (1996) Biosynthesis of triacylglycerols. Prog Lipid Res 35:169–201
Li K (2015) Phospholipids in Atlantic cod (*Gadus morhua* L.) larvae rearing: incorporation of DHA in live feed and larval phospholipids and the metabolic capabilities of larvae for de novo synthesis. Doctoral theses at NTNU, p 257
Li K, Olsen Y (2015) Effect of enrichment time and dietary DHA and non-highly unsaturated fatty acid composition on the efficiency of DHA enrichment in phospholipid of rotifer (*Brachionus* Cayman). Aquaculture 446:310–317
Li ZY, Vance DE (2008) Phosphatidylcholine and choline homeostasis. J Lipid Res 49:1187–1194
Li K, Kjorsvik E, Bergvik M et al (2015a) Manipulation of the fatty acid composition of phosphatidylcholine and phosphatidylethanolamine in rotifers *Brachionus* Nevada and *Brachionus* Cayman. Aquacult Nutr 21:85–97
Li K, Østensen M-A, Attramadal K et al (2015b) Gene regulation of lipid and phospholipid metabolism in Atlantic cod (*Gadus morhua*) larvae. Comp Biochem Physiol B 190:16–26
Li JY, Li XF, Xu WN et al (2016a) Effects of dietary choline supplementation on growth performance, lipid deposition and intestinal enzyme activities of blunt snout bream *Megalobrama amblycephal* fed high-lipid diet. Aquacult Nutr 22:181–190
Li K, Olsen RE, Ostensen MA et al (2016b) Atlantic cod (*Gadus morhua*) larvae can biosynthesis phospholipid de novo from 2-oleoyl-glycerol and glycerol precursors. Fish Physiol Biochem 42:137–147
Lie O, Lambertsen G (1985) Digestive lipolytic enzymes in cod (*Gadus morhua*)—fatty acid specificity. Comp Biochem Physiol B 80:447–450
Maehre HK, Hamre K, Elvevoll EO (2013) Nutrient evaluation of rotifers and zooplankton: feed for marine fish larvae. Aquacult Nutr 19:301–311
Mansbach CM (1977) Origin of chylomicron phosphatidylcholine in rat. J Clin Invest 60:411–420
Mantel CR, Schulz AR, Miyazawa K et al (1993) Kinetic selectivity of cholinephosphotransferase in mouse liver: the K_m for CDP-choline depends on diacylglycerol structure. Biochem J 289:815–820
McEvoy LA, Navarro JC, Hontoria F et al (1996) Two novel *Artemia* enrichment diets containing polar lipid. Aquaculture 144:339–352
Millikin MR (1982) Qualitative and quantitative nutrient requirements of fishes: a reveiw. Fish Bull US 80:655–686
Morimoto K, Kanoh H (1978) Acyl chain length dependency of diacylglycerol cholinephosphotransferase and diacylglycerol ethanolaminephosphotransferase. Effect of different saturated fatty acids at the C-1 or C-2 position of diacylglycerol on solubilized rat liver microsomal enzymes. J Biol Chem 253:5056
Navarro JC, Henderson RJ, McEvoy LA et al (1999) Lipid conversions during enrichment of *Artemia*. Aquaculture 174:155–166
Northfield TC, Hofmann AF (1975) Biliary lipid output during 3 meals and an overnight fast. 1. Relationship to bile acid pool size and cholesterol saturation of bile in gallstone and control subjects. Gut 16:1–12
O'Brien-MacDonald K, Brown JA, Parrish CC (2006) Growth, behaviour, and digestive enzyme activity in larval Atlantic cod (*Gadus morhua*) in relation to rotifer lipid. ICES J Mar Sci 63:275–284
Olsen Y (2004) Live food technology of cold-water marine fish larvae. In: Moksness E et al (eds) Culture of cold-water marine fish. Blackwell Publishing Ltd, Oxford, UK, pp 73–128
Olsen RE, Henderson RJ, Pedersen T (1991) The influence of dietary lipid classes on the fatty acid composition of small cod *Gadus morhua* L. juveniles reared in an enclosure in northern Norway. J Exp Mar Biol Ecol 148:59–76

Olsen Y, Reitan KI, Vadstein O (1993) Dependence of temperature on loss rates of rotifers, lipids, and 3 fatty acids in starved Brachionus plicatilis cultures. Hydrobiologia 255:13–20
Olsen RE, Myklebust R, Ringo E et al (2000) The influences of dietary linseed oil and saturated fatty acids on caecal enterocytes in Arctic char (*Salvelinus alpinus* L.): a quantitative ultrastructural study. Fish Physiol Biochem 22:207–216
Olsen RE, Dragnes BT, Myklebust R et al (2003) Effect of soybean oil and soybean lecithin on intestinal lipid composition and lipid droplet accumulation of rainbow trout, *Oncorhynchus mykiss* Walbaum. Fish Physiol Biochem 29:181–192
Olsen Y, Evjemo JO, Kjørsvik E et al (2014) DHA content in dietary phospholipids affects DHA content in phospholipids of cod larvae and larval performance. Aquaculture 428:203–214
Overrein I (2010) Copepod lipids in aquaculture. Doctoral theses at NTNU 2010:89
Oxley A, Torstensen BE, Rustan AC et al (2005) Enzyme activities of intestinal triacylglycerol and phosphatidylcholine biosynthesis in Atlantic salmon (*Salmo salar* L.). Comp Biochem Physiol B 141:77–87
Oxley A, Jutfelt F, Sundell K et al (2007) *Sn*-2-monoacylglycerol, not glycerol, is preferentially utilised for triacylglycerol and phosphatidylcholine biosynthesis in Atlantic salmon (*Salmo salar* L.) intestine. Comp Biochem Physiol B 146:115–123
Rainuzzo JR, Reitan KI, Jorgensen L et al (1994a) Lipid composition in turbot larvae fed live feed cultured by emulsions of different lipid classes. Comp Biochem Physiol A 107:699–710
Rainuzzo JR, Reitan KI, Olsen Y (1994b) Effect of short-and long-term lipid enrichment on total lipids, lipid class and fatty acid composition in rotifers. Aquacult Int 2:19–32
Rehberg-Haas S, Meyer S, Tielmann M et al (2015) Use of the microalga *Pavlova viridis* as enrichment product for the feeding of Atlantic cod larvae (*Gadus morhua*). Aquaculture 438:141–150
Reo NV, Adinehzadeh M, Foy BD (2002) Kinetic analyses of liver phosphatidylcholine and phosphatidylethanolamine biosynthesis using C-13 NMR spectroscopy. BBA-Mol Cell Biol Lipids 1580:171–188
Rodriguez C, Perez JA, Izquierdo MS et al (1996) Improvement of the nutritional value of rotifers by varying the type and concentration of oil and the enrichment period. Aquaculture 147:93–105
Rønnestad I, Rojas-Garcia CR, Tonheim SK et al (2001) In vivo studies of digestion and nutrient assimilation in marine fish larvae. Aquaculture 201:161–175
Rossmeisl M, Jilkova ZM, Kuda O et al (2012) Metabolic effects of n-3 PUFA as phospholipids are superior to triglycerides in mice fed a high-fat diet: possible role of endocannabinoids. PLoS One 7:e38834
Sæle O, Nordgreen A, Olsvik PA et al (2010) Characterization and expression of digestive neutral lipases during ontogeny of Atlantic cod (*Gadus morhua*). Comp Biochem Physiol A 157:252–259
Sæle O, Nordgreen A, Olsvik PA et al (2011) Characterisation and expression of secretory phospholipase A2 group IB during ontogeny of Atlantic cod (*Gadus morhua*). Br J Nutr 105:228–237
Sargent J, McEvoy L, Estevez A et al (1999) Lipid nutrition of marine fish during early development: current status and future directions. Aquaculture 179:217–229
Sargent JR, Tocher DR, Bell JG (2002) The lipids in fish nutrition. In: Halver JE, Hardy RW (eds) 3rd edn, pp 181–257
Schuchardt JP, Schneider I, Meyer H et al (2011) Incorporation of EPA and DHA into plasma phospholipids in response to different omega-3 fatty acid formulations—a comparative bioavailability study of fish oil vs. krill oil. Lipids Health Dis 10
Siddiqi S, Mansbach CM (2015) Dietary and biliary phosphatidylcholine activates PKC zeta in rat intestine. J Lipid Res 56:859–870
Strakova J, Demizieux L, Campenot RB et al (2011) Involvement of CTP: phosphocholine cytidylyltransferase-beta 2 in axonal phosphatidylcholine synthesis and branching of neurons. BBA-Mol Cell Biol Lipids 1811:617–625

Subbaiah PV, Kaufman D, Bagdade JD (1993) Incorporation of dietary n-3 fatty-acids into molecular-species of phosphatidyl choline and cholesteryl ester in normal human plasma. Am J Clin Nutr 58:360–368

Sugimoto H, Banchio C, Vance DE (2008) Transcriptional regulation of phosphatidylcholine biosynthesis. Prog Lipid Res 47:204–220

Sundler R, Akesson B (1975) Regulation of phospholipid biosynthesis in isolated rat hepatocytes-effect of different substrates. J Biol Chem 250:3359–3367

Tocher DR, Bendiksen EA, Campbell PJ et al (2008) The role of phospholipids in nutrition and metabolism of teleost fish. Aquaculture 280:21–34

Van Greevenbroek MMJ, Voorhout WF, Erkelens DW et al (1995) Palmitic acid and linoleic-acid Metabolism in Caco-2 cells—different triglyceride synthesis and lipoprotein secretion. J Lipid Res 36:13–24

Vance DE (2002) Phospholipid biosynthesis in eukaryotes. In: Vance DE, Vance JE (eds) Biochemistry of lipids, lipoproteins and membranes, 4th edn. Elsevier, Oxford, UK, pp 205–231

Vance DE, Vance JE (2009) Physiological consequences of disruption of mammalian phospholipid biosynthetic genes. J Lipid Res 50:S132–S137

Wold P, Hoehne-Reitan K, Cahu C et al (2007) Phospholipids vs. neutral lipids: effects on digestive enzymes in Atlantic cod (*Gadus morhua*) larvae. Aquaculture 272:502–513

Chapter 6
Fat-Soluble Vitamins in Fish: A Transcriptional Tissue-Specific Crosstalk that Remains to be Unveiled and Characterized

Ignacio Fernández, Paulo Gavaia, Maria J. Darias and Enric Gisbert

Abstract Fat-soluble vitamins play essential roles in vertebrate's development and homeostasis, and thus, an optimal, efficient and sustainable fish farming deeply depends on the optimization of their dietary levels provided. Subsequently, nutritional imbalances are considered one of the major causative factors of vertebrate's abnormal development. Although approaches such as nutritional-dose-response trials, gene knock-down and over-expression studies, have provided valuable knowledge on its metabolism and dietary requirements; this knowledge is still mostly based on studies with mammalian species. Even though nutritional approaches involving different (i) research tools (i.e. NGS, RT-qPCR, proteomics, histology, immunohistochemistry, in situ hybridization, etc.), (ii) experimental approaches (in vivo and in vitro), and (iii) developmental stages (larval, juvenile and adult stages), have been applied to different fish species, the biological roles, underlying mechanisms and nutritional requirements in farmed fish are not fully understood yet. Here, knowledge gained during the last decade for each of the fat-soluble vitamins is compiled, reviewed from a holistic point of view, and the potential points of convergence of fat-soluble vitamins signaling pathways at molecular, cellular and tissue levels identified for the proper development of nutritionally balanced diets, based on integrative, multifactorial and multidisciplinary nutritional studies to be conducted in the nearest future.

I. Fernández (✉) · P. Gavaia
Centre of Marine Sciences (CCMAR), University of Algarve, Faro, Portugal
e-mail: nacfm@hotmail.com; ivmonzon@ualg.pt

M. J. Darias
UNICAEN, UA, CNRS, Biologie des Organismes et Ecosystèmes Aquatiques (BOREA), IRD, Sorbonne Univ, MNHN, UPMC Univ Paris 06, Montpellier, France

E. Gisbert
IRTA, Centre de Sant Carles de la Ràpita (IRTA-SCR), Unitat de Cultius Experimentals, Sant Carles de la Ràpita, Spain

M. Yúfera (ed.), *Emerging Issues in Fish Larvae Research*,
https://doi.org/10.1007/978-3-319-73244-2_6

Keywords Fat-soluble vitamins · Retinoic acid · Calcitriol · α-tocopherol Phylloquinone · Menaquinone

6.1 Introduction

6.1.1 Nutrition and Nutritional Requirements for Proper Fish Larvae Development and Fry Quality

Fish larvae development and growth deeply depend on different environmental factors, genetic background, nutrition and rearing practices and conditions (feeding regime, density, etc.). While how each one of these factors determines fish larvae growth and development is still not fully understood, a holistic approach for understanding how these parameters interact might certainly warrant in the near future a successful and sustainable development of fish aquaculture.

From a practical (industrial) point of view, the definition of the optimal nutritional requirements for fish larvae is a major issue for aquaculture success and sustainability as an economic activity. Particular knowledge on the requirements for macronutrients (protein, lipid and/or carbohydrates) and micronutrients (vitamins and minerals) are limited to some of the most extensively farmed species—the salmonids—and is still far from being fine-tuned in marine fish larvae (Holt 2011). The amount and chemical form of each nutrient required by larvae depends on as many factors as the fish species, the physiological stage (stressed vs. non-stressed), developmental phase (pre-, pro- and/or post-metamorphic, juvenile, adult and/or sexually mature/active), the environmental conditions and the nutritional approach used (life prey carriers vs. formulated diets), among others. For years, the majority of studies determining the nutritional requirements have been performed in a unifactorial, fragmented and reductionist way, with a skewed view towards maximal growth and/or survival, lower cost of production or best fish health (Hamre et al. 2013). Nowadays, factors driving the consumers' decisions, and by extension those of the producers, have changed the above-mentioned premises, and not only the price but the quality of the product, the animal welfare and the sustainability of its production are important from a societal and economical points of view. For instance, quality perception of the consumers relies on how much the farmed fish is similar to the ones caught by wild fisheries and how this production implies a minimal (if not a total) lack of stress and/or suffering of animals being raised. Producing low quality fish showing skeletal deformities and/or pigmentation disorders, decrease their market price and increase production costs and discards (Boglione et al. 2013; Koumoundouros 2010). Furthermore, lack of welfare might represent an increased risk of pathological outbreaks due to reduced immunocompetence (Austin 2012). In this sense, are the nutritional requirements for a particular compound depending on the chemical form used and/or the physiological role (muscle growth, bone development, immunocompetence, etc.) considered?

Recent data suggested that different requirements of one given nutrient should be expected considering the biological process to be evaluated. The same dietary content of vitamin A that hampered skeletal development in Senegalese sole *Solea senegalensis* (Fernández et al. 2009) has been recently shown to stimulate fish immunocompetence in the same species (Fernández et al. 2015a). Therefore, future integrative research studies on fish nutritional requirements should provide knowledge regarding the different chemical forms in which one nutrient should be supplied in diets and the range of physiological and biological processes in which this nutrient might be required by the organism. In addition, different nutrients should be provided in equilibrated/balanced amounts to allow harmonic development and promote wellbeing. Thus, nutritional studies should focus on deciphering what is the "balanced requirement" of each nutrient, taking into account the amount of other nutrients that might interfere or compete in their regulatory pathways. In this sense, a balanced diet should be considered as the one where the different nutrients are provided in the needed ratios to each other, depending on the species, the developmental stage and the physiological condition of the animal. One already known example of balanced requirements are the levels established for essential fatty acids (Tocher et al. 2008) due to the extensive research effort done in different species and using different approaches, being fish oil inclusion on aquafeeds a hot topic in modern aquaculture (Tocher 2015). However, the nutritionally balanced requirements on other important nutrients on fish physiology such as fat-soluble vitamins are largely unknown, particularly in larvae of marine fish species, as only a few dose–response studies have been performed to obtain their quantitative requirements (Hamre et al. 2013).

A proper knowledge of nutritional requirements throughout larval development would allow fish nutrition and feeding practices optimization towards a better larval and juvenile performance and quality. However, an integrated understanding on how nutrients are uptaken, digested/metabolized, how and where they play key biological roles as well as if and how they interact with each other is fundamental. The present chapter provides a comprehensive compilation of the scientific knowledge gained through the last decade on nutrition, particularly focusing on fat-soluble vitamins (Cahu et al. 2003; Darias et al. 2011; Fernández and Gavaia 2016; Fernández and Gisbert 2011; Hamre et al. 2010; Lall and Lewis-McCrea 2007; Krossøy et al. 2011; Lock et al. 2010). Recent advances on vitamins metabolism, functions and requirements will be reviewed in detail, but most importantly highlighting the signaling crosstalk among them. This particular knowledge will help us to define the future research approaches needed to carry out in order to reach a balanced nutritional level for all the liposoluble vitamins, for meeting larval requirements and promoting a harmonic development and wellbeing.

6.1.2 *Fat-Soluble Vitamins as Key Morphogenetic Nutrients in Fish Larvae*

Within the general term of fat-soluble vitamins, the compounds known as vitamin A (VA), vitamin D (VD), vitamin E (VE) and vitamin K (VK) are included. These are essential micronutrients for development and homeostasis in fish that should be provided in a specific amount and chemical form within the diet, since fish are not able to de novo synthesize them. In the case of vitamin D (VD), in contrast to mammalians, which can photoconvert 7-dehydrocholesterol into pre-vitamin D_3 in the skin (Holick 1981), until now no strong evidences of VD photosynthesis have been documented in fish species (Lock et al. 2010).

All fat-soluble vitamins present a relatively heterogenic chemical structure, sharing some particularities, and have been the source of specific reviews in the last decade regarding their requirements and functions in fish species (Darias et al. 2011; Fernández and Gavaia 2016; Fernández and Gisbert 2011; Haga et al. 2011; Hamre et al. 2010; Krossøy et al. 2011; Lock et al. 2010). VA is a terpene and VE and VK have long terpene chains attached to an aromatic moiety, while in contrast, VD is a steroid derivative of ergosterol. Nevertheless, as its nomenclature stands for, they are soluble in lipids, in contrast to the water soluble vitamins like vitamins B and C. Also in contrast to the latest, fat-soluble vitamins are (i) absorbed through the small intestine with dietary fat, (ii) present a more complex metabolism (including the existence of specific nuclear receptors), (iii) can be easily stored in specific tissues (mainly in liver and adipose tissue) from which they can be mobilized when dietary intake is insufficient, and (iv) are excreted slowly, therefore lasting longer in the body. This is also the main reason why fat-soluble vitamins can exhibit toxicity, although not in the particular case of the natural metabolites of VK.

Retinoids, more generally known as VA, are a group of key morphogens including all compounds that possess the same biological activity of retinol (Blomhoff 1994), having numerous important functions in vision, pigmentation, body patterning, maintenance of epithelial surfaces, immune competence, nervous system, reproduction, and embryonic growth and development (Ross et al. 2000). Retinal, one of the two primary active forms of VA, is used as the chromophore of rhodopsin in the eye (Pepe 1999). Retinoic acid (RA), the main active metabolite of VA, acts as a ligand for the RA receptors and retinoid X receptors (RARs and RXRs, respectively) regulating cellular differentiation and proliferation processes through the control of expression of a large set of genes (Balmer and Blomhoff 2002). While all-*trans*-RA activate specifically RAR isoforms α, β, and γ, the 9-cis-RA could bind and activate both RAR and RXR isoforms α, β, and γ, although showing higher affinity for the RXRs isoforms (Germain et al. 2006a, b). Interestingly, among the chemical forms in which VA can be administered in aquafeeds, RA is not recommended due to its high activity and particularly its instability, which is also the reason why retinal and retinol are avoided in live prey (rotifers and *Artemia* metanauplii) enrichment (Haga et al. 2006). The most commonly used chemical source of VA in aquafeeds for marine fish species is the synthetic form retinol acetate, which is more stable and easily metabolized by fish (Fontagné-Dicharry et al. 2010; Villeneuve et al. 2005a). However, retinyl palmitate (the main form of VA storage in animals) has also been recommended as a VA form to be supplied (Fernández and Gisbert 2011). Independently of the source, the amount of VA for fulfilling fish nutritional requirements seems to be highly dependent on the species, developmental stage and organ/function considered (Fernández 2011). Initially, levels up to 750 IU kg^{-1} VA were considered the minimum requirement for fish (NRC 1993), although different optimal levels for normal skeletogenesis (main impact of dietary VA) were reported posteriorly. Dietary VA levels lower than 50,000 IU kg^{-1} for flatfish species (Dedi et al. 1995; Lewis-McCrea and Lall 2010), 35,000 IU kg^{-1} for marine round fish (Villeneuve et al. 2005a, b) and 37,000 IU kg^{-1} for Atlantic salmon *Salmo salar* post-smolts (Ørnsrud et al. 2013) have been recommended. Nevertheless, while Mazurais et al. (2009) showed that the optimal level of retinol for harmonious ontogenesis varies throughout European sea bass *Dicentrarchus labrax* larval development, the same dietary content of VA hampering skeletal development has been shown to stimulate fish immunocompetence in Senegalese sole (Fernández et al. 2009, 2015a). Although VA is the most studied fat-soluble vitamin in fish, the focus has often been on toxic effects, and the maximal non-toxic level of VA for fish larvae is still unknown (Hamre et al. 2013).

In zooplankton and phytoplankton, two forms of VD (VD_2 and VD_3 or calcitriol) have been found in high concentrations, possibly representing the most common source of VD for fish larvae (Mattila et al. 1997; Rao and Raghuramulu 1996). Nevertheless, VD_2 is almost absent in fish tissues, while for most fish species VD_3 is the primary VD metabolite for storage in liver and muscle among other tissues such as intestine, kidney, spleen, gills and skin (Lock et al. 2010). The best known action of the VD endocrine system is its role in calcium and phosphate homeostasis, controlling transcellular calcium uptake and inducing cytosolic calcium transport (Bronner 2009) as well as phosphorus uptake (Hamre et al. 2010). Nevertheless, VD has been found to play further roles in skeletogenesis and mineralization, muscle contraction, immunity and cardiovascular physiology through binding to its receptor (VDR; Cerezuela et al. 2009; Darias et al. 2010, 2011; Jiang et al. 2015; Kwon 2016; Lock et al. 2010).

VD requirements have been established based on the lack of incidence of clinical manifestations of tetany, the most common sign of VD_3 deficiency (Barnett et al. 1982). Thin epidermis, extensively necrotized underlying musculature and hypocalcaemia were also reported as signs of deficient levels of VD in fish diets (Taveekijakarn et al. 1996). In contrast, despite of the large set of studies that did not report negative effects of high dietary doses of VD_3, brook trout *Salvelinus fontinalis* juveniles fed 3750 IU VD_3 g^{-1} diet for 40 weeks were hypercalcaemic and showed high haematocrit values, although growth and survival were not affected (Poston 1968). More severe effects were reported when hypervitaminosis D occurred during early development, such as hypermelanosis on the blind side and higher frequency of vertebral deformities in Japanese flounder *Paralichthys olivaceus* fed a diet containing 20,000 IU VD_3 kg^{-1} (Haga et al. 2004), or an increased incidence of skeletal deformities in European sea bass larvae fed with dietary levels of 42,000 and 120,000 IU VD_3 kg^{-1} diet (Darias et al. 2010). The VD_3 content has been analysed in numerous wild fish species (reviewed in Schmid and Walther 2013). In cultured species, including Atlantic salmon and rainbow trout *Oncorhynchus mykiss* (Lu et al. 2007; Mattila et al. 1997), red sea bream *Pagrus major* and Japanese flounder (in Schmid and Walther 2013), and several carp, panga and tilapia species (Bogard et al. 2015), the estimated requirements range from 240 to 1,300,000 IU VD_3 kg^{-1}. It has been shown that VD_3 requirements are highest during the larval stage and should be determined for each species as they could vary greatly (Darias et al. 2010), being their requirements species- and developmental stage-specific (Darias et al. 2011). A recent study analysed the nutrient profiles of rotifers (*Brachionus* sp.) and rotifer diets from four different marine fish hatcheries and showed that VD content varied from 2.4 to 3.2 IU g^{-1} in non-enriched rotifers and from 2.8 to 15.2 IU g^{-1} in enriched rotifers (Hamre et al. 2016), which are in the range of the doses recommended by the NRC (2011). However, according to the suggested larval VD_3 requirements of 27.6 IU g^{-1} for European sea bass (Darias et al. 2010), the VD content of the rotifers used to feed fish larvae from these hatcheries might be deficient.

Tocopherols (α-, β-, γ -, and δ-tocopherol) and four tocotrienols (α-, β-, γ -, and δ-tocotrienol) are generally referred as VE, being α-tocopherol (α-TOH) the form showing the highest biological activity (Yoshida et al. 2003). This fat-soluble vitamin protects lipids from peroxidation, with the production of a tocopheryl radical, which can probably be regenerated to tocopherol (Frankel 1998). In addition to the antioxidant function, VE can also participate in non-oxidant roles as the modulation of eicosanoid synthesis and the immune responses, inhibit proliferation of smooth muscle cells and may have multiple signaling functions at the post-translational level, such as activation of protein phosphatase 2A, among other enzymes (reviewed in Hamre 2011). Most importantly, VE can bind to the pregnane X repector (PXR) and thus regulate the expression of downstream genes (Landes et al. 2003).

Since marine fish larvae are probably subjected to high levels of oxidative stress and high lipid content on fish diets can also favor oxidation, a high requirement on VE in fish species can be anticipated. However, since vitamin C is taking part on tocopheryl radical regeneration to tocopherol (Hamre et al. 1997), supplementation of marine fish larval diets with VE and VC is a common procedure. As α-TOH is preferentially retained in fish than the other tocopherols, it is the most commonly used source of VE in aquafeeds (Hamre 2011). Nevertheless, tocopherol requirements varies with fish species, size, metabolic functions, feeding patterns, environmental factors, as well as depending on the amount of polyunsaturated fatty acids and other interacting nutrients included in the diet. Thus, the VE requirement can range from 25 to 3,000 mg α-TOH kg^{-1} (Hamre 2011), although the contents of 110 mg kg^{-1} VE in copepods may provide a guideline to follow in larval diets.

VK compounds are derived from quinone, exhibiting a common 2-methyl-1,4-naphthoquinone ring but differing in the side chain at the C3-position (Lambert et al. 1992). Three different VK metabolites (or K vitamers) can be found depending on the source: phylloquinone (VK_1, produced by photosynthetic plants), menaquinones (VK_2 or MK, with microbial or animal origin) and menadiones (VK_3, synthetic water soluble salt; Booth and Suttie 1998; Fodor et al. 2010). Although VK_3 is easily excreted and shows lower bioavailability than the naturally occurring K vitamers, it is the most commonly used source of VK included in aquafeeds, either as menadione sodium bisulphite (MSB) or menadione nicotinamide bisulphite (MNB) (Krossøy et al. 2011).

Until the last decade, the major and unique role of VK was assumed to be the control of blood coagulation (Oldenburg et al. 2008). Now, a specific requirement in fish for prevention of soft tissue calcification, proper skeletogenesis and bone homeostasis has been reported (Fernández et al. 2014a). Further, Richard et al. (2014) by applying proteomic analysis presented evidences that VK might also be critical for a broader set of biological functions such as muscular contraction, resistance to osmotic stress, intracellular Ca^{2+} homeostasis or energetic metabolism. Furthermore, studies in mammalian species suggested an important role of VK in the synthesis of sphingolipids and thus, in the central nervous system development and cognition capacities (Ferland 2012). Furthermore, VK may play an important role in glucose metabolism, stimulating insulin secretion and β-cell proliferation in the pancreas, but also in reproduction by stimulating testosterone synthesis (Karsenty and Ferron 2012). Such roles are related with the activity of VK as a co-factor of γ-glutamyl carboxylase (GGCX) to convert glutamate into γ-carboxyl glutamate (Gla) residues in VK-dependent proteins (VKDPs; Oldenburg et al. 2008), but also as a ligand of PXR, modulating the expression of target genes (Tabb et al. 2003).

Since part of the requirement on VK seems to be covered by the intestinal microbiota production (Harshman et al. 2014), an equilibrated intake of VK from diverse dietary sources and a healthy intestinal microbiota seems to be necessary for fulfilling VK requirements. In contrast to the other above described fat-soluble vitamins, VK can be recycled in a way that one molecule of VK can be reused up to 50 times. Thus, nutritional requirements for this fat-soluble vitamin appear to be lower than for the other vitamins, as well as its toxicity. Upper tolerance of 100 mg kg^{-1} of VK_1, 2,500 mg kg^{-1} for MSB and 2,000 mg kg^{-1} for MNB have been reported in different fish species, being suggested that the dietary optimal levels for juveniles should be around 1.5–20 mg VK_3 kg^{-1} (reviewed in Krossøy et al. 2011). Recently, an improved osteological development in Senegalese sole larvae was achieved with a dietary supply of 250 mg VK_1 kg^{-1} (Richard et al. 2014). Thus, similarly to the other fat-soluble vitamins, the source and amounts of VK might have to be adapted to fish species and developmental stages.

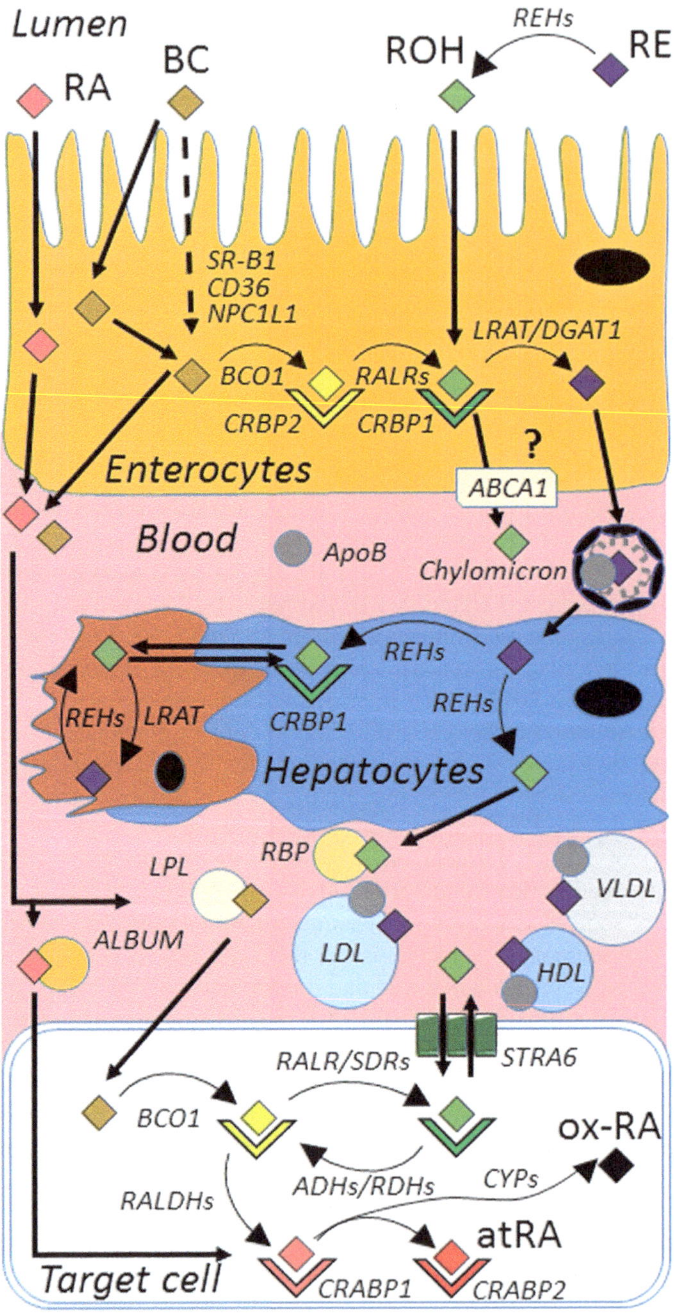
Lumen
RA
BC
ROH
REHs
RE
SR-B1
CD36
NPC1L1
LRAT/DGAT1
BCO1
RALRs
CRBP2
CRBP1
?
Enterocytes
ABCA1
Blood
ApoB
Chylomicron
REHs
REHs
LRAT
CRBP1
REHs
Hepatocytes
RBP
LPL
VLDL
ALBUM
LDL
HDL
STRA6
RALR/SDRs
BCO1
ox-RA
ADHs/RDHs
CYPs
RALDHs
atRA
Target cell
CRABP1
CRABP2

◂**Fig. 6.1** Overview of the metabolism of vitamin A (VA) and the retinoid signaling pathway (RSP) from absorption to target cells. *ABCA1*, ATP binding cassette A1; *ADHs*, alcohol dehydrogenases; *ALBUM*, albumin; *ApoB*, apolipoprotein B; *atRA*, all-trans retinoic acid; *BC*, carotenoids including mainly β-carotene but also lutein and lycopene; *BCO1*, β-carotene-15,150-oxygenase 1; *CD36*, cluster determinant 36; *CRBP1*, cellular retinol-binding protein 1; *CRBP2*, cellular retinol-binding protein 2; *CRABP1*, cellular retinoic acid binding protein 1; *CRABP2*, cellular retinoic acid binding protein 2; *CYPs*, cytochromes P450; *DGAT1*, diacylglycerol acyltransferase 1; *HDL*, high-density lipoprotein; *LDL*, low density lipoprotein; *LPL*, lipoprotein lipase; *LRAT*, lecithin:retinol acyltransferase; *NPC1l1*, Niemann-Pick C1-like 1; *RA*, retinoic acid; *RALDHs*, retinaldehyde dehydrogenases; *RALRs*, retinal reductases; *RBP*, retinol-binding protein; *RDHs*, retinol dehydrogenases; *RE*, retinyl ester; *REHs*, retinyl ester hydrolases, pancreatic triglyceride lipases and pancreatic lipase related protein 2; *ROH*, retinol; *SDRs*, short-chain dehydrogenase/reductases; *SR-B1*, scavenger receptor class B member 1; *STRA6*, stimulated by RA gene 6 protein homolog; *ox-RA*, oxidized retinoic acid metabolite; *VLDL*, very low density lipoprotein

6.2 Absorption, Transport, Metabolism and Function

6.2.1 Vitamin A

Vertebrates have a complex and tightly controlled metabolism for retinoids and their precursors (the carotenoids; Fig. 6.1). The retinoid signaling pathway (RSP) has been shown to be highly conserved through evolution, with only few exceptions such as in the tunicate *Oikopleura dioica* (Martí-Solans et al. 2016). Thus, the following RSP description is based on the studies with mammalian species, primarily characterized and reviewed in detail by D'Ambrosio et al. (2011), but updated with the published literature from the last 6 years. Particular differences with fish species will be given.

For optimal retinoid absorption, fat must be consumed along with the newly ingested retinoid in order to facilitate retinoid entry into enterocytes from the lumen through chylomicron formation. Carotenoids, like β-carotene, but also lutein and lycopene, are uptaken through a process involving scavenger receptor class B member 1 (SR-B1), cluster determinant 36 (CD36) and/or Niemann-Pick C1-like 1 (NPC1L1; Reboul 2013). β-carotene can be incorporated intact and unmodified into nascent chylomicrons and then transported by the bloodstream or can be converted to retinal by a central cleavage catalysed by β-carotene 15,15′ oxygenase 1 (BCO1) in the enterocytes that converts one molecule of β-carotene into two molecules of retinal (Chelstowska et al. 2016). The control of the expression of *bco1* at promoter level by the retinoic acid-inducible homeobox transcription factor ISX might represent the first negative feedback regulation step on RSP, which might also explain the large individual variability in intestinal β-carotene conversion (Lobo et al. 2013). In addition, ISX also regulates *sr-b1* expression in a RA dependent manner, meaning that the intestinal absorption of carotenoids is tightly regulated by a negative feedback loop, directly depending on the organism needs for VA (Chelstowska et al. 2016; Widjaja-Adhi et al. 2015). Independently of this regulatory step, the produced retinal must bind to cellular retinol-binding protein 2 (CRBP2) and then undergo reduction to retinol through short-chain dehydrogenase/reductase (SDR) enzymes and/or retinal reductases (RALR). In contrast, an eccentric cleavage of β-carotene through β,β-carotene 9′,10′-oxygenase (BCO2) will result in β-apocarotenoids that have been found to function as retinoid receptor antagonists (Eroglu et al. 2012).

Retinol can be also directly incorporated by enterocytes from retinyl esters hydrolysed in the intestine by pancreatic triglyceride lipase (PTL), pancreatic lipase related protein 2 (PLRP2) or retinyl ester hydrolases (REHs). Retinol formed from dietary carotenoids or directly absorbed from the intestine binds to CRBP2 and is esterified to retinyl ester (RE) by lecithin:retinol acyltransferase (LRAT, 90% of RE) or by diacylglycerol acyltransferase 1 (DGAT1, 10% of RE). The resulting RE, as the dietary carotenoid that has not undergone conversion to retinoid, is then packed into nascent chylomicrons, which are secreted into the lymphatic system. Retinol was also suggested to be secreted involving ATP binding cassette A1 (ABCA1; Reboul 2013).

Among the total amount of dietary retinoid included in chylomicrons from the blood stream, 66–75% is considered to be uptaken by the liver, with the remainder being cleared by peripheral tissues. The biological relevance of the clearance of this dietary retinoids in chylomicrons by peripheral tissues was uncovered by the general good health exhibited by mammals deficient in retinol-binding protein 4 (RBP4, generally known as RBP), which is the main protein responsible for the transport of retinol through the lymphatic system (Funkenstein 2001). The lipoprotein lipase (LPL) is the responsible for facilitating the uptake of postprandial retinoid into peripheral tissues.

Liver is the major site of retinoid storage and metabolism, taking place within the cells centrally involved in the uptake and processing of retinol in the liver, the hepatocytes (ca. 66% of cells), and the hepatic stellate cells (also known as fat-storing cells). Within the hepatocytes, retinyl ester undergoes rapid hydrolysis carried out by different REHs, carboxylesterases and/or lipases. Posteriorly, retinol is transferred through RBP or cellular retinol-binding protein 1 (CRBP1) to the stellate cells where it is re-esterified and stored in lipid droplets. When needed, lipid droplet retinyl ester stores are mobilized from hepatic stellate cells to supply peripheral tissues and maintaining the essential biological functions. The specific lipases involved in the hydrolysis of retinyl ester in hepatic stellate cells are not currently known, although hepatic carboxylesterases ES-4 and ES-10 and adipose triglyceride lipase are strong candidates.

Upon hydrolysis, retinol is thought to be transferred back to the hepatocyte, bound to RBP, and the retinol-RBP complex enters the blood stream and is transported to peripheral tissues, with RBP secretion being highly dependent on the retinoid status. Interestingly, in mammals, RBP is found in a 1:1 protein-protein complex with transthyretin (TTR), thought to prevent the filtration of the relatively small RBP molecule through the kidney glomeruli, although this process has not been reported in fish species (Zanotti et al. 2008). Nevertheless, under fasting conditions, >95% of retinoids in the circulation are found as retinol bound to RBP; while the rest are retinyl esters associated with lipoproteins (very low density lipoprotein (VLDL), low density lipoprotein (LDL), and high density lipoprotein

(HDL)), RA bound to albumin, and water-soluble β-glucuronides of retinol and retinoic acid (O'Byrne and Blaner 2013). In fish, in addition to RBP, retinol may also be transported by albumin (N'soukpoé-Kossi et al. 2007) and some carotenoids can be transported through LDL, HDL and very HDL (mainly astaxanthins); or retinal, being specifically transported through vitellogenin to oocytes (Lubzens et al. 2003).

The mechanisms mediating cellular uptake of retinyl ester and RA are not fully understood, but it has been established for certain tissues that LPL can hydrolyse retinyl ester to retinol, which can then be uptaken by cells through spontaneous transfer across the phospholipid bilayer. The stimulated by RA gene 6 protein homolog (STRA6) is able to bind RBP and facilitate its uptake into and efflux from the cells (Kawaguchi et al. 2015), although it is not expressed in all the cell types metabolizing retinoids. Remarkably, the cellular uptake of retinol/RA from RBP depends on functional coupling of STRA6 with intracellular LRAT (Amengual et al. 2012), CRBP1 (O'Byrne and Blaner 2013), cellular retinoic acid binding protein 1 (CRABP1) and CRBP2 (Kawaguchi et al. 2015). This suggests a well-coordinated uptake process at cellular level that may be responsible for the adipogenic differentiation potential of the cells (Muenzner et al. 2013). Loss-of-function studies revealed that RBP and STRA6 play a pivotal role in retinoid uptake in zebrafish *Danio rerio* (reviewed in Kelly and Von Lintig 2015). Indeed, gene knockdown of *stra6* induced embryonic malformations such as microphthalmia, curved body axis, edema of the pericardium, dysmorphic heart chambers, and peripheral bleedings, and were somehow rescued by *rbp* knockdown.

Within the target cell, retinoids are sequentially converted to RA (or just to retinaldehyde in the eye as the chromophore of rhodopsin; Pepe 1999) in two separate oxidation steps (reviewed in Gutierrez-Mazariegos et al. 2011). Retinol is first reversibly oxidized into retinaldehyde by alcohol dehydrogenase (ADH) or short-chain dehydrogenase/reductase (SDR) enzymes, mainly by retinol dehydrogenase 10 (RDH10) and retinoic acid-inducible dehydrogenase reductase 3 (DHRS3; Adams et al. 2014). Retinaldehyde is subsequently irreversibly oxidized into RA by retinaldehyde dehydrogenase (RALDH) enzymes, with RALDH2 representing the main RA synthesizing enzyme during early embryogenesis. In fact, the concerted action of RDH10 and RALDH2 have been suggested as the biosynthetic enzyme code required for axis formation and anterior-posterior patterning of the vertebrate embryo. In addition to the function of retinal as a chromophore and RA as a nuclear receptor (NR) ligand, retinol has been recently revealed to present its own physiological role as an electron carrier in mitochondria; thus, playing a key role in glycolytic energy generation (Hammerling 2016).

RA functions as a specific ligand for VA NRs (RARs and RXRs), and thus, controlling gene expression cascade for cell proliferation, differentiation and/or activity. Under RA excess, it is primarily hydroxylated by cytochromes from P450 family, the CYP26 enzymes in particular, into a wide variety of metabolites, such as 4-oxo RA, 4-OH RA and 18-OH RA that can be still biological active (Reijntjes et al. 2005). RA levels are also regulated by the presence of CRBPs and CRABPs, which can bind to retinol or RA molecules to store them, keeping away from the CYP enzymes. Furthermore, another step of tight control over the RSP is the autoregulation of RA signaling components. For example, while RA directly activates transcription of *rar* and *cyp26*, *raldh2* expression is repressed by RA. Such RA autoregulatory loop has been already demonstrated in different fish species under dietary VA imbalance at the transcriptional level for *rarα*, *rbp*, and *cyp26* (among others) from in vivo and in vitro approaches (Boglino et al. 2017; Fernández et al. 2011, 2014b; Lie et al. 2016; Oliveira et al. 2013). Nevertheless, and compared to the gained knowledge on VA metabolism in mammalian species, a great effort remains to be done in order to characterize VA metabolism in fish species (Fernández and Gisbert 2011). For example, although freshwater species are known to present both standard and didehydroretinoids, only one study reported its differential distribution in distinct tissues (Gesto et al. 2012). Didehydroretinoids were found to be the dominant VA form over retinoids in all analyzed tissues with the exception of plasma, although the biological relevance of these findings remains to be determined. Furthermore, building a model based on VA compartmentalization map among the different tissues—as the one already available for neonates from mammalian species (Tan et al. 2014)—will help to unveil the interaction among them and the modulation of its metabolism under different dietary VA conditions and at different developmental stages (larval, juvenile, adult and broodstock).

6.2.2 Vitamin D

From an evolutionary perspective, it was believed that VD_3 (cholecalciferol) was mainly a phosphate conserving hormone and thus, its significant role in calcium regulation occurred with the evolution of land animals (MacIntyre et al. 1976). However, nowadays it is known that VD_3 is also the active calcium regulatory hormone in fish (Larsson et al. 1995; Darias et al. 2010, 2011; Sundell and Björnsson 1990; Sundell et al. 1992, 1993). Studies made on fish indicate that the role of VD_3 rather seems to have evolved to a calcemic regulatory function in bony vertebrates (i.e., zebrafish, Lin et al. 2012) as it has been shown to have another role in early vertebrates lacking a calcified skeleton and teeth (i.e., lamprey *Petromyzon marinus*; Whitfield et al. 2003). Besides these main physiological functions, recent studies have demonstrated that the most active metabolite of VD_3, calcitriol, regulates multiple and biologically diverse pathways during fish development involved in nutrition, immune system and xenobiotic metabolism (Darias et al. 2010, 2011; Craig et al. 2012; Jiang et al. 2015; Qiu et al. 2007).

In mammals, VD is synthesized from a cholesterol derivative through photosynthesis in the skin under ultraviolet B (UVB) radiation exposure (Holick et al. 1977). Although UV light-induced synthesis of VD_3 from 7-dehydrocholesterol (7-DH) has been demonstrated in fish (Rao and Raghuramulu 1997), and a recent study suggested that some fish, such as rainbow trout, might depend upon solar radiation to induce VD_3 production in the skin to meet their VD requirements (Pierens and Fraser 2015); there has been a consensus that the contribution of this mode of synthesis is unlikely to be of any significance (at least for marine fish) in its natural habitat since most UVB irradiation is absorbed in the first meters of the water column (Lock et al. 2010; Rao and Raghuramulu 1997). Instead, fish likely obtain their required VD from the diet, being high concentrations of vitamin D_2 (VD_2, ergocalciferol) and VD_3 found in zoo- and phyto-plankton (Copping 1934; Mattila et al. 1997; Rao and Raghuramulu 1997; Takeuchi et al. 1986).

Although VD_2 and VD_3 are available from live food, the latter is almost the primary metabolite of storage in fish tissues (Mattila et al. 1997; Fig. 6.2), since it has a higher bioavailability and binding to vitamin D binding protein (DBP; from Lock et al. 2010). Nowadays, it is well known that fish have a VD endocrine system with similar functions to those reported in mammals (Darias et al. 2011; Hamre et al. 2010; Lock et al. 2010). After ingestion, VD_3 is mainly absorbed through passive diffusion, but also some transport proteins have been recently identified to be involved on its uptake such as SR-BI, CD36 and NPC1L1 (Reboul 2015). Then, VD is transported via DBP to the liver where it is hydroxylated to 25-hydroxyvitamin D_3 [25(OH)D_3, calcifediol] (Ponchon and DeLuca 1969) by vitamin D-25 hydroxylase (CYP2R1) (Lin et al. 2012). Then, calcifediol is hydroxylated to form the most active metabolite 1,25-dihydroxyvitamin D_3 [1,25 $(OH)_2D_3$, calcitriol] (Holick et al. 1971) by 1α-hydroxylase (CYP27B1) and this being finally degraded into calcitroic acid by 25-hydroxyvitamin D3-24-hydroxylase (CYP24A1) in the liver and other tissues, including the kidney (Bailly et al. 1988; Hayes et al. 1986; Holick et al. 1972; Lin et al. 2012; Sundh et al. 2007; Takeuchi et al. 1991, 1994). The three metabolites of VD_3 [calcifediol, 24,25$(OH)_2D_3$ and calcitriol] have been found in several organs like the liver, kidney, gills, spleen, skin, muscle and plasma of many fish species (Bailly et al. 1988; Horvli and Aksnes 1998; Graff et al. 1999; Takeuchi et al. 1987).

The most described VD_3 action in fish is the one mediated through binding to its NR, the vitamin D receptor (VDR; Lock et al. 2010), for which it has high affinity (DeLuca and Zierold 1998). Calcitriol can enter the cell and bind to VDR, forming a heterodimer with the RXR and activating the transcription of target genes through its binding to the VD responsive element (VDRE) on the gene promoter. Two isoforms of VDR exist in fish (VDRα and β), as the results of the independent genome duplication that occurred in ray-finned fish phylogeny (3R event; Suzuki et al. 2000). VDR expression has been detected in a wide range of tissues in fish (Craig et al. 2008; Lock et al. 2007; Suzuki et al. 2000) from early developmental stages to adults (Darias et al. 2011). Functional characterization of VDR paralogues in medaka *Oryzias latipes* showed that transactivation of VDRα with VD_3 was highly attenuated compared to that of VDRβ, suggesting a functional divergence between these two NRs (Howarth et al. 2008) that probably occurred before the 3R event (Kollitz et al. 2014). Similarly, VD_3-VDRβ signaling has been shown to be involved in Ca^{2+} uptake through the regulation of the transient receptor potential cation channel subfamily V member 6 (TRPV6) and in bone mineralization through the regulation of *bmp4* and *osteocalcin* expression in European sea bass larvae

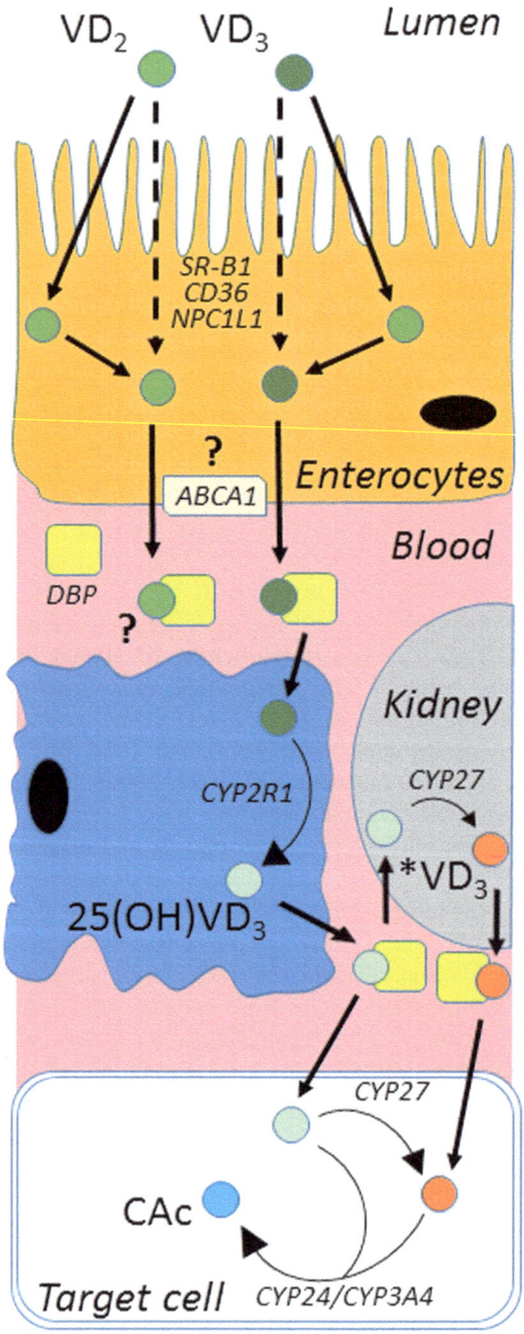

Fig. 6.2 Overview of the metabolism of vitamin D (VD) and the cholecalciferol signaling pathway (CSP) from absorption to target cells. *ABCA1*, ATP binding cassette A1; *CAc*, calcitroic acid; *CD36*, cluster determinant 36; *CYP2R1*, cytochrome P450 family 2 subfamily R member 1; *CYP24*, cytochrome P450 family 24; *CYP27*, cytochrome P450 family 27; *CYP3A4*, cytochrome P450 family 3 subfamily A member 4; *DBP*, vitamin D binding protein; *NPC1l1*, Niemann-Pick C1-like 1; *SR-B1*, scavenger receptor class B member 1; VD_2, ergocalciferol; VD_3, calcitriol; $*VD_3$; $1,25(OH)_2VD_3$ (1,25-dihydroxyvitamin D3); $25(OH)VD_3$, 25-dihydroxyvitamin D3 or colecalciferol

(Darias et al. 2010), whereas VDRα has been suggested to be functional in calcemic regulation in zebrafish (Lin et al. 2012). It has been reported that differences in protein-protein interactions between the VDR paralogs and essential co-regulators may drive the observed differential ligand sensitivities between VDRα and VDRβ (Kollitz et al. 2014, 2015).

The role of VD in calcium uptake regulation has been extensively studied in teleosts as they share similar mechanisms to those of mammals (Flik et al. 1996). VD is known to elevate the serum Ca^{2+} level in fish (Sundell et al. 1993; Swarup et al. 1991). Moreover, it has been demonstrated since the late 80s that plasma concentrations of VD metabolites in rainbow trout depend on environmental calcium concentrations in both fresh water (FW) and seawater (SW) environments (Hayes et al. 1986). Hypercalcemia and increased environmental calcium were associated with a greater transformation to the compound 25,26-dihydroxycholecalciferol, while hypocalcemia and reduced environmental calcium concentrations induced more conversion to the 1,25-dihydroxycholecalciferol-like compound (Hayes et al. 1986). Larsson et al. (1995) showed that $24,25(OH)_2D_3$ is a physiologically important calcium regulatory hormone in Atlantic cod *Gadus morhua*, involved in the short-term regulation of plasma calcium levels. This VD metabolite acts rapidly on intestinal calcium influx via a non-genomic pathway, whereas calcitriol regulates the intestinal calcium uptake in a slower, long-term manner, via a genome-mediated pathway. It was posteriorly demonstrated that $24R,25(OH)_2D_3$ and calcifediol are active regulators of intestinal Ca^{2+} uptake at physiological concentrations: $24R,25(OH)_2D_3$ decreases Ca^{2+} uptake through L-type Ca^{2+} channels, whereas calcifediol increases it, concurrent with an increase in Ca^{2+} extrusion by Na^+/Ca^{2+} exchangers in enterocytes (Larsson et al. 2002). However, the mechanisms of action of these metabolites and their physiological importance are quite unknown. Nemere et al. (2000) found differences in the expression of VDR in the FW common carp *Cyprinus carpio* compared with the SW Atlantic cod in response to calcitriol, and concluded that the different outcomes in rapid effects of the VD_3 metabolites may be due to an adaptation of the VD_3 endocrine system to the specific Ca^{2+} availability encountered by each animal group, suggesting a divergent evolution of the VD_3 endocrine system in SW fish compared to FW fish. Moreover, the evolutionarily conserved system that mediates rapid responses to calcitriol indicates that it is physiologically important for animals living in a low-Ca^{2+} environment (Nemere et al. 2000). Likewise, Lock et al. (2007) reported variations in VD_3 concentration and *vdr* expression in Atlantic salmon undergoing smoltification and migrating from FW to SW, revealing that VD_3 and VDR are regulated by water Ca^{2+} concentrations. Similarly, the transfer of rainbow trout from FW to SW induced an increase of $24,25(OH)_2D_3$ production and a decrease of calcitriol production in both liver and kidney, suggesting a physiological regulation and a differentiated importance of calcitriol and $24,25(OH)_2D_3$ in relation to environmental calcium concentrations (Sundh et al. 2007). The hormonal regulation of calcium is complex and several studies have shed light to the molecular mechanisms underlying this physiological process. VD_3 has been shown to modulate the expression of *vdr* to control intestinal Ca^{2+} absorption through the regulation of *trpv6* in European sea bass larvae (Darias et al. 2010), which is considered as the major transcellular mediator of Ca^{2+} uptake from the intestinal lumen (Hoenderop et al. 2005). Calcium absorption via TRPV6 has also been suggested to be regulated by estrogen and progesterone in view of the consensus cis-regulatory elements for the respective steroid hormone receptors found in the upstream regulatory region of the *trpv6* gene in pufferfish *Fugu rubripes* (Qiu and Hogstrand 2004). Stimulation of *trpv* expression by calcitriol has also been observed in zebrafish embryos (Hwang and Chou 2013). In addition to VD, parathyroid hormone (PTH) has also been demonstrated to be involved in Ca^{2+} regulation in fish (Lin et al. 2012; Sundell et al. 1993; Suzuki et al. 2011). Low Ca^{2+} FW stimulates *pth1* expression in zebrafish embryos, in turn, stimulates whole-body Ca^{2+} content and *trpv* expression (Lin et al. 2014). However, the corresponding receptor mediating PTH1 action and evidence for its role in mediating the effect of PTH1 on Ca^{2+} uptake is still unknown (Guh and Hwang 2016). Another member of the PTH family, the parathyroid hormone-related protein (PTHrP) is also considered as a hypercalcemic hormone in fish (Abbink and Flik 2007). Like VD_3 (Darias et al. 2010), PTHrP has been shown to be involved in skeletogenesis of zebrafish (Yan et al. 2012). Similarly, cortisol, a hypocalcemic hormone in mammals, was recently found to act as a stimulator of Ca^{2+} uptake in zebrafish through stimulation of *trpv* mRNA expression, as well as through modulation of the expression of the receptor and synthesizing enzyme of VD (Lin et al. 2011). These studies evidenced the complexity of the hormone system regulation in calcium homeostasis in fish.

Little is known about the VD_3 mediation of phosphorous (P) homeostasis in fish. The influence of VD in P metabolism has been demonstrated in several fish species, such as the American eel *Anguilla rostrata* (Fenwick et al. 1984), the catfish *Clarias batrachus* (Swarup et al. 1984), the common carp (Swarup et al. 1991), the Atlantic cod (Sundell et al. 1993) or the rainbow trout (Avila et al. 1999). VD_3 has been shown to induce hyperphosphatemia in some fish species (Fenwick et al. 1984; Swarup et al. 1984, 1991). In the case of rainbow trout, it was observed that increasing dietary levels of VD_3 under sufficient dietary P conditions did not enhance inorganic phosphate absorption by the intestine, but increased inorganic phosphate plasma concentration, suggesting alterations in the liver function (Avila et al. 1999). Vielma and Lall (1998) suggested an inverse relationship between the hepatic accumulation of VD_3 and the bone mineralization degree in Atlantic salmon, the latter being correlated with dietary P utilization. P deficiency has been suggested to have several indirect effects on immune functions via VD-mediated pathways and via depression of leukocyte function associated with decreased ATP content (Sugiura and Ferraris 2004). Indeed, VD_3 has been recently shown to play a role in the immune system regulation. The anti-inflammatory effect of VD_3 may be associated with decreasing the expression of pro-inflammatory cytokines (*tnf-α*, *il-1β*, *il-6* and *il-8*) by down-regulating *toll-like receptor 4* (*tlr4*), *myeloid differentiation primary response gene 88* (*myd88*) and *nf-κbp65* mRNA abundance in juvenile Jian carp *Cyprinus carpio* var. Jian in vivo and in enterocytes in vitro (Jiang et al. 2015). No information about the regulation of phosphorous homeostasis by VD_3 exists during the fish larval stages.

Finally, VD_3 also mediates body mineral balance in fish other than P and Ca^{2+}. For instance, VD_3 enhances zinc uptake across rainbow trout gills and this effect is associated with an increased expression of transporters that mediate zinc uptake, such as *slc39a1* and *trpv* (Qiu et al. 2007). Nevertheless, functions of VD are not only restricted to mineral homeostasis through orchestrated actions on the intestine, kidney, and bone, but also in immune cells, the skin and tissues of the cardiovascular system, muscle, liver, and brain (Pike et al. 2017). Advances in our understanding of the multiple roles of VD have emerged through either clinical discovery or through studies of the numerous genetically modified mouse models. With the advent of new tools for developing mutant lines in species like zebrafish and medaka, further insights on the mechanistically consequences of having two VDR paralogs will be elucidated. Since recent studies suggested the presence of VDR in activated hepatic stellate cells (Ding et al. 2013), the possible interactions between VD and VA signaling pathways through their respective NRs and the physiological consequences, deserves to be elucidated. Further research efforts should be performed for a more integrative understanding of VD metabolism and physiology on the specific activities of VD on central nervous system (CNS) neurons and the proposed roles through non-genomic mechanisms, recently summarized in Hii and Ferrante (2016).

As several studies have shown that VD_3 levels for optimal larval development are within a restricted range (reviewed in Darias et al. 2011), variations can cause severe physiological disruptions in cascade. Moreover, VD_3 has been shown to dramatically alter the expression of genes involved in fatty acid, amino acid, and xenobiotic metabolism pathways, as well as transcription factors, leptin, peptide hormones, RANKL, and calcitonin-like ligand receptor pathways in developing zebrafish embryos/larvae, demonstrating the multiple targets of this vitamin and its involvement in multiple biologically diverse pathways during development (Craig et al. 2012). Moreover, it has been recently demonstrated in zebrafish the complex regulation of mRNA expression by VD_3 through its control at miRNA expression level (Craig et al. 2014). These studies show the relevance of this nutrient in morphogenesis and growth in fish, pointing out the need for further research to better understand the role of this hormone in fish larval development and physiology in order to determine their specific requirements for each species.

6.2.3 Vitamin E

There is limited knowledge about the specific absorption and transport of vitamin E (VE) in fish (Hamre 2011), but should be similar to the ones reported in mammalian species (Fig. 6.3). In higher vertebrates, the primary site of VE absorption is the medial small intestine, where esterified forms of the VE are hydrolysed and absorbed as free alcohols (reviewed in Rigotti 2007). Most studies have shown no appreciable differences in the efficiency of absorption of the acetate ester (α-TOAc) and free alcohol forms, nor differences in the absorption of the various tocopherol

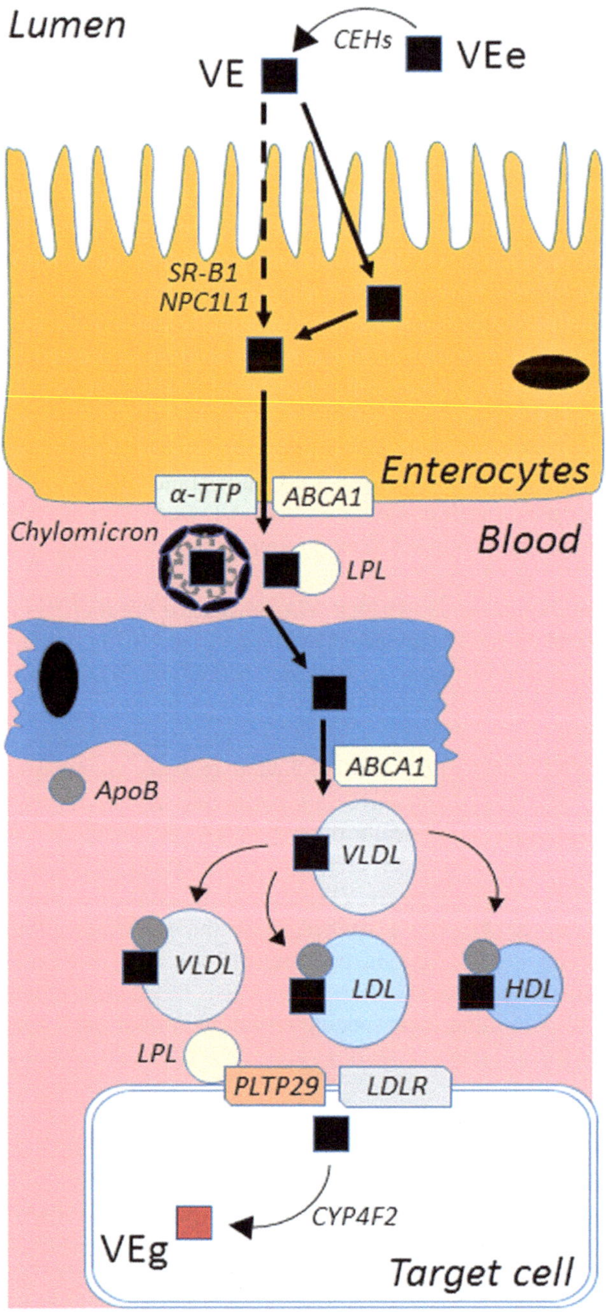
Lumen
CEHs
VE
VEe
SR-B1
NPC1L1
Enterocytes
α-TTP
ABCA1
Chylomicron
Blood
LPL
ABCA1
ApoB
VLDL
VLDL
LDL
HDL
LPL
PLTP29
LDLR
CYP4F2
VEg
Target cell

◀**Fig. 6.3** Overview of the metabolism of vitamin E (VE) and the tocopherol signaling pathway (TSP) from absorption to target cells. *ABCA1*, ATP binding cassette A1; *ApoB*, apolipoprotein B; *α-TTP*, alpha-tocopherol transfer protein; *CEHs*, carboxyl ester hydrolases; *CYP4F2*, cytochrome P450 family 4 subfamily F member 2; *HDL*, high-density lipoprotein; *LDL*, low density lipoprotein; *LDLR*, low-density lipoprotein receptor; *LPL*, lipoprotein lipase; *NPC1l1*, Niemann-Pick C1-like 1; *PLTP29*, phospholipid transfer protein 29; *SR-B1*, scavenger receptor class B member 1; *VE*, vitamin E; *VEe*, vitamin E esters; *VEg*, vitamin E glucunorides; *VLDL*, very low density lipoprotein

and tocotrienol vitamers. Regardless of the form, higher VE intakes generally lead to higher absorption, but with lower efficiencies (i.e., fractional absorption; Combs and McClung 2017). In a similar way to other hydrophobic substances, VE appears to be absorbed by (i) non-saturable passive diffusion dependent on the formation of micelles in the intestinal lumen (Borel et al. 2013), and (ii) receptor-mediated transporter (Schmölz et al. 2016).

The proper absorption of VE is dependent on the adequate absorption of lipids. This process requires the presence of fat in the lumen of the gut, and the secretion of pancreatic esterases for the release of free fatty acids from dietary triglycerides, bile acids for the formation of mixed micelles, and esterases for the hydrolytic cleavage of tocopheryl esters when those forms are provided in the diet (Combs and McClung 2017). Tocopheryl esters are hydrolyzed, at least partly, by carboxyl ester hydrolase, also called bile salt-dependent lipase, carboxyl ester lipase, and bile salt-stimulated lipase, which is secreted by the exocrine pancreas and whose activity requires bile salts. However, by analogy with retinol esters, some other candidate enzymes exist: pancreatic lipase, PLRP2 and phospholipase B. This hydrolysis could also be carried out by some brush border enzymes of enterocytes. Candidates for brush border enzymes could include an esterase localized in the membrane or in the endoplasmic reticulum. The relative percentage of tocopheryl esters hydrolyzed by the candidate esterases is not known, but it is suggested that most tocopheryl esters are hydrolyzed in the lumen of the intestine (Borel et al. 2013). As tocopherols can interact with PUFAs in the intestinal lumen, this can result in absorption being stimulated by medium-chain triglycerides and inhibited by linoleic acid, although VE absorption is not as efficiently absorbed as triacylglycerols (Combs and McClung 2017). Since the uptake of VE into enterocytes is less efficient compared to other types of lipids, this may explain the relatively low bioavailability of VE (Schmölz et al. 2016). The kinetics of VE absorption are biphasic, reflecting the initial uptake of the vitamin by existing chylomicrons followed by a lag phase due to the assembly of new chylomicrons (Combs and McClung 2017).

Although it is hypothesized that most VE released from food matrices is localized in mixed micelles in the intestinal lumen, the possibility that some VE is incorporated in other lipid structures during digestion, i.e., lipid droplets and vesicles, cannot be excluded. Vesicles, like liposomes, are constituted of either single bilayers of phospholipids (unilamellar vesicles) or multiple bilayers of phospholipids (multilamellar vesicles). The assumption that VE could be incorporated in vesicles during digestion is supported by the fact that VE is incorporated in phospholipid bilayers in vitro. Furthermore, α-TOH facilitates the assembly of phospholipid bilayers. Evidences have been presented for roles of cholesterol and lipid transporters in the uptake of α-TOH by enterocytes. Several transporters have been suggested: the SR-BI, CD36, NPC1L1, and ABCA1. In particular, ABCA1 has also been found to be involved in the export of tocopherols from enterocytes into the lymphatic circulation. Its interaction with α-tocopherol transfer protein (α-TTP) promotes the preferential trafficking of α-TOH over non-α-vitamers. The competitive binding to these receptors seems also to involve carotenoids (Chung et al. 2015). Absorbed VE, alike other hydrophobic substances, enters the lymphatic circulation in association with nascent triglyceride-rich chylomicrons (Combs and McClung 2017).

As Borel et al. (2013) more recently reviewed, the absorption efficiency of VE depends on several factors including: (i) the food matrix; (ii) the nature and amount of macronutrients, especially dietary lipids; (iii) the activity of digestive enzymes, and (iv) differences in transport efficiency across the intestinal cell. Recent studies in mice have shown that differences exist in VE absorption along the intestine. There are several hypotheses explaining why the efficiency of VE absorption is not similar along the intestine: (i) proteins involved in VE absorption (SR-B1, NPC1L1, CD36, ABCA1) are not distributed equally along the intestine; thus, the efficiency of VE absorption is maximal where VE intestinal transporters are highly expressed; (ii) differences in the repartition of VE transporters between the basolateral and apical membranes of enterocytes; and (iii) the major sites of absorption might be those where the bioaccessibility of VE is the highest, i.e., where the concentration of VE in micelles, and possibly vesicles, is the highest. Since vitamins are essential and that large amounts of fat-soluble vitamins can be toxic, it has been suggested that VE absorption could be regulated by the VE status in the organism. This mechanism could involve a VE-mediated regulation of the synthesis of membrane proteins implicated in VE absorption (e.g., ABCA1 and SR-BI). This is supported by a reduction of transcriptional activity of liver X receptor alpha (LXRα) and, thereby, a reduction of the expression level of *abca1* by α-TOH. Similarly, the expression of *sr-b1* is regulated post-transcriptionally by VE as a negative feedback regulation mechanism. Nevertheless, there is no study dedicated to the effect of VE status on the efficiency of VE absorption (Borel et al. 2013).

Regarding its transport, VE combines with other lipids and apolipoproteins (apo) to form chylomicrons within the enterocytes, which are released into the lymphatic system in mammals or the portal circulation in birds and reptiles. Triglyceride-rich chylomicrons containing VE are transported mainly to the liver, although some transfer of VE to the muscle, adipose tissue and perhaps other peripheral tissues also takes place during catabolism of chylomicrons by lipoprotein lipase (LPL; Rigotti 2007). Then, VE is transported from the liver to peripheral tissues by VLDL synthesized by hepatic parenchymal cells. Although the majority of the triglyceride-rich VLDL remnants are returned to the liver, some are converted by LPL to LDL. During this process, VE may also be spontaneously transferred to lipoproteins containing apoB, including the VLDLs, LDLs, and HDLs.

The above-mentioned exchanges between lipoproteins are mediated by the phospholipid transfer protein (PLTP) 27. Thus, plasma tocopherols are distributed among these three lipoprotein classes, with LDL and HDL classes comprising the major VE carriers (Combs and McClung 2017).

The cellular uptake of VE takes place similarly to other lipids that are transferred between lipoproteins and cells and was reviewed by Combs and McClung (2017). The lipase-mediated lipid transfer of α-TOH uptake from the amphipathic lipoprotein outer layer is unidirectional and mediated by the PLTP 29 (PLTP29) and the lipoprotein lipase-mediated exchange from chylomicrons. This route is considered of importance in cells that express lipase such as the adipose tissue, muscle and brain, and particularly important in the transport of α-TOH across the blood–brain barrier into the CNS.

The metabolism of VE is one of the most complex among liposoluble vitamins. Most α-TOH is transported to body tissues without metabolic transformation, and is subsequently metabolized through one cycle of ω-hydroxylation, followed by five cycles of β-oxidation (Combs and McClung 2017). The chromanol hydroxyl group renders VE metabolites like tocopherols and tocotrienols capable of undergoing both one- and two-electron oxidations. As a result, α-TOH is converted to α-tocopheryl quinine and (5,6- or 2,3-)-epoxy-α-tocopheryl quinones and thus, enabling it to scavenge free radicals such as peroxynitrate and lipid peroxyl radicals. In addition, oxidation of the chromanol ring is the basis of the in vivo antioxidant function of VE. It involves oxidation primarily to tocopherylquinone, which proceeds through the tocopheroxyl radical intermediate. A significant portion of VE may be recycled in vivo by reduction of tocopheroxyl radical back to tocopherol. Because tocopherylquinone lacks VE activity, its production represents the loss of the vitamin from the organism. It can be reduced to α-tocopherylhydroquinone, which can be conjugated with glucuronic acid and secreted in the bile, thus making excretion with the faeces the major route of elimination of VE. The catabolism of VE to water-soluble metabolites begins with a ω-hydroxylation of the side-chain, which is catalyzed by cytochrome P450 (CYP) enzymes, namely CYP4F2 or CYP3A4 (Schmölz et al. 2016). CYP4F2-initiated ω-oxidation metabolism is estimated to account for generation of >70% whole body VE metabolites (Jiang 2014). This oxidation is the rate-limiting step in VE metabolism. How all this absorption, transport and metabolism of VE is evolutionarily conserved remains to be deciphered, and might offer key knowledge for fish farming. The understanding of VE metabolic pathways will help to determine the better VE sources, its bioavailability, and their requirements along production cycle (larval development, juvenile out-growth, and adult reproduction). For instance, Miller et al. (2012) found that α-TTP is essential for larval development in zebrafish, being expressed in the developing brain, eyes and tail bud and which translational blocking using specific oligonucleotide morpholinos resulted in severe malformations of the head and eyes.

The main function of VE is as a biological antioxidant, preventing the propagation of free radical reactions in cell membranes and lipoproteins. It is distributed in membranes and serves as a lipid-soluble biological antioxidant as it scavenges lipid peroxyl radicals by donating hydrogen from the phenolic group on the chromanol ring. Examples of its antioxidant function are preventing the conversion of PUFAs to fatty hydroperoxides, or the conversion of free or protein-bound sulfhydryls to disulfides, and thus, having a key role in maintaining membrane integrity. Since the co-transport of VE co-occurs with polyunsaturated lipids, it ensures protection of the latter from free-radical attack, being circulating tocopherol levels correlated with those of total lipids and cholesterol (reviewed in Combs and McClung 2017). In scavenging free radicals, tocopherols and tocotrienols undergo oxidation of their respective alcohol forms to semistable radical intermediates, tocopheroxyl (or chromanoxyl) radicals. Unlike free radicals formed from PUFAs, the tocopheroxyl radical is relatively unreactive, thus blocking the destructive propagation of the cycle of lipid peroxidation. Because α-TOH can compete for peroxyl radicals much faster than PUFAs, small amounts of VE are required for the antioxidant protection of relatively large amounts of PUFAs. In contrast, VE deficiency reduces the radical scavenging ability in the organism, which may be partly related to disturbance of the antioxidant system. SODs, CAT, GPx, GST, GR and GSH are important components of the antioxidant defence system and play a vital role in attenuating the potential toxicity of free radicals (Pan et al. 2017). For further reading on the VE-mechanisms preventing lipid oxidation, readers are recommended to consult the works of Combs and McClung (2017), and particularly Hamre (2011) regarding VE in fish species and Hamre et al. (2016) regarding the general role of VE as antioxidant.

The available literature on dietary inclusion of VE in aquafeeds as an antioxidant nutrient is mostly focused on juveniles. Recommended levels of VE inclusion in diets for antioxidant function vary: 31 mg VE kg^{-1} for red drum *Sciaenops ocellatus* (Peng and Gatlin 2009); 500 mg kg^{-1} dietary VE supplementation in rainbow trout under high rearing density conditions (Naderi et al. 2017); 40–1,000 mg VE kg^{-1} in Nile tilapia *Oreochromis niloticus* (Wu et al. 2017); or 100 mg VE kg^{-1} in cobia *Rachycentron canadum* juveniles (Ding et al. 2017). The nutritional requirements of VE in marine fish larvae have not been properly identified. Two different studies in gilthead sea bream (Atalah et al. 2012) and European sea bass (Betancor et al. 2011) recommended an optimal level of 3 g VE kg^{-1} dry diet in feeds containing high levels of PUFAs, while lower dietary VE levels (>1.5 g VE kg^{-1} dry diet) were reported for meagre *Argyrosomus regius* larvae (El Kertaoui et al. 2017). However, as previously indicated by Hamre (2011), data between different studies are not directly comparable for extracting a general recommendation of VE inclusion levels in aquafeeds due to inherent differences in the experimental design (i.e. fish species, dietary lipid levels and other antioxidative nutrients) and the parameters measured for evaluating the antioxidative properties of VE (i.e., MDA levels, activity of antioxidative enzymes, fillet quality, etc.). Although the dietary recommendations for VE are 25–100 mg kg^{-1} for promoting adequate growth in most fish species (NRC 2011), Hamre (2011) recommended including a safety margin when including VE in fish diets due to the potential effects of interactions between VE and other feed components. Since α-TOH can also promote lipid peroxidation in LDLs in the absence of other antioxidants (i.e., ascorbic acid, coenzyme Q10, urate), those secondary antioxidants need to be included in the diets to prevent LDL oxidation (Combs and McClung 2017).

In addition, several studies have revealed that VE may also have a function in improving fish health (Li et al. 2013; Lim et al. 2009; Pan et al. 2017; Zhou et al. 2013), resistance to stress (Li et al. 2013; Montero et al. 2001; Naderi et al. 2017; Trushenski and Kohler 2008) and disease prevention (Liu et al. 2014; Trichet 2010; Trushenski and Kohler 2008; Yildirim-Aksoy et al. 2008). VE has also a key role during embryogenesis since VE deficiency has been reported to cause developmental abnormalities (cranial-facial malformations, bent anterior-posterior axis, pericardial oedema, swim bladder malformations, and yolk-sac oedema) and early death in zebrafish embryos (Miller et al. 2012). Last, but not the least, VE act as a ligand for PXR and thus, control the transcription of downstream genes (Landes et al. 2003). In this sense, further research should focus on the different pathways where VE might have an impact, either through genomic (via PXR) or non-genomic actions.

6.2.4 *Vitamin K*

Vitamin K, or the coagulation vitamin, was discovered by Henrik Dam, meanwhile studying the role of cholesterol (Dam 1935). The current knowledge regarding VK fish requirements has been compiled by Krossøy et al. (2011) and more recently updated by Fernández and Gavaia (2016). Nevertheless, and similarly to the other fat-soluble vitamins, most of knowledge on VK metabolism in vertebrates comes from studies in mammals, extensively reviewed in Shearer and Newman (2008, 2014) and Card et al. (2014).

The main sources of VK are the dietary composition and the intestinal microbiota, although contribution of colonic flora to VK requirements remains controversial (Harshman et al. 2014), some bacterial-derived menaquinones have been found stored in the human liver (Suttie 1995). Nevertheless, since antibiotics are known to affect the intestinal microbiota (Mathers et al. 1990), the relevance and the impact (if any) of antibiotics on intestinal production of VK might deserve specific investigation. In general, it is assumed that absorption of VK mainly takes place in the jejunum and ileum in the form of mixed micelle complexes with bile salts, and less likely from the colon (Suttie 1995). First comparative experiments demonstrated that the composition of the micelle and the presence or absence of other fat-soluble compounds (e.g. lipid and bile salt composition) have profound effects on the absorption rates of VK_1 (Hollander and Rim 1976a). In contrast, since the presence of oleic and linoleic acids, VK_1 and VK_3 did not change the absorption of VK_2, it was initially suggested that VK_2 is absorbed by a passive non-carrier-mediated diffusion process (Hollander and Rim 1976b). More recently, it was shown that the absorption of vitamers K from the diet was low (between 5–20% for VK_1 and 55% for VK_2; Gijsbers et al. 1996; Koivu-Tikkanen et al. 2000), suggesting a very high efficiency of hepatic VK recycling during the

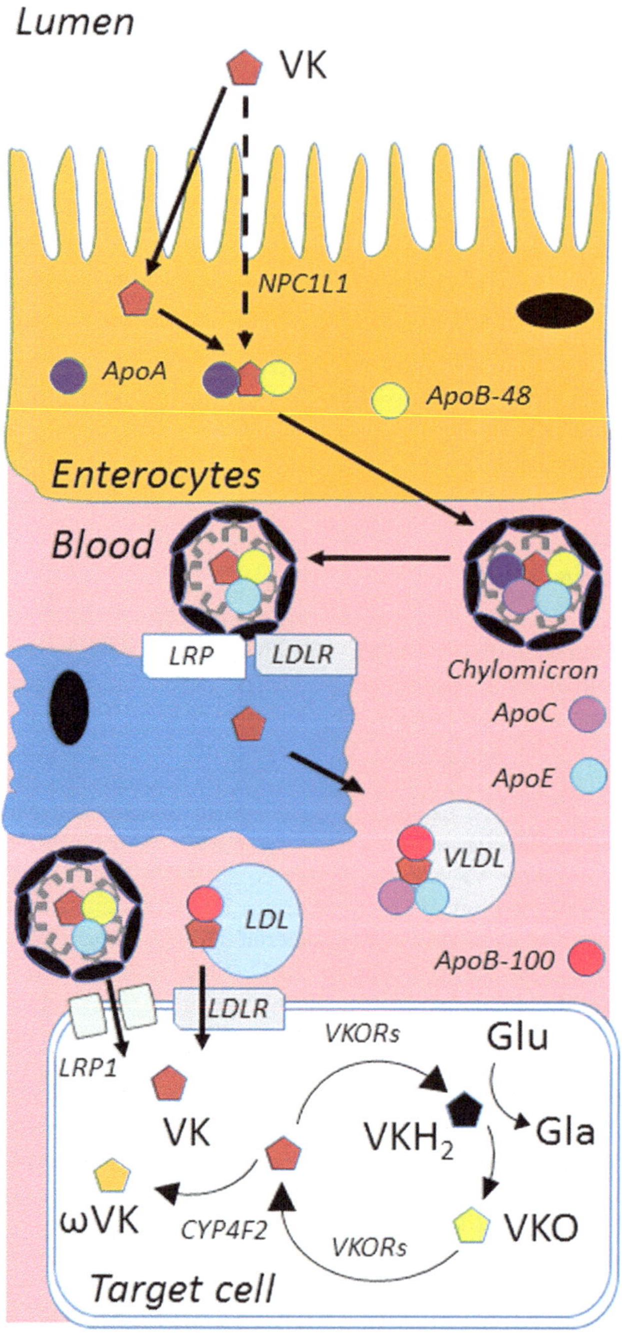
Lumen
VK
NPC1L1
ApoA
ApoB-48
Enterocytes
Blood
LRP
LDLR
Chylomicron
ApoC
ApoE
VLDL
LDL
ApoB-100
LDLR
LRP1
VKORs
Glu
Gla
VK
VKH2
ωVK
CYP4F2
VKORs
VKO
Target cell

◂**Fig. 6.4** Overview of the metabolism of vitamin K (VK) and the (phyto-/mena-) quinone signaling pathway (QSP) from absorption to target cells. *ApoA*, apolipoprotein A; *ApoB-48*, apolipoprotein B 48; *ApoB-100*, apolipoprotein B 100; *ApoC*, apolipoprotein C; *ApoE*, apolipoprotein E; *CYP4F2*, cytochrome P450 family 4 subfamily F member 2; *Gla*, γ-carboxyl glutamate; *Glu*, glutamic acid; *LDL*, low density lipoprotein; *LDLR*, low-density lipoprotein receptor; *LRP*, low-density lipoprotein receptor-related protein; *LRP1*, low-density lipoprotein receptor-related protein 1; *NPC1l1*, Niemann-Pick C1-like 1; *VK*, vitamin K; *VKH$_2$*, reduced vitamin K; *VKO*, vitamin K epoxide; *VKORs*, vitamin K epoxide reductases; *VLDL*, very low density lipoproteins; ωVK: hydroxyvitamin K1

synthesis of blood clotting factors, and raising the question of whether a similar efficiency is obtained in other tissues (e.g. bone and vessel wall). Nevertheless, while high intakes of phylloquinone suppress the colonic production of all higher menaquinones, high menaquinone-4 intake specifically induces very high menaquinone-8 concentrations (Koivu-Tikkanen et al. 2000). These results suggest a tight control of VK metabolism (Fig. 6.4).

Until recently, no protein has been associated to intestinal uptake of VK. Takada et al. (2015), recently demonstrated that NPC1L1 protein, a cholesterol transporter, plays a central role in intestinal VK uptake using in vitro (NPC1L1-overexpressing intestinal cells) and in vivo (*npc1l1*-knockout mice) studies, revealing that intestinal VK absorption is NPC1L1-dependent. So far, no specific proteins are known for VK transport unless for menadione, being transported by lipoproteins. Depending on the physiological state of the organisms (fasting state vs. postpandrial phase), vitamers K are transported in triacylglycerol-rich lipoproteins (TRLs), comprising chylomicron remnants (CRs) and VLDLs or distributed between LDLs and HDLs (reviewed in Shearer and Newman 2014). Interestingly, there is a clear evidence that the length and degree of saturation of the isoprene side chain significantly influences their kinetics of appearance and clearance from the circulation (Schurgers and Vermeer 2002), being the clearance of MK-7 very slow, with a terminal plasma half-life of around 3 days and having a much greater efficiency for γ-carboxylation of osteocalcin in bone than equimolar doses of VK$_1$ (Schurgers et al. 2007). In brief, after intestinal absorption, nascent chylomicrons containing apoA and apoB-48 are secreted into lymphatic vessels and eventually ending into the blood circulation. There, chylomicrons acquire apoC and apoE from HDL. In capillaries, chylomicrons are stripped of their triglyceride (TG) content by lipoprotein lipase (LPL) lining the capillary walls. The smaller CR re-enter the circulation loosing much of their apoA and apoC, but retaining the VK in the lipophilic core. At the liver, CRs enter hepatocytes by binding to low-density lipoprotein receptor-related protein (LRP) and low-density lipoprotein receptor (LDLR) followed by receptor mediated endocytosis. Lipids are then repackaged in VLDL containing apoB-100 and returning to the circulation where they acquire apoC and apoE. After, TG is removed by LPL in the capillaries and a subsequent loss of apoC and apoE occurs. Then, CRs contain almost exclusively apoB-100. In target cells like bone cells, circulating CRs and LDL can deliver lipids to these cells through LDLR and LRP1. While it is suggested that osteoblasts obtain most of their VK$_1$ via the CRs pathway, most of their MK-7 seems to be via the LDL pathway.

Plasma concentrations of VK are really low, of the order of 0.5 nM in healthy fasting humans, which represents one, three, and four orders of magnitude lower than concentrations encountered for 25-hydroxyvitamin D, retinol, and α-TOH, respectively. This is in line with the low tissue reserves. In mammals, while the liver comprise the majority of VK reserves of long chain forms (MK-7 to MK-13) and with only low concentrations of MK-4, extrahepatic tissues can also store VK, although it contains mainly VK_1 and/or MK-4. Furthermore, since around 60–70% of the amounts of VK_1 absorbed from each meal will be lost through excretion (McBurney et al. 1980), it was suggested that the body stores of VK_1 are being constantly replenished (Shearer and Newman 2014). Interestingly, dietary VK_1 releases VK_3 by the cleavage of the side chain in the intestine, followed by the delivery of VK_3 to tissues via blood circulation, where it is finally converted to MK-4 by the prenylating enzyme UBIAD1 (Hirota et al. 2013). UBIAD1 catalyses the non-mitochondrial biosynthesis of CoQ10 in zebrafish and its mutations have been reported to cause cardiac oedema and cranial haemorrhages (Hegarty et al. 2013; Mugoni et al. 2013). In humans, mutations in *UBIAD1* cause a rare eye disease (Schnyder corneal dystrophy) characterised by abnormal deposition of cholesterol and phospholipids in the cornea, resulting in progressive corneal opacification and vision loss (Orr et al. 2007).

From a functional point of view, VK can have two different roles in addition to the ancient antioxidant activity. In one hand, VK acts as a co-factor of the γ-glutamyl carboxylase (GGCX) to perform the γ-carboxylation of the VK dependent proteins (VKDPs; Oldenburg et al. 2008). The γ-carboxylation of VKDPs is essential for the biological activity of these proteins in Ca^{2+} homeostasis regulation, and thus controlling the blood clotting process (reviewed in Brenner et al. 2009) as well as the mineralization of bony and cartilaginous structures (Oldenburg et al. 2008; Viegas et al. 2008; Neacsu et al. 2011). From this reaction results an VK 2,3 epoxide (VKO) that can be recycled to VK hydroquinone through the enzymes known as VK epoxide reductases (Presnell and Stafford 2002). Up to date, the two different VKOR proteins (vitamin K epoxide reductase complex subunit 1 (VKORC1) and VKORC1-like 1 (VKORC1L1)), commonly found in vertebrates (Oldenburg et al. 2015), have been functionally studied in mammals. While VKORC1 may be more related to VK recycling for VKDP γ-carboxylation in particular tissues, VKORC1L1 was suggested as more efficient in VK recycling towards an antioxidant role (Westhofen et al. 2011), but also with a VK recycling action of VKDP γ-carboxylation in specific tissues (Hammed et al. 2013). The presence of these two VKORs and their molecular regulation in vivo and in vitro under different physiological conditions on VK has been recently described in fish, and the evolutionary conservation of their expression pattern confirmed (Richard et al. 2014; Fernández et al. 2015b). These and previous results are evidencing a particular differential role of both VKORs depending on the tissues/cell types considered.

Besides its canonical action as a co-factor for the γ-carboxylation, VK also has a role in the transcriptional regulation through its binding to the PXR [also known as steroid and xenobiotic receptor (SXR); Tabb et al. 2003]. In this sense, VK has been shown to control bone development and homeostasis through the transcription of bone related genes through PXR activation (Ichikawa et al. 2006). PXR was largely known as a master regulator of xenobiotic (Chen et al. 2012), cholesterol and bile acid metabolisms (Makishima 2005). More recently, the structural analysis of heterodimers formed by PXR to drive the transcription of target genes revealed that PXR bind to RXR in a heterotetramer conformation (Wallace et al. 2013), evidencing a crosstalk point at the nucleus where the different fat-soluble signaling pathways might interact (see below). Due to the different and wider roles (e.g. reproduction, brain development and cognitive capacities) of VK suggested in studies with mammalian species (Ferland 2012; Ito et al. 2011), future studies using high throughput technologies might clearly identify these different roles of VK in fish. The first results observed applying a proteomic approach in flatfish larvae evidenced that VK might have a role on muscle contraction, resistance to osmotic stress, intracellular Ca^{2+} homeostasis or energetic metabolism (Richard et al. 2014). The use of Next Generation Sequencing (NGS) technologies like RNA- or chromatin immunoprecipitation-(ChIP-)seq might reveal further insights on the biological functions and gene networks involved in VK metabolism. This knowledge will help to determine the nutritional requirements for VK in fish during early development, growth and/or reproduction.

As above mentioned, the use of VK_3 as the main source of VK in most formulations for fish is widespread, mostly menadione salts such as MSB or MNB. This salts are water soluble, contrary to the liposoluble nature of VK_1 and VK_2, and some concerns have been already reported such as inducing bone deformities in mummichog *Fundulus heteroclitus* larvae when fed with high doses of MSB (2500 mg kg^{-1}; Udagawa 2001) or the depression of weight gain in Atlantic salmon fry and fingerlings when fed in a long-term period (Grisdale-Helland et al. 1991). Those effects might be due to its metabolism, forming superoxide anions, hydrogen peroxide, and hydroxyl radicals causing cell death by the activation of caspase 3 in hepatocytes (de la Rosa et al. 2015). The toxicity of pure menadione was proven to be extremely high in eggs and larvae of sheepshead minnow *Cyprinodon variegates*, being killed in less than 24 h when exposed to doses as low as 1 mg l^{-1} (Wright et al. 2007). Moreover, menadione itself has no biological activity as a co-factor for the GGCX, requiring a biological activation through enzymatic alkylation to MK-4 in animal tissues (Buitenhuis et al. 1990; Graff et al. 2002, 2010; Krossøy et al. 2009; Udagawa 2000). Thus, it is recommended that formulations for larval diets incorporate the natural forms of VK (phylloquinone and menadiones) in order to prevent VK-induced hepatotoxicity. The differential accumulation of VK_1 and VK_2 in tissues, being the levels of VK_1 in the liver 10-fold higher than those of MK-4, and the preferential accumulation of MK-4 in extrahepatic tissues (e.g. bone and arterial vessels; Spronk et al. 2003), suggested a differential use of both vitamers K, and thus, the need of both being included in aquafeeds.

6.3 The Multilevel Fat-Soluble Vitamins Crosstalk

The detailed knowledge gained on fat-soluble vitamins in the last decades uncovered several crosstalk points between them along their signaling pathways: from intestinal absorption to detoxification mechanisms. The latest has been recently pointed out when the different CYPs involved in the detoxification process of hypervitaminosis conditions were identified. In this sense, CYP4F2 promotes the ω-hydroxylation of both VK and VE, while CYP4F11 only acts for VK (McDonald et al. 2009; Traber 2013). Nevertheless, ω-hydroxylation of VK_1 by CYP4F2 seems not to be increased by α-TOH levels (Farley et al. 2013).

Although the interaction between fat-soluble vitamins in fish species has been already reported, particularly between RA and $1,25(OH)_2D_3$ in Atlantic salmon where RA injections reduce plasma $1,25(OH)D_3$ values, in line with a direct involvement of RA in $1,25(OH)_2D_3$ metabolism (Ørnsrud et al. 2009); most known interactions of fat-soluble vitamins were identified in mammalian species and are centred at the level of their absorption in the intestine, their transport from the intestine to the liver (and other target tissues) through the blood stream, and at the nucleus of target cells.

6.3.1 *The Intestinal Border: A Starting Point for Competition?*

The first step to fulfil the nutritional requirements for optimal growth, development and homeostasis maintenance is the absorption of all the different metabolites of fat-soluble vitamins in the amounts needed. Although their uptake mostly relies on (allegedly) a passive diffusion mechanism and thus, fulfilling their requirements might depend on the amount provided for each vitamin in the diet in a unifactorial perception, several of them also use the same proteins at the enterocyte's cell membrane. In this sense, β-carotenes (precursors of VA), VD, VE and VK have NCP1L1 as a common transporter (Borel et al. 2013; Reboul 2013, 2015; Takada et al. 2015) and might represent the main convergent point in fat-soluble vitamins uptake at the intestine. Furthermore, except for the case of VK, the other fat-soluble vitamins might also compete for SR-B1 (Borel et al. 2013; Reboul 2013, 2015), while VA and VD further compete for CD36 (Reboul 2013, 2015). However, these interactions at the intestinal level are poorly documented, even in mammalian species. Although each fat-soluble vitamin has their own preferential intestinal absorption region (proximal intestine for VA, median intestine for VD, and distal intestine for both VE and VK), recent data showed that significant competitive interactions for uptake were found, supporting the hypothesis of common absorption pathways (see above). While VA significantly decreased the uptake of the other fat-soluble vitamins, its uptake was not impaired by VD and VK, but promoted by VE (Gonçalves et al. 2015). In this sense, even low VA dietary contents might hamper fish development and/or wellbeing if a high dose of VE is provided at the same time. Following the same line of reasoning, the negative effects in fish physiology of hypervitaminosis A (reviewed in the previous section of present chapter) might be not only due to high levels of VA on the diet, but also to the supposedly relatively low levels of VD and/or VK. Likewise, it has been reported in farmed growing mink *Mustela vison* that increased dietary levels of VA (as retinol) or VE (as RR-α-tocopherol isomer) decreased the 25-hydroxycholecalciferol concentration in plasma (Hymøller et al. 2016). Then, a higher requirement of VD is expected when increased levels of VA and some particular VE isomers are included in the diets. These results in a mammalian species highlight the need of similar experiments to confirm the evolutionary conservation of fat-soluble vitamins competition at intestinal uptake in fish species.

In addition to "competitive" uptake of fat-soluble vitamins at the intestine, and previously to the incorporation of VA and VE metabolites by enterocytes, PLRP2 has been identified as one of the enzymes responsible for the intestinal hydrolysis of both retinyl and tocopheryl esters (reviewed in D'Ambrosio et al. 2011; Borel et al. 2013). Although differences in hydrolysis efficiency of this enzyme when dealing with VA or VE esters remains to be experimentally deciphered, it might be an unanticipated crosstalk point between both vitamins. Nevertheless, since ABCA1 has been found to be responsible for the VE efflux from enterocytes (Chung et al. 2015), but also suggested in the case of retinol (Reboul 2013) and VD (Reboul 2015), a more complex interaction of fat-soluble vitamins at the intestine might take place than the one previously thought. Further information may be found in Card et al. (2014) who reviewed the specific interactions between VE and VK. Bleeding events have been known to occur due to a VE overdose and associated with a prolonged prothrombin time, activated partial thromboplastin time and lowered VK dependent coagulation factors (Booth et al. 2004; Glynn et al. 2007; Helson 1984). Rats under high-dose of α-TOH exhibited a MK-4 depletion in brain, lung, kidney and heart, as well as VK_1 depletion in lung. Since depletion of VK metabolites seemed not to be due to an increase in VK urinary excretion, this might be through biliary excretion, as a consequence of an alteration in ABC transporters (Farley et al. 2012), like the increased expression of *multidrug resistance protein 1* (*mdr1*) transporter (Mustacich et al. 2006).

Nutritional strategies for future aquaculture sustainability, as fish oil and fish meal substitution by alternative sources, might also represent a new scenario for reanalyzing fat-soluble vitamin requirements since absorption of all fat-soluble vitamins depends on the dietary lipid content. Both content and (most importantly) fatty acid profile have been found to be different when aquafeeds are formulated with fish oil or vegetable oil sources (Turchini et al. 2011). Thus, further increases in fat-soluble vitamins content should be considered in those new fish-free diets or at least, specific attention should be paid to the altered lipid contents. Previous studies already demonstrated that the composition of the micelle and the presence or absence of other fat-soluble compounds, have profound effects on the absorption rates of VK_1 (Hollander and Rim 1976a); while PUFAs have been shown to interfere with VE absorption (Horwitt 1962). Another supporting fact that fish-free aquafeed might represent a nutritional concern on fulfilling fat-soluble vitamins requirements is the inhibition of VE absorption by linoleic acid or the stimulated absorption by medium-chain triglycerides (Combs and McClung 2017). Furthermore, the presence of sterols in vegetable oils (as main alternative sources of fish oil), and particularly their ester forms, needs to be investigated. Decreased absorption of α-TOH has been found with an increased level of plant sterol esters (Richelle et al. 2004). Although it remains to be demonstrated, it might be one of the underlying mechanisms of gastrointestinal disturbance found in European sea bass juveniles fed diets containing phytosterols (Couto et al. 2015). Taking into account the ongoing search for alternative sources of fish oil and fish meal, an intense research work might be expected in the next years to specifically study the case of fat-soluble vitamins crosstalk at the intestine on farmed fish species in order to avoid nutritional deficiencies in such essential compounds.

6.3.2 A Race for Protein's Cargo at the Blood Stream

Another interacting/convergent point between fat-soluble vitamins was already anticipated: the blood protein's cargo. HDLs are dealing with the transport of retinyl esters, but also with VE (Figs. 6.1 and 6.3). For years, HDLs were considered as central for cholesterol transport. Nevertheless, the greater understanding of the complexity of HDL composition and biology has prompted researchers to redefine HDL away from a cholesterol-centric view (reviewed in Vickers and Remaley 2014). Many of the recently discovered functions of HDL are, in fact, the transport of other molecules, including a diverse set of proteins, small RNAs, hormones, carotenoids, bioactive lipids and vitamins. Thus, HDLs have the remarkable capacity to affect a wide variety of endocrine-like systems. It is not known whether this protein cargo has differential preference for different fat molecules and particularly between carotenoids, VA and VE. Nevertheless, the association of dietary intakes of folate, iron and magnesium, but also dietary levels of saturated fat, fiber and protein, with different levels of HDLs in plasma (Kim et al. 2014) translates in the need for further integrative studies on this issue.

Similarly to HDLs, VLDLs and LDLs are other important lipoprotein cargoes of cholesterol and other lipophilic substances such as VA, VE and VK, and both have been linked to coagulation and lipid metabolism (Dashty et al. 2014), with these two biological functions being related with VK and PXR signaling respectively. In fact, prothrombin, protein S, and fibrinogen γ (all related with blood clothing process) were identified to be present in VLDLs, but not in LDLs (Dashty et al. 2014). Thus, in addition to the important delivery of fat-soluble vitamins, these cargo proteins are also essential for the transport of proteins involved on the same process where fat-soluble vitamins act (e.g. blood coagulation). Might the transport efficiency of prothrombin be somehow influenced by dietary VK level through γ-carboxylation statement? Or might the intake of other fat-soluble vitamins have an impact on VKDPs γ-carboxylation? In this last case, it is known that expression of VKDPs is under the control of NR activated by different fat-soluble vitamins (e.g. VA and VD; please see last subheading). Further specific research is needed to uncover the relevance of this interacting point of fat-soluble vitamins.

6.3.3 At the Nucleus: The Main Hotspot that Needs Further Characterization

NRs comprise a superfamily of ligand-activated transcription factors, comprising 6 subfamilies. Among them, the NRs ligand-activated by fat-soluble vitamins are within subfamily 1: RARs from the Group B, and VDR and PXR from the Group I; and subfamily 2: RXRs. While *all-trans*-RA activates RARs (α, β, γ), 9-cis RA can activate both RXRs (α, β, γ) and RARs, although showing higher affinity by the firsts (Germain et al. 2006a, b). VDR is specifically ligand-activated by VD (Moore et al. 2006), while PXR can be activated by VK and VE, as well as by a large quantity of different xenobiotic compounds (Landes et al. 2003; Tabb et al. 2003; Ekins et al. 2008). The biological relevance of all these NRs relies on the fact that they are the responsible for driving the genomic actions of each fat-soluble vitamin.

Remarkably, all of these NRs need to form heterodimers (or heterotetramers in the particular case of PXR; Wallace et al. 2013) with the same NR: the RXR (Fig. 6.5). Another relevant NRs, activated by other nutrients or hormones, playing central roles in vertebrate development and homeostasis that have RXR as heterodimer partner, are the peroxisome proliferator-activated receptors (PPARs) α, β, γ that are activated by fatty acids, among other molecules (Michalik et al. 2006); thyroid hormone receptors (TRs α, β; Flamant et al. 2006); and/or liver X receptors (LXRs α, β; Moore et al. 2006). Taking into account whether each NR is ligand activated by its respective fat-soluble vitamin or not, the dimers formed in the nucleus may sustain basal (normal), repress or enhance transcription of target genes. Thus, considering the presence of fat-soluble vitamins and their efficiency on activating NRs, RXR is one of the most critical crosstalk points between all fat-soluble vitamins, since any perturbation of the natural equilibrium formed between all those NRs in the nucleus might induce a great modification of the downstream signaling. In this sense, variations in RXR homo-/heterodimer equilibrium have been shown to cause severe abnormalities in zebrafish embryos (Minucci et al. 1997). Nevertheless, the key question still remains to be elucidated: is there a higher affinity of any NRs for RXR as a partner? Here, instead of reviewing the immense literature regarding the mechanisms and actions of the different NRs activated by the fat-soluble vitamins and other molecules, we will highlight the particularities and differences in those NRs between mammalian and fish species as well as the future research needs to identify the different genes targeted by each NR (and thus, each fat-soluble vitamin) to unveil the biological roles of each fat-soluble vitamin and their related gene networks, fundamental for determining their nutritional requirements accurately.

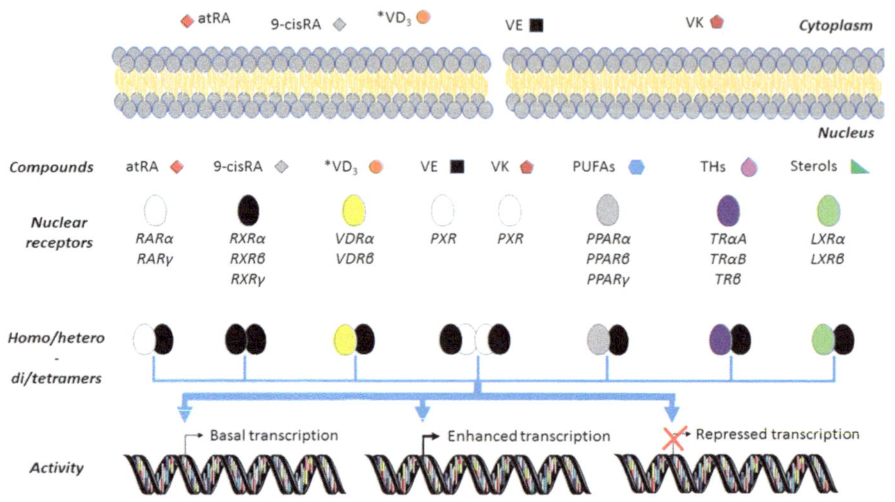

Fig. 6.5 Fat-soluble vitamin metabolites (and others) and their respective nuclear receptors. From up to down: Compounds, their specific nuclear receptors and their isomers, possible combinations of nuclear receptors regulating gene transcription, and potential action on the transcriptional regulation of target genes at promoter region. *atRA*, *all-trans* retinoic acid; *9-cis-RA*, 9-cis retinoic acid; *$*VD_3$; $1,25(OH)_2VD_3$* (1,25-dihydroxyvitamin D3); *VE*, vitamin E; *VK*, vitamin K; *PUFAs*, polyunsaturated fatty acids; *THs*, thyroid hormones; *LXRα*, liver X receptor alpha; *LXRβ*, liver X receptor beta; *PPARα*, peroxisome proliferator-activated receptor alpha; *PPARβ*, peroxisome proliferator-activated receptor beta; *PPARγ*, peroxisome proliferator-activated receptor gamma; *PXR*, pregnane X receptor; *RARα*, retinoic acid receptor alpha; *RARγ*, retinoic acid receptor gamma; *RXRα*, retinoid X receptor beta; *RXRβ*, retinoid X receptor beta; *RXRγ*, retinoid X receptor gamma; *TRαA*, thyroid hormone receptor alpha A; *TRαB*, thyroid hormone receptor alpha B; *TRβ*, thyroid hormone receptor beta; *VDRα*, vitamin D receptor alpha; *VDRβ*, vitamin D receptor beta

Any NR is generally characterized by a DNA binding domain (DBD), involved in the binding to specific DNA sequences; a H (or hinge) region, enabling flexible conformations of the DBD and the ligand binding domain (LBD); and a C-terminal LBD, where the ligands are bound (Bertrand et al. 2004). From an evolutionary perspective, the DBD sequence is the most highly conserved region of each NR, while LBD is the second, with these high degrees of conservation being representative of their essential biological role in vertebrates (Sluder et al. 1999). Nevertheless, some exceptions exist, like the case of PXR, exhibiting the lowest degree of conservation on its LBD among the NRs (Ekins et al. 2007). Furthermore, PXR has a LBD highly flexible and forms a binding cavity substantially larger than that of other NRs, enabling the binding of a wide range of small endogenous and exogenous molecules (Pavek 2016). Even when a high conservation of particular features of NRs occurred during evolution, several differences among mammals and fish species are still relevant to be considered before transposing the knowledge gained in mammalian model species to farmed fish species.

Comparative analysis is a powerful approach to extract functional or evolutionary information from biological sequences (reviewed in O'Brien and Fraser

2005). Although DBD is highly conserved along evolution, implying that the same NR from a mammalian species will specifically bind to the same DNA sequence as does the NR from another vertebrate (fish) species, the conservation of DNA sequences in non-coding regions (e.g. gene promoters) is not so high, even when close-related mammalian species are compared (Chiba et al. 2008). Thus, the sequence of a gene promoter might be quite different between mammalian and fish species and therefore, while a particular gene could be transcriptionally regulated by a specific NR in mammals, the same gene promoter might not be controlled by this NR in a fish species (Meisler 2001). In this sense, the recent and rapid advance on Next-Generation Sequencing (NGS) technologies has made DNA sequencing broadly available, allowing the sequencing of important farmed fish species, like Atlantic salmon, rainbow trout or turbot *Scophthalmus maximus* (Berthelot et al. 2014; Lien et al. 2016; Figueras et al. 2016). This fact opens new possibilities to such comparative analysis, in order to confirm the conservation of downstream signaling cascades of each NR activated by fat-soluble vitamins, generating knowledge that will increase our understanding on the biological function of each fat-soluble vitamin. Moreover, by combining NGS technologies, such as chromatin immunoprecipitation (ChIP-seq) and RNA sequencing (RNA-seq), a deeper knowledge can be gained. Although to the best of our knowledge this kind of approach has still not been applied to NRs in fish species, similar approaches have already been applied to histones (e.g. H3) in model species like medaka *Oryzias latipes* and zebrafish (Tena et al. 2014). More research efforts have been applied using ChIP-seq technologies regarding the identification of *cis*-regulatory sequences of particular NRs ligand activated by fat-soluble vitamins (e.g. RARs, VDR and PXR; Meyer et al. 2010; Cui et al. 2010; He et al. 2014) in mammalian species. In contrast to ChIP-seq, RNA-seq has already been applied to a wide range of issues in fish species (reviewed in Li and Li 2014) like the identification of NRs (Bain et al. 2015; Cheng et al. 2015), or the identification of more than 90 biological pathways altered by VK-induced deficiency at transcriptional level (Fernández et al. 2015c). Similarly, wide transcriptomic assays were already applied using microarrays to identify differentially expressed genes under different VA dietary regimes or exposure in fish species (Oliveira et al. 2013; Lie et al. 2016). Comparatively, since microarray technology is limited towards the amount of RNA, the quantification of transcript levels and the sequence information, RNA-seq provides nearly unlimited possibilities in modern bioanalysis. RNA-seq analyses, not only the expression level of mRNA, but also splice variants, non-coding RNA, and microRNA on a genome-wide scale. Further, the combination of ChIP-seq and RNA-seq data will allow the discovery of novel transcriptional mechanisms.

Sequence divergence in LBD between mammals and fish species is responsible that not all the described ligands for a NR in mammals also function as specific ligand for the same NR in fish. A particular example is rifampicin, an effective specific ligand for human and mouse PXR, but not for zebrafish PXR (Ekins et al. 2008). Thus, while genomic inference from model (mammalian) species is normally applied when studying genes with no functional characterization in farmed fish species, extra caution should be taken when the known ligands of NR are considered in fish physiology. Cell based assays have been useful tools to identify NR agonists and antagonists in pharmacology (Raucy and Lasker 2013). Those in vitro systems and/or in silico analysis, as those performed in green spotted pufferfish *Tetraodon nigriviridis* (Krasowski et al. 2011), based on LBD sequences from farmed species might be developed to confirm if known mammalian NR agonists and antagonists are also effective in fish species.

The scenario of NRs ligand activated by fat-soluble vitamins interaction is further complicated by two aspects. Firstly, in fish species some NRs are not present or might have suffered a sub- or neo-functionalization process, due to the extra genome duplication event in teleosts (Glasauer and Neuhauss 2014). For instance, constitutive androstane receptor (CAR) is a NR with overlapping roles with PXR in humans, but it is not present in teleosts since it was originated posteriorly in the evolution (Zhao et al. 2015). Another relevant example is the absence of RARβ in fish species (Linville et al. 2009). On the other hand, teleosts present more genes or isoforms of particular NRs than mammalian organisms. One particular case is the duplicated isoforms of RARα and RARγ in zebrafish (Linville et al. 2009), the TRαA and TRαB in flatfish species like Senegalese sole (Manchado et al. 2009) or the two VDR genes (α and β) in all teleosts. In this particular case, while 1α,25 $(OH)_2D_3$ significantly increased transactivation activity of VDRβ, the one of VDRα was highly attenuated, suggesting a functional divergence between these two NRs paralogs (Howarth et al. 2008). Second, due to the evolutionary pressure, subtler differences have been found between mammals and fish species regarding NRs ligand activated by fat-soluble vitamins. For example, domains of expression in time (along development) and/or space (tissues and cell types) need to be characterized in order to get a complete picture on where a certain fat-soluble vitamin might exert a role in a particular biological process, and when and where the interaction of fat-soluble vitamins at nuclear level through their respective NRs might be expected. In this sense, different transcriptomic approaches (PCR, qPCR and ISH) have been applied, not only in fish model species but also in important farmed fish species such as European sea bass, gilthead sea bream, Atlantic salmon or Senegalese sole (Fernández et al. 2011, 2014a, b; Howarth et al. 2008; Lock et al. 2007; Marques et al. 2017; Villeneuve et al. 2004). This knowledge will provide more accurate information regarding NRs than the one extracted by comparative/inference approaches from databases of mammalian species, as it could be the case of the Nuclear Receptor Signaling Atlas (NURSA; https://www.nursa.org/nursa/index.jsf; McKenna et al. 2009).

Finally, another level of interaction between fat-soluble vitamins subsequent to NRs crosstalk, is the fact that those NRs may regulate the transcription of genes involved on their own metabolism. For instance, *abca1* is up-regulated by the heterodimer LXR-RXR and thus, VA signaling pathway through RXR might increase VE intestinal efflux (reviewed in Traber 2004). Furthermore, VA and VD through their respective NRs control the expression of VKDPs (Conceição et al. 2008; Darias et al. 2010; Lian et al. 1989), which might increase VK requirements for their γ-carboxylation.

6.4 The Significance of the Research on Liposoluble Vitamin Requirements in the Context of a Sustainable Aquaculture Development

The stagnation of fish catches from fisheries and the demand for fish production to fulfil the increased fish consumption and human population were the drivers for aquaculture development on its first decades. Nowadays, low profit margins in largest produced fish species in Europe (European sea bass and gilthead seabream, among others) and higher sustainability standards push the industry to three different directions while demanding nutrition innovation to face new challenges. Aquaculture diversification deals with new/emerging finfish species to be farmed, requiring optimization of the environmental and biotic factors, as well as husbandry practices to achieve species domestication. Thus, new and species-specific diets are expected to be developed to fulfil their own nutritional requirements. Further, some new/emerging species are fast growing species that might have special nutritional requirements, and particularly on fat-soluble vitamins such as the case of meagre (El Kertaoui et al. 2017). Besides, sustainability of aquaculture has been highly discussed and a source of controversy for many years. The dependence of aquafeeds on fish oil and fish meal from wild caught fisheries was one of its major drawback on consumer's perception of aquaculture sustainability. Although, improvements on feed conversion efficiency as well as an increasing utilization of fisheries by-products (35% of the global raw material) has been achieved, increasing farmed fish demand forced the sector to look for a more sustainable sources of lipids and proteins in aquafeeds, and particularly to launch the F3 (Fish-Free Feed) challenge (www.iffo.net/). Fish-free diets based in vegetable oil and meal sources already exist for many farmed fish like tilapia, but its use in carnivorous farmed fish (e.g. salmon) causes gut inflammation and growth impairment, as well as lower quality flesh, particularly in terms of PUFAs content. Importantly, meal and oil from alternative vegetable sources, like soybean or canola oils, may contain higher levels of natural VK_1 compared to marine ingredients (Krossøy et al. 2011). Nevertheless, the same alternative sources contain some anti-nutritional factors hampering the assimilation/use of nutritional components like the fat-soluble vitamins. One example could be the presence of phytoestrogens such as genistein and daidzein, abundantly found in the major alternative source of vegetal proteins, the soybean (Chung et al. 2014; Robaina et al. 1995). Genistein has been recently shown to alter thyroid, estrogenic and metabolic biomarkers in early life stages of the Senegalese sole (Sarasquete et al. 2017). Those phytoestrogens are known to bind PXR and regulate the transcription of downstream genes (Li et al. 2009), and thus might interact with VE and VK signaling. Furthermore, a specific design and assessment of these and new alternative raw materials for aquafeeds (e.g. meal from feather, insects or from byproducts from other industries, and oil from krill or marine algae) might contribute to a successful implementation of these innovative sustainable feeds, while keeping the high nutritional value and quality of farmed seafood. In this regard, lipid content in seafood, and particularly that of PUFAs, is the major responsible of the health outcomes of consuming fish by human population (Calder 2012). In contrast, one of the main drawbacks of using vegetable oil sources is its low PUFA content (Turchini et al. 2010). Therefore, using vegetable oil sources will change fat-soluble vitamin requirements such as the one for VE since different enzymes involved in lipid metabolism depend on their interaction with this vitamin (Lebold and Traber 2014; Zingg 2007). New approaches have been applied to overcome these limitations (low content of n-3 and -6 fatty acids), like the production of nutritionally-enhanced oil from transgenic *Camelina sativa* (Betancor et al. 2015).

Finally, improvement of efficiency is the third issue to fill the increasing demand for farmed fish. In this sense, the development of more and better breeding programs, which represents the basis for these major improvements, has been recently addressed in European project FISHBOOST (www.fishboost.eu/). Farmed fish from inbreeding programs might have specific nutritional requirements or digestive capacities, being feed formulation able to differentially affect the fish depending on its genomic background, as already demonstrated by Geay et al. (2011) in European seabass. A big effort on determining genetic background for improved nutrients absorption, metabolism and utilization through the development of nutrigenomics is expected in the future years.

Altogether, the current knowledge points out that if a balanced dietary requirement for each fat-soluble vitamin has not been already efficiently proven to be achieved for largely cultivated species; the success on domestication of new species, breeding selection, and the implementation of new raw materials in aquafeeds might urge new and integrative approaches to get key knowledge in this regard.

6.5 Conclusions

Fat-soluble vitamins are essential micronutrients for fish development and homeostasis. Since fish are not able to synthesize them de novo, they should be provided in a specific amount and chemical form within the diet. The identification of their biological roles and metabolism allowed the definition of optimal nutritional requirements in fish. However, this knowledge has been gained through unifactorial, fragmented and in a skewed manner. Nutritional requirements should be defined taking into account the quality of the product, the animal welfare and the sustainability of its production since the consumers perception on these issues is nowadays a strong driver of aquaculture production; but also from an integrated/holistic point of view, since the amount and chemical form of each nutrient required by fish—mainly during larval development—depends on as many factors as the fish species, the environmental rearing conditions, the physiological stage (stressed vs. non stressed), developmental phase (pre-, pro- and/or post-metamorphic, juvenile, adult and/or sexually mature/active), the nutritional approach used (life prey carriers vs. formulated diets), among others. The detailed knowledge gained on the different fat-soluble vitamins in the last decades uncovered several crosstalk points between them along its signaling pathways: from intestinal absorption to detoxification mechanisms. Thus, nutritional studies should focus on deciphering which is the "balanced requirement" for each nutrient, taking into account the amount of other nutrients.

Most known interactions of fat-soluble vitamins were identified in studies using mammalian species, raising the need for specific research efforts on this issue in fish species, preferentially the farmed ones, since nutritional requirements identified in model fish species would not be potentially extrapolated to them. Nevertheless, knowledge, biological features, and the development of biotechnological tools in model fish species, such as zebrafish (reviewed in Goldsmith 2004), proposed as a suitable model for nutritional research in aquaculture (Ulloa et al. 2014), as well as medaka, might represent an invaluable model to get further insights on fat-soluble vitamins metabolism. Furthermore, the recent advancement and the implementation of OMIC tools such as those provided by NGS technologies will enable a wider and integrative analysis in fish physiology on this matter.

Finally, renewed research efforts on fat-soluble vitamins as well as new dietary recommendations are expected to result from the implementation of the latest research lines, boosting future aquaculture suitability. The application of nutritional strategies replacing fish oil and fish meal by alternative sources might imply reanalyzing requirements in fat-soluble vitamins, as their absorption depends on the dietary lipid content. Both content and fatty acid profile have been found to be different when aquafeeds are formulated with fish oil or vegetable oil sources, and thus correction or supplementation strategies should be adopted in particular cases, although it might be not the case of VK, since vegetable sources contain higher levels of VK (Krossøy et al. 2011). On the other hand, the development of new species for aquaculture will require to determine their specific nutritional requirements in these (and other) nutrients, while the breeding selection procedures in largely farmed fish species might also require the identification of genes and gene networks involved in the metabolism of fat-soluble vitamins.

Acknowledgements This work was co-funded by the Portuguese Foundation for Science and Technology (FCT) through the European Commission (ERDF-COMPETE) and through PEst-C/MAR/LA0015/2011 project. In addition, this review was partially funded by the following grants AGL2005-02478, AGL2008-03897-C04-01/ACU, AGL2014-51839-C5-5-R from the Ministerio de Economía, Industria y Competitividad (MINECO) of the Spanish Government + FEDER; the FINEFISH project, a collective research project of the sixth framework programme of the European Union (contract 012451); and the IberoAmerican network LARVAplus "Estrategias de desarrollo y mejora de la producción de larvas de peces en Iberoamérica" (117RT0521) funded by the Programa Iberoamericano de Ciencia y Tecnología para el Desarrollo (CYTED). IF was financed by the FCT through a postdoctoral fellowship (SFRH/BPD/82049/2011).

References

Abbink W, Flik G (2007) Parathyroid hormone-related protein in teleost fish. Gen Comp Endocrinol 152:243–251

Adams MK, Belyaeva OV, Wu L et al (2014) The retinaldehyde reductase activity of DHRS3 is reciprocally activated by retinol dehydrogenase 10 to control retinoid homeostasis. J Biol Chem 289:14868–14880

Austin B (2012) Infectious disease in aquaculture. Woodhead Publishing, UK, p. 560.

D'Ambrosio DN, Clugston RD, Blaner WS (2011) Vitamin A metabolism: an update. Nutrients 3:63–103

Amengual J, Golczak M, Palczewski K et al (2012) Lecithin: retinol acyltransferase is critical for cellular uptake of vitamin A from serum retinol-binding protein. J Biol Chem 287:24216–24227

Atalah E, Hernández-Cruz C, Ganga R et al (2012) Enhancement of gilthead seabream (*Sparus aurata*) larval growth by dietary vitamin E in relation to two different levels of essential fatty acids. Aquac Res 43:1816–1827

Avila EM, Basantes SP, Ferraris RP (1999) Cholecalciferol modulates plasma phosphate but not plasma vitamin D levels and intestinal phosphate absorption in rainbow trout (*Oncorhynchus mykiss*). Gen Comp Endocrinol 469:460–469

Bailly M, Milet C, Garabedian M et al (1988) Calcium-dependent metabolism of 25-hydroxycholecalciferol in silver eel tissues. Gen Comp Endocrinol 71:1–9

Bain PA, Papanicolaou A, Kumar A (2015) Identification of putative nuclear receptors and steroidogenic enzymes in Murray-darling rainbowfish (*Melanotaenia fluviatilis*) using RNA-Seq and de novo transcriptome assembly. PLoS ONE 10:1–25

Balmer JE, Blomhoff R (2002) Gene expression regulation by retinoic acid. J Lipid Res 43: 1773–1808

Barnett BJ, Cho CY, Slinger SJ (1982) Relative biopotency of dietary ergocalciferol and cholecalciferol and the role of and requirement for vitamin D in rainbow trout (*Salmo gairdneri*). J Nutr 112:2011–2019

Berthelot C, Brunet F, Chalopin D et al (2014) The rainbow trout genome provides novel insights into evolution after whole-genome duplication in vertebrates. Nature 5:1–10

Bertrand S, Brunet F, Escriva H et al (2004) Evolutionary genomics of nuclear receptors: from twenty-five ancestral genes to derived endocrine systems. Mol Biol Evol 21:1923–1937

Betancor M, Atalah E, Caballero M et al (2011) α-Tocopherol in weaning diets for European sea bass (*Dicentrarchus labrax*) improves survival and reduces tissue damage caused by excess dietary DHA contents. Aquac Nutr 17:112–122

Betancor MB, Sprague M, Usher S et al (2015) A nutritionally-enhanced oil from transgenic *Camelina sativa* effectively replaces fish oil as a source of eicosapentanoic acid for fish. Sci Rep 5:1–10

Blomhoff R (1994) Vitamin A in health and disease. Marcel Dekker, New York

Bogard JR, Thilsted SH, Marks GC et al (2015) Nutrient composition of important fish species in Bangladesh and potential contribution to recommended nutrient intakes. J Food Compos Anal 42:120–133

Boglino A, Ponce M, Cousin X et al (2017) Transcriptional regulation of genes involved in retinoic acid metabolism in Senegalese sole larvae. Comp Biochem Physiol Part B 203:35–46

Boglione C, Gisbert E, Gavaia P et al (2013) Skeletal anomalies in reared European fish larvae and juveniles. Part 2: main typologies, occurrences and causative factors. Rev Aquac 5(Suppl. 1): S121–S167

Booth SL, Suttie JW (1998) Dietary intake and adequacy of vitamin K1. J Nutr 128:785–788

Booth SL, Golly I, Sacheck JM et al (2004) Effect of vitamin E supplementation on vitamin K status in adults with normal coagulation status. Am J Clin Nutr 80:143–148

Borel P, Preveraud D, Desmarchelier C (2013) Bioavailability of vitamin E in humans: an update. Nutr Rev 71:319–331

Brenner B, Kuperman A, Watzka M et al (2009) Vitamin K-dependent coagulation factors deficiency. Semin Thromb Hemost 35:439–446

Bronner F (2009) Recent developments in intestinal calcium absorption. Nutr Rev 67:109–113

Buitenhuis HC, Soute BAM, Vermeer C (1990) Comparison of the vitamins K1, K2 and K3 as cofactors for the hepatic vitamin K-dependent carboxylase. Biochim Biophys Acta 1034:170–175

Cahu C, Infante Z, Takeuchi T (2003) Nutritional components affecting skeletal development in fish larvae. Aquaculture 227:245–258

Calder PC (2012) Omega-3 polyunsaturated fatty acids and inflammatory processes: nutrition or pharmacology? Br J Clin Pharmacol 75:645–662

Card DJ, Gorska R, Cutler J et al (2014) Vitamin K metabolism: current knowledge and future research. Mol Nutr Food Res 58:1590–1600

Cerezuela R, Cuesta A, Meseguer J et al (2009) Effects of dietary vitamin D3 administration on innate immune parameters of seabream (*Sparus aurata* L.). Fish Shellfish Immunol 26:243–248

Chelstowska S, Widjaja-adhi MAK, Silvaroli JA et al (2016) Molecular basis for vitamin A uptake and storage in vertebrates. Nutrients 8:1–24

Chen Y, Tang Y, Guo C et al (2012) Nuclear receptors in the multidrug resistance through the regulation of drug-metabolizing enzymes and drug transporters. Biochem Pharmacol 83:1112–1126

Cheng Y, Tao W, Chen J et al (2015) Genome-wide identification, evolution and expression analysis of nuclear receptor superfamily in Nile tilapia, *Oreochromis niloticus*. Gene 569:141–152

Chiba H, Yamashita R, Kinoshita K et al (2008) Weak correlation between sequence conservation in promoter regions and in protein-coding regions of human-mouse orthologous gene pairs. BMC Genom 15:1–15

Chung I, Yu B, Park I et al (2014) Isoflavone content and profile comparisons of cooked soybean–rice mixtures: electric rice cooker versus electric pressure rice cooker. J Agric Food Chem 62:11862–11868

Chung S, Atkinson J, Parker R et al (2015) Intracellular trafficking of vitamin E by the α-tocopherol transfer protein. FASEB J 29:886

Combs DF, McClung JP (2017) The vitamins, fundamental aspects in nutrition and health. Academic Press, London

Conceição N, Laizé V, Simões B et al (2008) Retinoic acid is a negative regulator of matrix Gla protein gene expression in teleost fish *Sparus aurata*. Biochim Biophys Acta 1779:28–39

Copping MA (1934) Origin of vitamin d in cod-liver oil: vitamin D content of zooplankton. Biochem J 28:1516–1520

Couto A, Kortner T, Penn M et al (2015) Saponins and phytosterols in diets for Europeansea bass (*Dicentrarchus labrax*) juveniles: effects on growth, intestinal morphology and physiology. Aquac Nutr 21:180–193

Craig TA, Sommer S, Sussman CR et al (2008) Expression and regulation of the vitamin D receptor in the zebrafish, *Danio rerio*. J Bone Mineral Res 23:1486–1496

Craig TA, Zhang Y, Magis AT et al (2014) Detection of 1α,25-dihydroxyvitamin D-regulated miRNAs in zebrafish by whole transcriptome sequencing. Zebrafish 11:207–218

Craig TA, Zhang Y, Mcnulty MS et al (2012) Whole transcriptome RNA sequencing detects multiple 1α,25-dihydroxyvitamin D3-sensitive metabolic pathways in developing zebrafish. Mol Endocrinol 26:1630–1642

Cui JY, Gunewardena SS, Rockwell CE et al (2010) ChIPing the cistrome of PXR in mouse liver. Nucleic Acids Res 38:7943–7963

Dam H (1935) The antihaemorrhagic vitamin of the chick. Biochem J 29:1273–1285

Darias MJ, Mazurais D, Koumoundouros G et al (2010) Dietary vitamin D 3 affects digestive system ontogenesis and ossification in European. Aquaculture 298:300–307

Darias MJ, Mazurais D, Koumoundouros G et al (2011) Overview of vitamin D and C requirements in fish and their influence on the skeletal system. Aquaculture 315:49–60

Dashty M, Motazacker MM, Levels J et al (2014) Proteome of human plasma very low-density lipoprotein and low-density lipoprotein exhibits a link with coagulation and lipid metabolism. J Thromb Haemost 111:381–564
de la Rosa LC, Vrenken TE, Buist-Homan M et al (2015) Metformin protects primary rat hepatocytes against oxidative stress-induced apoptosis. Pharmacol Res Perspect 3:e00125
Dedi J, Takeuchi T, Seikai T et al (1995) Hypervitaminosis and safe levels of vitamin A for larval flounder (*Paralichthys olivaceus*) fed *Artemia* nauplii. Aquaculture 133:135–146
Deluca HF, Zierold C (1998) Mechanisms and functions of vitamin D. Nutr Rev 56:S4–S10
Ding N, Yu RT, Subramaniam N et al (2013) A vitamin D receptor/SMAD genomic circuit gates hepatic fibrotic response. Cell 153:601–613
Ding Z, Li W, Huang J et al (2017) Dietary alanyl-glutamine and vitamin E supplements could considerably promote the expression of GPx and PPARα genes, antioxidation, feed utilization, growth, and improve composition of juvenile cobia. Aquaculture 470:95–102
Ekins S, Chang C, Mani S et al (2007) Human pregnane X receptor antagonists and agonists define molecular requirements for different binding sites. Mol Pharmacol 72:592–603
Ekins S, Reschly EJ, Hagey LR et al (2008) Evolution of pharmacologic specificity in the pregnane X receptor. BMC Evol Biol 8:103
El Kertaoui N, Hernández-Cruz M, Montero D et al (2017) The importance of dietary HUFA for meagre larvae (*Argyrosomus regius*; Asso, 1801) and its relation with antioxidant vitamins E and C. Aquac Res 48:419–433
Eroglu A, Hruszkewycz DP, Sena C et al (2012) Naturally occurring eccentric cleavage products of provitamin A B-carotene function as antagonists of retinoic acid receptors. J Biol Chem 287:15886–15895
Farley SM, Leonard SW, Labut EM et al (2012) Vitamin E decreases extra-hepatic menaquinone-4 concentrations in rats fed menadione or phylloquinone. Mol Nutr Food Res 56:912–922
Farley SM, Leonard SW, Taylor AW et al (2013) ω-Hydroxylation of phylloquinone by CYP4F2 is not increased by a-tocopherol. Mol Nutr Food Res 57:1785–1793
Fenwick JC, Smith K, Smith J et al (1984) Effect of various vitamin D analogs on plasma calcium and phosphorus and intestinal calcium absorption in fed and unfed American eels, *Anguilla rostrata*. Gen Comp Endocrinol 55:398–404
Ferland G (2012) Vitamin K, an emerging nutrient in brain function. BioFactors 38:151–157
Fernández I (2011) Hypervitaminosis A effects on *Sparus aurata* and *Solea senegalensis*: characterization of larval performance and the underlying gene expression of abnormal skeletogenesis. PhD thesis. University of Barcelona
Fernández I, Gisbert E (2011) The effect of vitamin A on flatfish development and skeletogenesis: a review. Aquaculture 315:34–48
Fernández I, Gavaia PJ (2016) Vitamin K, much more than blood clotting: roles, metabolism and specific requirements: a mini-review. In: Cruz Suárez LE, Ricque Marie D, Tapia Salazar M, Nieto López M, Villarreal Cavazos DA, Gamboa Delgado J, Rivas Vega ME (eds) Nutrición Acuícola: Investigación y Desarrollo. Universidad Autónoma de Nuevo León, San Nicolás de los Garza, Nuevo León, México, pp 361–376
Fernández I, Pimentel MS, Ortiz-Delgado JB et al (2009) Effect of dietary vitamin A on Senegalese sole (*Solea senegalensis*) skeletogenesis and larval quality. Aquaculture 295:250–265
Fernández I, Darias M, Andree KB et al (2011) Coordinated gene expression during gilthead sea bream skeletogenesis and its disruption by nutritional hypervitaminosis A. BMC Dev Biol 11:7
Fernández I, Santos A, Cancela ML et al (2014a) Warfarin, a potential pollutant in aquatic environment acting through Pxr signaling pathway and γ-glutamyl carboxylation of vitamin K-dependent proteins. Environ Pollut 194:86–95
Fernández I, Tiago DM, Laizé V et al (2014b) Retinoic acid differentially affects in vitro proliferation, differentiation and mineralization of two fish bone-derived cell lines: different gene expression of nuclear receptors and ECM proteins. J Steroid Biochem Mol Biol 140:34–43

Fernández I, Carmen L, Andree KB et al (2015a) Vitamin A supplementation enhances Senegalese sole (*Solea senegalensis*) early juvenile's immunocompetence: new insights on potential underlying pathways. Fish Shellfish Immunol 46:703–709

Fernández I, Vijayakumar P, Marques C et al (2015b) Zebrafish vitamin K epoxide reductases: expression *in vivo*, along extracellular matrix mineralization and under phylloquinone and warfarin *in vitro* exposure. Fish Physiol Biochem 41:745–759

Fernández I, Granadeiro L, Gavaia PJ et al (2015c) Zebrafish as a model for unveiling underlying signaling pathways of warfarin embryopathy. FASEB—Molecular, structural & clinical aspects of vitamin k & vitamin k-dependent proteins. USA, Chicago

Figueras A, Robledo D, Corvelo A et al (2016) Whole genome sequencing of turbot (*Scophthalmus maximus*; Pleuronectiformes): a fish adapted to demersal life. DNA Res 23:181–192

Flamant F, Baxter J, Forrest D et al (2006) International Union of Pharmacology. LIX. The pharmacology and classification of the nuclear receptor superfamily: thyroid hormone receptors. Pharmacol Rev 58:705–711

Flik G, Klaren P, Schoenmakers T et al (1996) Cellular calcium transport in fish: unique and universal mechanisms. Physiol Zool 69:403–417

Fodor D, Albu A, Poantă L et al (2010) Vitamin K and vascular calcifications. Acta Physiol Hung 97:256–266

Fontagné-Dicharry S, Lataillade E, Surget A et al (2010) Effects of dietary vitamin A on broodstock performance, egg quality, early growth and retinoid nuclear receptor expression in rainbow trout (*Oncorhynchus mykiss*). Aquaculture 303:40–49

Frankel EN (1998) Antioxidants. In: Frankel EN (ed) Lipid oxidation. The Oily Press, Dundee, pp 129–160

Funkenstein B (2001) Developmental expression, tissue distribution and hormonal regulation of fish (*Sparus aurata*) serum retinol-binding protein. Comp Biochem Physiol Part B 129:613–622

Geay F, Ferraresso S, Zambonino-Infante JL et al (2011) Effects of the total replacement of fish-based diet with plant-based diet on the hepatic transcriptome of two European sea bass (*Dicentrarchus labrax*) half-sibfamilies showing different growth rates with the plant-based diet. BMC Genom 12:522

Germain P, Chambon P, Eichele G et al (2006a) International Union of Pharmacology. LX. Retinoic acid receptors. Pharmacol Rev 58:712–725

Germain P, Chambon P, Eichele G et al (2006b) International Union of Pharmacology. LXIII. Retinoid X receptors. Pharmacol Rev 58:760–772

Gesto M, Castro LFC, Reis-Henriques MA et al (2012) Tissue-specific distribution patterns of retinoids and didehydroretinoids in rainbow trout *Oncorhynchus mykiss*. Comp Biochem Physiol Part B 161:69–78

Gijsbers B, Jie K, Vermeer C (1996) Effect of food composition on vitamin K absorption in human volunteers. Br J Nutr 76:223–229

Glasauer SMK, Neuhauss SCF (2014) Whole-genome duplication in teleost fishes and its evolutionary consequences. Mol Genet Genomics 289:1045–1060

Glynn RJ, Ridker PM, Goldhaber SZ et al (2007) Effects of random allocation to vitamin E supplementation on the occurrence of venous thromboembolism: report from the Women's Health Study. Circulation 116:1497–1503

Goldsmith P (2004) Zebrafish as a pharmacological tool: the how, why and when. Curr Opin Pharmacol 4:504–512

Gonçalves A, Roi S, Nowicki M et al (2015) Fat-soluble vitamin intestinal absorption: absorption sites in the intestine and interactions for absorption. Food Chem 172:155–160

Graff IE, Lie O, Aknes L (1999) *In vitro* hydroxylation of vitamin D3 and 25-hydroxy vitamin D3 in tissues of Atlantic salmon *Salmo salar*, Atlantic mackerel *Scomber scombrus*, Atlantic halibut *Hippoglossus hippoglossus* and Atlantic cod *Gadus morhua*. Aquac Nutr 5:23–32

Graff IE, Waagbo R, Fivelstad S et al (2002) A multivariate study on the effects of dietary vitamin K, vitamin D3 and calcium, and dissolved carbon dioxide on growth, bone minerals,

vitamin status and health performance in smolting Atlantic salmon *Salmo salar* L. J Fish Dis 25:599–614
Graff IE, Krossoy C, Gjerdevik K et al (2010) Influence of dietary menadione nicotinamide bisulphite (vitamin K3) and phylloquinone (vitamin K1) on Atlantic salmon (*Salmo salar* L.) tissue levels, determined by high- performance liquid chromatography with fluorescence detection. Aquac Nutr 16:637–647
Grisdale-Helland B, Helland SJ, Åsgård T (1991) Problems associated with the present use of menadione sodium bisulfite and vitamin A in diets for Atlantic salmon. Aquaculture 92:351–358
Guh Y, Hwang P (2016) Insights into molecular and cellular mechanisms of hormonal actions on fish ion regulation derived from the zebrafish model. Gen Comp Endocrinol 251:12–20
Gutierrez-Mazariegos J, Theodosiou M, Campo-Paysaa F et al (2011) Vitamin A: a multifunctional tool for development. Semin Cell Dev Biol 22:603–610
Haga Y, Takeuchi T, Murayama Y et al (2004) Vitamin D3 compounds induce hypermelanosis on the blind side and vertebral deformity in juvenile Japanese flounder *Paralichthys olivaceus*. Fish Sci 70:16–20
Haga Y, Tarui F, Ohta K et al (2006) Effect of light irradiation on dynamics of vitamin A compounds in rotifers and *Artemia*. Fish Sci 72:1020–1026
Haga Y, Du S, Satoh S et al (2011) Analysis of the mechanism of skeletal deformity in fish larvae using a vitamin A-induced bone deformity model. Aquaculture 315:26–33
Hammed A, Matagrin B, Spohn G et al (2013) VKORC1L1, an enzyme rescuing the vitamin K 2,3-epoxide reductase activity in some extrahepatic tissues during anticoagulation therapy. J Biol Chem 288:28733–28742
Hammerling U (2016) Retinol as electron carrier in redox signaling, a new frontier in vitamin A research. Hepatobiliary Surg Nutr 5:15–28
Hamre K (2011) Metabolism, interactions, requirements and functions of vitamin E in fish. Aquac Nutr 17:98–115
Hamre K, Waagbø R, Berge RK et al (1997) Vitamins C and E interact in juvenile Atlantic salmon. Free Radical Biol Med 22:137–149
Hamre K, Krossöy C, Lock E et al (2010) Roles of lipid-soluble vitamins during ontogeny of marine fish larvae. Aquac Nutr 41:745–750
Hamre K, Yúfera M, Rønnestad I et al (2013) Fish larval nutrition and feed formulation: knowledge gaps and bottlenecks for advances in larval rearing. Rev Aquac 5(Suppl. 1):S26–S58
Hamre K, Sissener NH, Lock EJ et al (2016) Antioxidant nutrition in Atlantic salmon (*Salmo salar*) parr and post-smolt, fed diets with high inclusion of plant ingredients and graded levels of micronutrients and selected amino acids. PeerJ 4:e2688
Harshman SG, Saltzman E, Booth SL (2014) Vitamin K: dietary intake and requirements in different clinical conditions. Curr Opin Clin Nutr Metab Care 17:531–538
Hayes ME, Guilland-Cumming D, Russell R et al (1986) Metabolism of 25-Hydroxycholecalciferol in a teleost fish the rainbow trout (*Salmo gairdneri*). Gen Comp Endocrinol 64:143–150
He Y, Tsuei J, Wan YY (2014) Biological functional annotation of retinoic acid alpha and beta in mouse liver based on genome-wide binding. Am J Physiol Gastrointest Liver Physiol 307:205–218
Hegarty JM, Yang H, Chi NC (2013) UBIAD1-mediated vitamin K_2 synthesis is required for vascular endothelial cell survival and development. Development 140:1713–1719
Helson L (1984) The effect of intravenous vitamin E and menadiol sodium diphosphate on vitamin K dependent clotting factors. Thromb Res 35:11–18
Hii CS, Ferrante A (2016) The non-genomic actions of vitamin D. Nutrients 8:135
Hirota Y, Tsugawa N, Nakagawa K et al (2013) Menadione (vitamin K_3) is a catabolic product of oral phylloquinone (vitamin K_1) in the intestine and a circulating precursor of tissue menaquinone-4 (vitamin K_2) in rats. J Biol Chem 288:33071–33080
Hoenderop JGJ, Nilius B, Bindels RJM (2005) Calcium absorption across epithelia. Physiol Rev 85:373–422

Holick MF (1981) The cutaneous photosynthesis of previtamin D3: a unique photoendocrine system. J Invest Dermatol 77:51–58
Holick MF, Schonoes H, Deluca HF (1971) Identification of 1,25-dihydroxycholecalciferol, a form of vitamin D3 metabolically active in the intestine. Proc Natl Acad Sci U S A 68:803–804
Holick MF, Schnoes HK, DeLuca HF et al (1972) Isolation and identification of 24,25-dihydroxycholecalciferol, a metabolite of vitamin D3 made in the kidney. Biochem 11:4251–4255
Holick M, Frommer H, McNeill S et al (1977) Photometabolism of 7-dehydrocholesterol to previtamin D3 in skin. Biochem Biophys Res Commun 76:107–114
Hollander D, Rim E (1976a) Factors affecting the absorption of vitamin K-1 *in vitro*. Gut 17:450–455
Hollander D, Rim E (1976b) Vitamin K_2 absorption by rat everted small intestinal sacs. Am J Physiol 231:415–419
Holt GJ (2011) Larval fish nutrition. Wiley-Blackwell, West Sussex
Horvli OL, Aksnes L (1998) Tissue distribution of vitamin D3 in Atlantic salmon *Salmo salar*: effect of dietary level. Aquac Nutr 4:127–131
Horwitt MK (1962) Interrelations between vitamin E and poly-unsaturated fatty acids in adult men. Vitam Horm 20:541–558
Howarth DL, Law SHW, Barnes B et al (2008) Paralogous vitamin D receptors in teleosts: transition of nuclear receptor function. Endocrinology 149:2411–2422
Hwang P, Chou M (2013) Zebrafish as an animal model to study ion homeostasis. Pflugers Arch Eur J Physiol 465:1233–1247
Hymøller L, Clausen TN, Jensen SK (2016) Interactions between retinol, α-tocopherol and cholecalciferol need consideration in diets for farmed mink (*Mustela vison*). Br J Nutr 115:751–758
Ichikawa T, Horie-Inoue K, Ikeda K et al (2006) Steroid and xenobiotic receptor SXR mediates vitamin K2-activated transcription of extracellular matrix-related genes and collagen accumulation in osteoblastic cells. J Biol Chem 281:16927–16934
Ito A, Shirakawa H, Takumi N et al (2011) Menaquinone-4 enhances testosterone production in rats and testis-derived tumor cells. Lipids Health Dis 10:158
Jiang Q (2014) Natural forms of vitamin E: metabolism, antioxidant, and anti-inflammatory activities and their role in disease prevention and therapy. Free Radic Biol and Med 72:76–90
Jiang J, Shi D, Zhou X et al (2015) Vitamin D inhibits lipopolysaccharide-induced in flammatory response potentially through the Toll-like receptor 4 signalling pathway in the intestine and enterocytes of juvenile Jian carp (*Cyprinus carpio* var. Jian). Br J Nutr 114:1560–1568
Karsenty G, Ferron M (2012) The contribution of bone to whole-organism physiology. Nature 481:314–320
Kawaguchi R, Zhong M, Kassai M et al (2015) Vitamin A transport mechanism of the multitransmembrane cell-surface receptor STRA6. Membranes 5:425–453
Kelly M, Von Lintig J (2015) STRA6: role in cellular retinol uptake and efflux. Hepatobiliary Surg Nutr 4:229–242
Kim DS, Burt AA, Ranchalis JE et al (2014) Effects of dietary components on high-density lipoprotein measures in a cohort of 1,566 participants. Nutr Metab 11:44
Koivu-Tikkanen TJ, Schurgers LJ, Thijssen HHW et al (2000) Intestinal, hepatic, and circulating vitamin K levels at low and high intakes of vitamin K in rats. Br J Nutr 83:185–190
Kollitz EM, Hawkins MB, Whitfield GK et al (2014) Functional diversification of vitamin D receptor paralogs in Teleost fish after a whole genome duplication event. Endocrinology 155:4641–4654
Kollitz EM, Zhang G, Hawkins MB et al (2015) Molecular cloning, functional characterization, and evolutionary analysis of vitamin D receptors isolated from basal vertebrates. PLoS ONE 1:27
Koumoundouros G (2010) Morpho-anatomical abnormalities in Mediterranean marine aquaculture. In: Koumoundouros G (ed) Recent advances in aquaculture research. Transworld Research Network, Kerala, pp 125–148

Krasowski MD, Ai N, Hagey LR et al (2011) The evolution of farnesoid X, vitamin D, and pregnane X receptors: insights from the green-spotted pufferfish (*Tetraodon nigriviridis*) and other non-mammalian species. BMC Biochem 12:5

Krossøy C, Waagbø R, Fjelldal PG et al (2009) Dietary menadione nicotinamide bisulphite (vitamin K3) does not affect growth or bone health in first-feeding fry of Atlantic salmon (*Salmo salar* L.). Aquac Nutr 15:638–649

Krossøy C, Waagbø R, Ørnsrud R (2011) Vitamin K in fish nutrition. Aquac Nutr 17:585–594

Kwon H (2016) Vitamin D receptor signaling is required for heart development in zebrafish embryo. Biochem Biophys Res Commun 470:575–578

Lall SP, Lewis-Mccrea LM (2007) Role of nutrients in skeletal metabolism and pathology in fish—an overview. Aquaculture 267:3–19

Lambert WE, Vanneste L, De Leenheer AP (1992) Enzymatic sample hydrolysis and HPLC in a study of phylloquinone concentration in human milk. Clin Chem 38:1743–1748

Landes N, Pfluger P, Kluth D, Birringer M et al (2003) Vitamin E activates gene expression via the pregnane X receptor. Biochem Pharmacol 65:269–273

Larsson D, Aksnes L, Björnsson BT et al (2002) Antagonistic effects of 24R,25-dihydroxyvitamin D3 and 25-hydroxyvitamin D3 on L-type Ca^{2+} channels and Na^{+}/Ca^{2+} exchange in enterocytes from Atlantic cod (*Gadus morhua*). J Mol Endocrinol 28:53–68

Larsson D, Bjrnsson BTH, Sundell K (1995) Physiological concentrations of 24,25-dihydroxyvitamin D3 rapidly decrease the in vitro intestinal calcium uptake in the Atlantic cod, *Gadus morhua*. Gen Comp Endocrinol 100:211–217

Lebold KM, Traber MG (2014) Free radical biology and medicine interactions between α-tocopherol, polyunsaturated fatty acids, and lipoxygenases during embryogenesis. Free Radical Biol Med 66:13–19

Lewis-Mccrea BLM, Lall SP (2010) Effects of phosphorus and vitamin C deficiency, vitamin A toxicity, and lipid peroxidation on skeletal abnormalities in Atlantic halibut (*Hippoglossus hippoglossus*). J Appl Ichthyol 26:334–343

Li Y, Ross-viola JS, Shay NF et al (2009) Human CYP3A4 and murine Cyp3A11 are regulated by equol and genistein via the pregnane X receptor in a species-specific manner. J Nutr 139:898–904

Li M, Chen L, Qin JG et al (2013) Growth performance, antioxidant status and immune response in darkbarbel catfish *Pelteobagrus vachelli* fed different PUFA/vitamin E dietary levels and exposed to high or low ammonia. Aquaculture 406–407:18–27

Li E, Li C (2014) Use of RNA-seq in aquaculture research. Poult Fish Wild Sci 2:e108

Lian J, Stewart C, Puchacz E et al (1989) Structure of the rat osteocalcin gene and regulation of vitamin D-dependent expression. Proc Natl Acad Sci U S A 86:1143–1147

Lie KK, Kvalheim K, Rasinger JD et al (2016) Vitamin A and arachidonic acid altered the skeletal mineralization in Atlantic cod (*Gadus morhua*) larvae without any interactions on the transcriptional level. Comp Biochem Physiol Part A 191:80–88

Lien S, Koop BF, Sandve SR et al (2016) The Atlantic salmon genome provides insights into rediploidization. Nature 533:200–205

Lim C, Yildirim-Aksoy M, Li MH et al (2009) Influence of dietary levels of lipid and vitamin E on growth and resistance of Nile tilapia to *Streptococcus iniae* challenge. Aquaculture 298:76–82

Lin C, Su C, Hwang P (2014) Calcium-sensing receptor mediates Ca^{2+} homeostasis by modulating expression of PTH and stanniocalcin. Endocrinology 155:56–67

Lin C, Su C, Tseng D et al (2012) Action of vitamin D and the receptor, VDRa, in calcium handling in zebrafish (*Danio rerio*). PLoS ONE 7:e45650

Lin C, Tsai I, Su C et al (2011) Reverse effect of mammalian hypocalcemic cortisol in fish: cortisol stimulates Ca^{2+} uptake via glucocorticoid receptor-mediated vitamin D3 metabolism. PLoS ONE 6:e23689

Linville A, Radtke K, Waxman JS et al (2009) Combinatorial roles for zebrafish retinoic acid receptors in the hindbrain, limbs and pharyngeal arches. Dev Biol 325:60–70

Liu B, Xu P, Xie J et al (2014) Effects of emodin and vitamin E on the growth and crowding stress of Wuchang bream (*Megalobrama amblycephala*). Fish Shellfish Immunol 40:595–602

Lobo GP, Amengual J, Baus D et al (2013) Genetics and diet regulate vitamin A production via the homeobox transcription factor ISX. J Biol Chem 288:9017–9027
Lock EJ, Ørnsrud R, Aksnes L et al (2007) The vitamin D receptor and its ligand 1α,25-dihydroxyvitamin D3 in Atlantic salmon (*Salmo salar*). J Endocrinol 193:459–471
Lock E, Waagbo R, Wendelaar Bonga S et al (2010) The significance of vitamin D for fish: a review. Aquac Nutr 16:100–116
Lu Z, Chen TC, Zhang A et al (2007) An evaluation of the vitamin D3 content in fish: is the vitamin D content adequate to satisfy the dietary requirement for vitamin D? J Steroid Biochem Mol Biol 103:642–654
Lubzens E, Lissauer L, Levavi-Sivan B et al (2003) Carotenoid and retinoid transport to fish oocytes and eggs: what is the role of retinol binding protein? Mol Aspects Med 24:441–457
MacIntyre I, Colston K, Evans I et al (1976) Regulation of vitamin D: an evolutionary view. Clin Endocrinol 5:85–95
Makishima M (2005) Nuclear receptors as targets for drug development: regulation of cholesterol and bile acid metabolism by nuclear receptors. J Pharmacol Sci 97(2):177–183.
Manchado M, Infante C, Rebordinos L et al (2009) General and comparative endocrinology molecular characterization, gene expression and transcriptional regulation of thyroid hormone receptors in Senegalese sole. Gen Comp Endocrinol 160:139–147
Marques C, Roberto VP, Granadeiro L et al (2017) The xenobiotic sensor PXR in a marine flatfish species (*Solea senegalensis*): gene expression patterns and its regulation under different physiological conditions. Mar Environ Res 130:187–199
Martí-Solans J, Belyaeva OV, Torres-Aguila NP et al (2016) Co-elimination and survival in gene network evolution: dismantling the RA-signaling in a chordate. Mol Biol Evol 33:2401–2416
Mathers J, Fernández F, Hill M et al (1990) Dietary modification of potential vitamin K supply from enteric bacterial menaquinones in rats. Br J Nutr 63:639–652
Mattila P, Piironen V, Haapala R (1997) Possible factors responsible for the high variation in the cholecalciferol contents of fish. J Agric Food Chem 45:3891–3896
Mazurais D, Glynatsi N, Darias MJ et al (2009) Optimal levels of dietary vitamin A for reduced deformity incidence during development of European sea bass larvae (*Dicentrarchus labrax*) depend on malformation type. Aquaculture 294:262–270
McBurney A, Shearer MJ, Barkhan P (1980) Preparative isolation and characterization of the urinary aglycones of vitamin K1 (phylloquinone) in man. Biochem Med 24:250–267
McDonald MG, Rieder MJ, Nakano M et al (2009) CYP4F2 is a vitamin K1 oxidase: an explanation for altered warfarin dose in carriers of the V433M variant. Mol Pharmacol 75:1337–1346
McKenna NJ, Cooney AJ, DeMayo FJ et al (2009) Minireview: evolution of NURSA, the nuclear receptor signaling atlas. Mol Endocrinol 23:740–746
Meisler MH (2001) Evolutionarily conserved noncoding DNA in the human genome: how much and what for? Genome Res 11:1617–1618
Meyer MB, Goetsch PD, Pike JW (2010) Genome-wide analysis of the VDR/RXR cistrome in osteoblast cells provides new mechanistic insight into the actions of the vitamin D hormone. J Steroid Mol Biol 121:136–141
Michalik L, Auwerx J, Berger JP et al (2006) International Union of Pharmacology. LXI. Peroxisome proliferator-activated receptors. Pharmacol Rev 58:726–741
Miller GW, Ulatowski L, Labut EM et al (2012) The a-Tocopherol transfer protein is essential for vertebrate embryogenesis. PLoS ONE 7:e47402
Minucci S, Leid M, Toyama R et al (1997) Retinoid × Receptor (RXR) within the RXR-retinoic acid receptor heterodimer binds its ligand and enhances retinoid-dependent gene expression. Mol Cell Biol 17:644–655
Montero D, Tort L, Robaina L et al (2001) Low vitamin E in diet reduces stress resistance of gilthead seabream (*Sparus aurata*) juveniles. Fish Shellfish Immunol 11:473–490
Moore DD, Kato S, Xie WEN et al (2006) International Union of Pharmacology. LXII. The NR1H and NR1I receptors: constitutive androstane receptor, pregnene X receptor, farnesoid X

receptor α, farnesoid X receptor B, liver X receptor α, liver X receptor B, and vitamin D receptor. Pharmacol Rev 58:742–759
Muenzner M, Tuvia N, Deutschmann C et al (2013) Retinol-binding protein 4 and its membrane receptor STRA6 control adipogenesis by regulating cellular retinoid homeostasis and retinoic acid receptor a activity. Mol Cell Biol 33:4068–4082
Mugoni V, Postel R, Catanzaro V et al (2013) Ubiad1 is an antioxidant enzyme that regulates eNOS activity by CoQ10 synthesis. Cell 152:504–518
Mustacich DJ, Leonard SW, Devereaux MW et al (2006) α-tocopherol regulation of hepatic cytochrome P450s and ABC transporters in rats. Free Radical Biol Med 41:1069–1078
Naderi M, Keyvanshokooh S, Salati AP et al (2017) Proteomic analysis of liver tissue from rainbow trout (*Oncorhynchus mykiss*) under high rearing density after administration of dietary vitamin E and selenium nanoparticles. Comp Biochem Physiol Part D Genom Proteome 22:10–19
Neacsu CD, Grosch M, Tejada M et al (2011) Ucmaa (Grp-2) is required for zebrafish skeletal development. Evidence for a functional role of its glutamate γ-carboxylation. Matrix Biol 30:369–378
Nemere I, Larsson D, Sundell K (2000) A specific binding moiety for 1,25-dihydroxyvitamin D3 in basal lateral membranes of carp enterocytes. Am J Physiol Endocrinol Metab 279:614–621
National Research Council (1993) Nutrient requirements of fish. The National Academies Press, Washington, USA
National Research Council (2011) Nutrient Requirements of Fish and Shrimp. The National Academies Press, Washington, USA
N'soukpoé-Kossi C, Sedaghat-Herati R, Ragi C et al (2007) Retinol and retinoic acid bind human serum albumin: stability and structural features. Int J Biol Macromol 40:484–490
O'Brien SJ, Fraser CM (2005) Genomes and evolution: the power of comparative genomics. Curr Opin Genet Dev 15:569–571
O'Byrne SMO, Blaner WS (2013) Retinol and retinyl esters: biochemistry and physiology. J Lipid Res 54:1731–1743
Oldenburg J, Marinova M, Müller-Reible C et al (2008) The Vitamin K cycle. Vitam Horm 78:35–62
Oldenburg J, Watzka M, Bevans CG (2015) VKORC1 and VKORC1L1: why do vertebrates have two vitamin K 2,3-epoxide reductases? Nutrients 7:6250–6280
Oliveira E, Casado M, Raldúa D et al (2013) Retinoic acid receptors' expression and function during zebrafish early development. J Steroid Biochem Mol Biol 138:143–151
Ørnsrud R, Lock EJ, Glover CN et al (2009) Retinoic acid cross-talk with calcitriol activity in Atlantic salmon (*Salmo salar*). J Endocrinol 202:473–482
Ørnsrud R, Lock EJ, Waagbø R et al (2013) Establishing an upper level of intake for vitamin A in Atlantic salmon (*Salmo Salar* L.) postsmolts. Aquac Nutr 19:651–664
Orr A, Dubé MP, Marcadier J et al (2007) Mutations in the UBIAD1 gene, encoding a potential prenyltransferase, are causal for Schnyder crystalline corneal dystrophy. PLoS ONE 2:e685
Pan JH, Feng L, Jiang WD et al (2017) Vitamin E deficiency depressed fish growth, disease resistance, and the immunity and structural integrity of immune organs in grass carp (*Ctenopharyngodon idella*): referring to NF-kB, TOR and Nrf2 signaling. Fish Shellfish Immunol 60:219–236
Pavek P (2016) Pregnane X receptor (PXR)-mediated gene repression and cross-talk of PXR with other nuclear receptors via coactivator interactions. Front Pharmacol 7:456
Peng LI, Gatlin DM III (2009) Dietary vitamin E requirement of the red drum *Sciaenops ocellatus*. Aquac Nutr 15:313–319
Pepe IM (1999) Rhodopsin and phototransduction. J Photochem Photobiol 48:1–10
Pierens SL, Fraser DR (2015) The origin and metabolism of vitamin D in rainbow trout. J Steroid Biochem Mol Biol 145:58–64
Pike JW, Meyer MB, Lee S et al (2017) The vitamin D receptor: contemporary genomic approaches reveal new basic and translational insights. J Clin Invest 127:1146–1154

Poston HA (1968) Effects of massive doses of vitamin D3 on fingerling brook trout. Fish Res Bull 32:48–50
Ponchon G, Deluca HF (1969) The role of the liver in the metabolism of vitamin D. J Clin Invest 48:1273–1279
Presnell SR, Stafford DW (2002) The vitamin K-dependent carboxylase. Thromb Haemost 87:937–946
Qiu A, Glover CN, Hogstrand C (2007) Regulation of branchial zinc uptake by 1α, 25-(OH)2D3 in rainbow trout and associated changes in expression of ZIP1 and ECaC. Aquat Toxicol 84:142–152
Qiu A, Hogstrand C (2004) Functional characterisation and genomic analysis of an epithelial calcium channel (ECaC) from pufferfish, *Fugu rubripes*. Gene 342:113–123
Rao DS, Raghuramulu N (1996) Food chain as origin of vitamin D in fish. Comp Biochem Physiol 114:15–19
Rao DS, Raghuramulu N (1997) Vitamin D3 in Tilapia mossambica: relevance of photochemical synthesis. J Nutr Sci Vitaminol 43:425–433
Raucy JL, Lasker JM (2013) Cell-based systems to assess nuclear receptor activation and their use in drug development. Drug Metab Rev 45:101–109
Reboul E (2013) Absorption of vitamin A and carotenoids by the enterocyte: focus on transport proteins. Nutrients 5:3563–3581
Reboul E (2015) Intestinal absorption of vitamin D: from the meal to the enterocyte. Food Funct 6:356–362
Reijntjes S, Blentic A, Gale E et al (2005) The control of morphogen signalling: regulation of the synthesis and catabolism of retinoic acid in the developing embryo. Dev Biol 285:224–237
Richard N, Fernández I, Wulff T et al (2014) Dietary supplementation with vitamin K affects transcriptome and proteome of Senegalese sole, improving larval performance and quality. Mar Biotechnol 16:522–537
Richelle M, Enslen M, Hager C et al (2004) Both free and esterified plant sterols reduce cholesterol absorption and the bioavailability of β-carotene and α-tocopherol in normocholesterolemic humans. Am J Clin Nutr 80:171–177
Rigotti A (2007) Absorption, transport and tissue delivery of vitamin E. Mol Aspects Med 28:423–436
Robaina L, Izquierdo MS, Moyano FJ et al (1995) Soybean and lupin seed meals as protein sources in diets for gilthead seabream (*Sparus aurata*): nutritional and histological implications. Aquaculture 130:219–233
Ross SA, Caffery PJMC, Drager UC et al (2000) Retinoids in embryonal development. Physiol Rev 80:1021–1055
Sarasquete C, Úbeda-Manzanaro M, Ortiz-Delgado JB (2017) Effects of the soya isoflavone genistein in early life stages of the Senegalese. Gen Comp Endocrinol 250:136–151
Schmid A, Walther B (2013) Natural vitamin D content in animal products. Adv Nutr 4:453–462
Schmölz L, Birringer M, Lorkowski S et al (2016) Complexity of vitamin E metabolism. World J Biol Chem 26:14–43
Schurgers LJ, Teunissen KJF, Hamulyák K et al (2007) Vitamin K–containing dietary supplements: comparison of synthetic vitamin K1 and natto-derived menaquinone-7. Blood 109:3279–3284
Schurgers LJ, Vermeer C (2002) Differential lipoprotein transport pathways of K-vitamins in healthy subjects. Biochim Biophys Acta 1570:27–32
Shearer MJ, Newman P (2008) Metabolism and cell biology of vitamin K. Thromb Haemost 100:530–547
Shearer MJ, Newman P (2014) Recent trends in the metabolism and cell biology of vitamin K with special reference to vitamin K cycling and MK-4 biosynthesis. J Lipid Res 55:345–362
Sluder A, Mathews SW, Hough D et al (1999) The nuclear receptor superfamily has undergone extensive proliferation and diversification in nematodes. Genome Res 9:103–120
Spronk H, Soute B, Schurgers L et al (2003) Tissue-specific utilization of menaquinone-4 results in the prevention of arterial calcification in warfarin-treated rats. J Vasc Res 40:531–537

Sugiura SH, Ferraris RP (2004) Dietary phosphorus-responsive genes in the intestine, pyloric ceca, and kidney of rainbow trout. Am J Physiol Regul Integr Comp Physiol 287:541–550

Sundell K, Bishop JE, Bt Bjornsson et al (1992) 1,25-dihydroxyvitamin D3 in the Atlantic cod: plasma levels, a plasma binding of a high affinity receptor. Endocrinology 131:2279–2286

Sundell K, Bt Björnsson (1990) Effects of vitamin D3, 25(OH) vitamin D3, 24,25(OH)$_2$ vitamin D3, and 1,25(OH)2 vitamin D3 on the in vitro intestinal calcium absorption in the marine Teleost, Atlantic cod (*Gadus morhua*). Gen Comp Endocrinol 78:74–79

Sundell K, Norman A, Bt Bjornsson (1993) 1,25(OH)$_2$ vitamin D3 increases ionized plasma calcium concentrations in the immature Atlantic cod *Gadus morhua*. Gen Comp Endocrinol 91:344–351

Sundh H, Larsson D, Sundell K (2007) Environmental salinity regulates the in vitro production of [3H]-1,25-dihydroxyvitamin D3 and [3H]-24,25 dihydroxyvitamin D3 in rainbow trout (*Oncorhynchus mykiss*). Gen Comp Endocrinol 152:252–258

Suttie JW (1995) The importance of menaquinones in human-nutrition. Ann Rev Nutr 15:399–417

Suzuki N, Danks JA, Maruyama Y et al (2011) Parathyroid hormone 1 (1–34) acts on the scales and involves calcium metabolism in gold fish. Bone 48:1186–1193

Suzuki T, Suzuki N, Srivastava AS et al (2000) Identification of cDNAs encoding two subtypes of vitamin D receptor in flounder, *Paralichthys olivaceus*. Biochem Biophys Res Commun 270:40–45

Swarup K, Dash VK, Norman AW (1991) Dose-dependent vitamin D3 and 1,25dihydroxyvitamin D3-induced hypercalcemia and hyperphosphatemia in male cyprinoid *Cyprinus carpio*. Comp Biochem Physiol 100:445–447

Swarup K, Norman AW, Srivastav AK (1984) Dose-dependent vitamin D3 and 1,25-dihydroxyvitamin D3-induced hypercalcemia and hyperphosphatemia in male catfish *Clarias batrachus*. Comp Biochem Physiol 78:553–555

Tabb MM, Sun A, Zhou C et al (2003) Vitamin K2 regulation of bone homeostasis is mediated by the steroid and xenobiotic receptor SXR. J Biol Chem 278:43919–43927

Takada T, Yamanashi Y, Konishi K et al (2015) NPC1L1 is a key regulator of intestinal vitamin K absorption and a modulator of warfarin therapy. Sci Transl Med 7:275ra23

Takeuchi A, Okano T, Kobayashi T (1991) The existence of 25-hydroxyvitamin D3-1α-hydroxylase in the liver of carp and bastard halibut. Life Sci 48:275–282

Takeuchi A, Okano T, Sayamoto M et al (1986) Tissue distribution of 7-dehydrocholesterol, vitamin D3 and 25-dydroxyvitamin D3 in several species of fishes. J Nutr Sci Vitaminol 32:13–22

Takeuchi A, Okano T, Sekimoto H et al (1994) The enzymatic formation of 1α,25-dihydroxyvitamin D3 from 25-hydroxyvitamin D3 in the liver of fetal rats. Comp Biochem Physiol 109:1–7

Takeuchi A, Torii M, Hatanaka Y et al (1987) Comparative studies on the contents of vitamin D3, 25-hydroxy vitamin D3 and 7-dehydrocholesterol in fish liver. Comp Biochem Physiol 88:569–573

Tan L, Wray AE, Green MH et al (2014) Compartmental modeling of whole-body vitamin A kinetics in unsupplemented and vitamin A-retinoic acid-supplemented neonatal rats. J Lipid Res 55:1738–1749

Taveekijakarn P, Miyazakp T, Matsumoto M et al (1996) Histopathological and haematological changes in amago salmon, *Oncorhynchus rhodurus* (Jordan & McGregor), fed a vitamin-D-free diet. J Fish Dis 19:289–294

Tena JJ, González-Aguilera C, Fernández-Miñan A et al (2014) Comparative epigenomics in distantly related teleost species identifies conserved cis-regulatory nodes active during the vertebrate phylotypic period. Genome Res 24:1075–1085

Tocher DR (2015) Omega-3 long-chain polyunsaturated fatty acids and aquaculture in perspective. Aquaculture 449:94–107

Tocher DR, Bendiksen EÅ, Campbell PJ et al (2008) The role of phospholipids in nutrition and metabolism of teleost fish. Aquaculture 280:21–34

Traber MG (2004) The ABCs of vitamin E and B-carotene absorption. Am J Clin Nutr 80:3–4

Traber MG (2013) Mechanisms for the prevention of vitamin E excess. J Lipid Res 54:2295–2306
Trichet V (2010) Nutrition and immunity: an update. Aquac Res 41:356–372
Trushenski JT, Kohler CC (2008) Influence of vitamin E source and dietary supplementation level on production performance of sunshine bass, *Morone chrysops* ♀ × *Morone saxatilis* ♂, fillet tocopherol content, and immunocompetency during stress and bacterial challenge. J World Aquac Soc 39:454–466
Turchini GM, Ng WK, Tocher DR (2010) Fish oil replacement and alternative lipid sources in aquaculture feeds. CRC Press, Boca Raton, USA, Florida
Turchini GM, Francis DS, Keast RSJ et al (2011) Transforming salmonid aquaculture from a consumer to a producer of long chain omega-3 fatty acids. Food Chem 124:609–614
Udagawa M (2001) The effect of dietary vitamin K (phylloquinone and menadione) levels on the vertebral formation in mummichog *Fundulus heteroclitus*. Fish Sci 67:104–109
Udagawa M (2000) Physiological role of vitamin K in fish—review. Japan Agric Res Q 34:279–284
Ulloa PE, Medrano JF, Feijoo CG (2014) Zebrafish as animal model for aquaculture nutrition research. Front Genet 5:313
Vickers KC, Remaley AT (2014) HDL and cholesterol: life after the divorce? J Lipid Res 55:4–12
Viegas CSB, Simes DC, Laize V et al (2008) Gla-rich protein (GRP), a new vitamin K-dependent protein identified from sturgeon cartilage and highly conserved in vertebrates. J Biol Chem 283:36655–36664
Vielma J, Lall SP (1998) Phosphorus utilization by Atlantic salmon (*Salmo salar*) reared in freshwater is not influenced by higher dietary calcium intake. Aquaculture 160:117–128
Villeneuve L, Gisbert E, Cahu CL et al (2004) Expression and localization of some retinoid receptors during European sea bass (*Dicentrarchus labrax*) larvae development. Aquaculture 242:537–551
Villeneuve L, Gisbert E, Delliou L et al (2005a) Dietary levels of all-trans retinol affect retinoid nuclear receptor expression and skeletal development in European sea bass larvae. Br J Nutr 93:791–801
Villeneuve L, Gisbert E, Zambonino-Infante JL et al (2005b) Effect of nature of dietary lipids on European sea bass morphogenesis: implication of retinoid receptors. Br J Nutr 94:877–884
Wallace BD, Betts L, Talmage G et al (2013) Structural and functional analysis of the human nuclear xenobiotic receptor PXR in complex with RXRα. J Mol Biol 425:2561–2577
Westhofen P, Watzka M, Marinova M et al (2011) Human vitamin K 2,3-epoxide reductase complex subunit 1-like 1 (VKORC1L1) mediates vitamin K-dependent intracellular antioxidant function. J Biol Chem 286:15085–15094
Whitfield GK, Dang HTL, Schluter SF et al (2003) Cloning of a functional vitamin D receptor from the lamprey (*Petromyzon marinus*), an ancient vertebrate lacking a calcified skeleton and teeth. Endocrinology 144:2704–2716
Widjaja-Adhi MAK, Lobo GP, Golczak M et al (2015) A genetic dissection of intestinal fat-soluble vitamin and carotenoid absorption. Hum Mol Genet 24:3206–3219
Wright DA, Dawson R, Cutler SJ et al (2007) Screening of natural product biocides for control of non-indigenous species. Environ Technol 28:309–319
Wu F, Jiang M, Wen H et al (2017) Dietary vitamin E effects on growth, fillet textural parameters, and antioxidant capacity of genetically improved farmed tilapia (GIFT), *Oreochromis niloticus*. Aquac Int 25:991–1003
Yan Y, Bhattacharya P, He XJ et al (2012) Duplicated zebrafish co-orthologs of parathyroid hormone-related peptide (PTHrP, Pthlh) play different roles in craniofacial skeletogenesis. J Endocrinol 214:421–435
Yildirim-Aksoy M, Lim C, Li MH et al (2008) Interaction between dietary levels of vitamins C and E on growth and immune responses in channel catfish, *Ictalurus punctatus* (Rafinesque). Aquac Res 39:1198–1209
Yoshida Y, Niki E, Noguchi N (2003) Comparative study on the action of tocopherols and tocotrienols as antioxidant: chemical and physical effects. Chem Phys Lipids 123:63–75

Zanotti G, Folli C, Cendron L et al (2008) Structural and mutational analyses of protein-protein interactions between transthyretin and retinol-binding protein. FEBS J 275:5841–5854
Zhao Y, Zhang K, Giesy JP et al (2015) Families of nuclear receptors in vertebrate models: characteristic and comparative toxicological perspective. Sci Rep 5:8554
Zhou Q, Wang L, Wang H et al (2013) Dietary vitamin E could improve growth performance, lipid peroxidation and non-specific immune responses for juvenile cobia (*Rachycentron canadum*). Aquac Nutr 19:421–429
Zingg J-M (2007) Modulation of signal transduction by vitamin E. Mol Aspects Med 28:481–506

Chapter 7
Nutritional Modulation of Marine Fish Larvae Performance

Sofia Engrola, Cláudia Aragão, Luisa M. P. Valente and Luís E. C. Conceição

Abstract Nutrition shapes the individual physiological pathways prior to hatching, resulting in long-term effects on postnatal growth and physiological functions. The environment, including nutrition, determines the rate of myogenesis, the number and size of muscle fibres, the composition of sub-cellular organelles, the patterns of gene expression, influences protein turnover and the efficiency of protein deposition, among others. Moreover, protein retention efficiency and fish adaptive fitness have been said to be negatively correlated. In addition, high mortalities are normally observed in the marine larval stages in optimized farming conditions (70–80%) affecting fish production. The consequences and implications of early nutrition for aquaculture production is an important challenge for the future. Likewise, optimized feeds need to be provided in sufficient quantities and in a manner adequate to feeding behaviour, to fully express growth potential, while avoiding deterioration of water quality or disease problems caused by excessive feeding. Understanding the mechanisms that control early development and growth and their relation with nutrition are critical for the identification of time windows in development that introduce growth variation, impact growth potential, and affect viability and quality of juveniles.

S. Engrola (✉) · C. Aragão
Centro de Ciências do Mar do Algarve (CCMAR), Universidade do Algarve, Campus de Gambelas, 8005-139 Faro, Portugal
e-mail: sengrola@ualg.pt

L. M. P. Valente
CIIMAR – Centro Interdisciplinar de Investigação Marinha e Ambiental, Universidade do Porto, Terminal de Cruzeiros do Porto de Leixões, Avenida General Norton de Matos, S/N, 4450-208 Matosinhos, Portugal

L. M. P. Valente
ICBAS – Instituto de Ciências Biomédicas de Abel Salazar, Universidade do Porto, Rua dos Bragas, 289, 4050-123 Porto, Portugal

L. E. C. Conceição
Sparos Lda, Área Empresarial de Marim, Lote C, 8700-221 Olhão, Portugal

M. Yúfera (ed.), *Emerging Issues in Fish Larvae Research*,
https://doi.org/10.1007/978-3-319-73244-2_7

Keywords Fish larvae · Nutrition · Programming · Microdiets
Growth performance

7.1 The Role of Nutrients in Promoting Larval Growth

7.1.1 Protein and Amino Acids Larval Nutrition—Modulating Growth Performance

Growth is essentially muscle protein deposition and, dependent of the balance between protein synthesis and degradation (Carter and Houlihan 2001). Since protein synthesis requires that all indispensable amino acids (IAA) occur at an optimal ratio with respect to each other, dietary amino acids imbalances will compromise protein deposition and consequently affect negatively the high growth potential of fish larvae. Due to the methodological difficulties in determining amino acid requirements in fish larvae, indispensable amino acid profiles of whole-larvae or egg protein have been used as an indicator (Conceição et al. 2011). Using this approach, several studies have shown that live feeds used in marine larviculture present imbalanced amino acid profiles (e.g. Aragão et al. 2004; Hamre 2016; Saavedra et al. 2015). Therefore, the use of microdiets designed for a species according to the whole-larvae amino acid profile of the species has been suggested to enhance growth performance.

However, studies performed on large yellow croaker *Larimichthys crocea* and Senegalese sole *Solea senegalensis* larvae did not entirely support this hypothesis. Li et al. (2013) used a Control diet formulated using intact protein (a mixture of white fish meal, shrimp meal and squid meal) as the only protein source and experimental microdiets where crystalline amino acids replaced 40% of fish meal protein-bound nitrogen, and simulated the overall amino acid pattern of white fish meal (WFM) or of large yellow croaker egg, whole-larvae or muscle protein. Both Control and WFM microdiets resulted in higher larval growth and higher protein content than the other microdiets. Therefore, microdiets formulated according to egg or larval amino acid profile were less suitable for large yellow croaker larvae compared with microdiets based on white fish meal amino acid pattern (Li et al. 2013). Studies in Senegalese sole larvae also showed that microdiets with a balanced amino acid profile do not necessarily translate enhanced larval growth performances (Canada et al. 2016a, b). By feeding Senegalese sole with a practical microdiet supplemented with encapsulated crystalline amino acids in order to balance the dietary amino acid profile, a positive impact on the larvae capacity to retain either 1.0 or 6.8 kDa peptides during the metamorphosis climax was observed. However, the positive short-term impact om protein retention observed at metamorphosis did not translate into increased growth (Canada et al. 2016b). A similar study using higher levels of crystalline amino acids to balance the dietary amino acid profile resulted in reduced growth performances of Senegalese sole larvae fed the Balanced diet (Canada et al. 2016a). In the latter study, the expression pattern of key genes regulating myogenesis was affected by the microdiet amino acid profile at metamorphosis, with the expression of myogenin and myosin heavy chain transcript levels being significantly reduced in larvae fed the Balanced diet.

These results were surprising, but several hypotheses have been raised. Besides methodological difficulties in obtaining an effectively amino acid balanced microdiet using crystalline amino acids due to leaching losses, the different rates of absorption between free amino acids (FAA) and proteins (Rønnestad et al. 2000) was suggested to impair results using amino acid balanced microdiets (Canada et al. 2016a, b). However, Li et al. (2013) showed that there were no significant differences on growth and survival of large yellow croaker larvae when fed microdiets with an amino acid profile resembling that of fish meal either using supplementation with crystalline amino acids or intact protein. Another possible explanation raised, is that the whole-larvae amino acid profile changes during ontogeny (Aragão et al. 2004; Saavedra et al. 2006, 2015), thus using a diet which referenced the whole-larvae amino acid composition in a certain period, may not meet the amino acid requirements over the entire larval stage (Li et al. 2013). An increase in voluntary feed intake to compensate the dietary amino acid imbalances has also been proposed (Canada et al. 2016a, b). Another alternative explanation is that even when apparently the amino acid requirements for growth are being covered by amino acid balanced microdiets, these may be insufficient to cover metabolic processes other than growth and energy supply.

Although the major fate of amino acids is towards protein synthesis, it is has been recognized that amino acids regulate key metabolic pathways important not only for growth (Li et al. 2009). Saavedra et al. (2010) showed that further supplementation of amino acid balanced microdiets (based on the larvae whole-body amino acid profile) with aromatic amino acids did not increase growth or survival of white seabream *Diplodus sargus* larvae, but reduced skeletal deformities and mortalities caused by stress. The importance of aromatic amino acids for physiological processes other than growth has also been demonstrated by Pinto et al. (2009) using nutrient flux trials with ^{14}C-phenylalanine and ^{14}C-tyrosine. The authors found that aromatic amino acids were highly retained by Senegalese sole (a species with a very marked metamorphosis process) during pre-metamorphosis and metamorphosis climax, while no significant differences were found for gilthead seabream *Sparus aurata* larvae along its smooth metamorphosis process. Further nutrient flux trials with ^{14}C-tyrosine indicated potential benefits of dietary aromatic amino acid supplementation during metamorphosis climax of Senegalese sole, while no apparent benefit was found for gilthead seabream along the metamorphosis process (Pinto et al. 2010b). Furthermore, even for some dispensable amino acids (DAA), the regulatory role in key metabolic pathways may be translated in growth enhancement. Microdiets supplemented with glutamine promoted antioxidant status and stress resistance capacities of half-smooth tongue sole *Cynoglossus semilaevis* post-larvae through modulation of activities and mRNA expression of antioxidant enzymes (such as glutathione peroxidase and catalase), decreased malondialdehyde contents, and increased mRNA levels of the 70 kDa heat shock protein (*Hsp70*) expression after hypoxia stress. The enhanced antioxidant abilities and increased hypoxia stress resistance were ultimately translated in better survival and growth performance (Liu et al. 2015).

Amino acids that are not incorporated in proteins are often not considered in feed formulations. Taurine, a β-sulfonic-amino acid that only exists in free form, is the most abundant amino acid in the FAA pool from animal blood and tissues. In the recent past, taurine has been considered a required nutrient for normal growth of marine fish (NRC 2011). Several studies have demonstrated that dietary taurine supplementation improved growth performance of several species of marine fish larvae, such as Senegalese sole (Pinto et al. 2010a), northern rock sole *Lepidopsetta polyxystra* (Hawkyard et al. 2014, 2015), red seabream *Pagrus major* (Kim et al. 2016), cobia *Rachycentron canadum* (Salze et al. 2011), greater amberjack *Seriola dumerili* (Matsunari et al. 2013), California yellowtail *Seriola lalandi* (Hawkyard et al. 2016; Rotman et al. 2017), Pacific bluefin tuna *Thunnus orientalis* and yellowfin tuna *Thunnus albacares* (Katagiri et al. 2017). However, for some marine species, dietary taurine supplementation did not improve growth performance, as is the case of gilthead seabream (Pinto et al. 2013) and white seabass *Atractoscion nobilis* (Rotman et al. 2017). This is not entirely surprisingly, since the capacity to biosynthesize taurine has been shown to be highly species-specific (Goto et al. 2003; Yokoyama et al. 2001) and even stage-dependent (Kim et al. 2008).

Taurine is not used by the cells for protein synthesis and the mechanisms by which dietary supplementation may improve larval growth are still largely unknown. Studies using nutrient flux trials showed an increase in amino acid retention concomitant with a higher body taurine content in Senegalese sole larvae fed taurine supplemented microdiets (Pinto et al. 2010a), which may explain the increased growth performance. The larval growth improvement by dietary taurine supplementation was also suggested to be due to an increase in protein synthesis efficiency (Katagiri et al. 2017). Furthermore, Salze et al. (2012) showed that dietary taurine supplementation directed protein translation towards digestive enzymes in cobia larvae during the first two weeks after hatching. The authors suggested that the heightened enzymatic activities may lead to enhanced nutrient availability, thus providing some explanation for the growth improvement in taurine-supplemented larvae.

In line with the improved growth performance, dietary taurine supplements also increased development rates and enhanced metamorphosis of flatfish species (Hawkyard et al. 2014, 2015; Pinto et al. 2010a). This effect may be due to the correlation between metamorphosis and larval size, though Hawkyard et al. (2014) suggested some degree of developmental plasticity associated with dietary taurine concentrations. Dietary taurine supplementation has been shown to affect more than larval growth. Nutrient flux studies showed a higher retention of ^{14}C-methionine in gilthead seabream larvae receiving a taurine supplement (Pinto et al. 2013). These results indicated that although increased methionine retention was not translated into higher growth performance, dietary taurine supplementation may ultimately affect larval metabolism by increasing methionine availability for several important physiological purposes.

For most altricial fish species it has been generally assumed that early-stage larvae have a limited capacity to digest and absorb the native protein sources commonly used in commercial fish feed formulations. Since dietary protein is mainly absorbed as FAA and di- or tri-peptides (Conceição et al. 2011), pre-digested proteins have been introduced in larval feed formulations in order to ease the dietary protein digestion, with the expectation of promoting absorption and further protein synthesis. This formulation strategy has been tested since the 90s, but even in more recent years the results are still unsatisfactory (Cai et al. 2015; Gisbert et al. 2012; Srichanun et al. 2014). A moderate inclusions of hydrolysed protein (<25% inclusion on protein basis) in larval microdiets is still the advisable strategy in order to promote growth and survival. In more recent years the formulations that have included higher levels of proteins hydrolysates presented similar results to previous studies regarding larval performance, like in Atlantic halibut *Hippoglossus hippoglossus* (Kvåle et al. 2009), Asian sea bass *Lates calcarifer* (Srichanun et al. 2014), gilthead seabream (de Vareilles Sommières 2013), large yellow croaker (Cai et al. 2015), spotted wolfish *Anarhichas minor* (Savoie et al. 2011), and white seabream (de Vareilles et al. 2012). This lower larval performance when fed microdiets containing protein hydrolysates has been attributed to a saturation of the peptide transport system in the intestinal brush-border membrane due to overloading of short peptides, and/or to impaired utilization of the fast absorbed FAA and/or di or tri-peptides, with these being used for energy production rather than growth. However, growth of larval Senegalese sole was augmented by a diet with high inclusion level of a protein hydrolysate with a moderate hydrolysis, but not with a diet with a high degree of protein hydrolysis (Canada et al. 2017). Moreover, this diet with moderate protein hydrolysis lead to decreased growth in post-larvae in comparison to intact protein. Therefore, the degree of protein hydrolysis is important, and the optimal dietary protein complexity will likely change throughout larval development.

The impact of dietary protein complexity on the efficiency of protein utilization by Senegalese sole has been studied at short-term in a nutrient flux study (Richard et al. 2015). In this study it was shown that, whereas 1.0 kDa oligopeptides are highly digestible and its retention efficiency is constant throughout development, the digestibility and body retention of larger polypeptides (6.8 kDa) are low in pre-metamorphic larvae, but improve throughout development. Also in Senegalese sole the nutrient flux method was used to assess the microdiet fed larvae digestive capacity to utilize polypeptides with different molecular weight (MW), 1.0 and 7.2 kDa. The PartH microdiet (target peptide MW 5–70 kDa) stimulated growth in metamorphosing larvae, whereas the Intact microdiet (target peptide molecular weight (MW) > 70 kDa) stimulated growth after weaning. The Intact microdiet stimulated the larvae absorption capacity for 1.0 kDa peptides at metamorphosis climax stage, which may have contributed for enhanced growth in later stages (Canada et al. 2017). The conclusion seems quite similar among the different experiments and fish species: a moderate rate of inclusion of protein hydrolysates is advisable, but a higher inclusion may be beneficial when hydrolysates with a lower degree of hydrolysis are used. A more complex protein should be incorporated into the larval microdiets as the maturation of the digestive system progresses and a higher proteolytic capacity is acquired by the fish.

7.1.2 Improving Lipid Utilization

Most research on larval lipid nutrition has been centred on essential fatty acid requirements, since its importance to larval growth and quality has been long-recognized. However, recent studies have shown that total lipid content in microdiets should also be considered to understand larval nutritional requirements. The effects of dietary lipid level on growth and some lipid metabolism related genes were investigated in orange-spotted grouper larvae *Epinephelus coioides* and half-smooth tongue sole, using isonitrogenous microdiets with graded contents of lipid (6–22% dry weight approximately) and fish oil as main lipid source (Li et al. 2016; Yuan et al. 2017). On both studies, microdiets with lipid contents close to the estimated requirement (13.56 and 15.99% dry weight, respectively for sole and grouper) improved larval growth performances. Moreover, dietary lipid content influenced both de novo lipogenesis and lipolysis at transcriptional level. It was shown that larvae may cope with high dietary lipid contents mainly through down-regulating lipogenesis-related gene expression of fatty acid synthesis (FAS) and acetyl-CoA carboxylase beta (ACC2) (Li et al. 2016; Yuan et al. 2017). Furthermore, in orange-spotted grouper larvae, the expression of lipolysis-related genes, lipoprotein lipase (LPL) and hormone sensitive lipase (HSL), significantly increased first and then decreased in relation to the dietary lipid content (Li et al. 2016).

Through genomic approaches it has also been demonstrated that Senegalese sole larvae modulate metabolism to manage dietary lipid levels and obtain the energy and molecules that warrant optimal growth rates (Hachero-Cruzado et al. 2014; Román-Padilla et al. 2017). Senegalese sole larvae were fed live prey enriched with different emulsions resulting in high and low triacylglycerols (TAG) levels (Hachero-Cruzado et al. 2014). Larvae fed high TAG microdiets activated co-ordinately the transcription of apolipoproteins and other related transcripts involved in chylomicron formation, likely to facilitate proper lipid absorption and delivery. In contrast, larvae fed low TAG microdiets showed higher mRNA levels of several pancreatic enzymes and appetite modulators and some intra- and extra-cellular lipases. Senegalese sole fed live preys enriched in oil emulsions differing in fatty acid composition showed also different expression of genes involved in lipid metabolism, absorption and transport (Bonacic et al. 2016). Live preys were enriched in cod liver oil (rich in long-chain polyunsaturated fatty acids, LC-PUFA), linseed oil (rich in *n*-3 PUFA), soybean oil (rich in *n*-6 PUFA), and olive oil (rich in monounsaturated fatty acids). Larval performance was higher when larvae was fed preys containing higher levels of LC-PUFA and *n*-3 PUFA and this was partly explained by an up-regulation of phospholipid metabolism and apolipoprotein synthesis, which resulted in enhanced lipid transport and mobilization, as well as tissue growth and remodelling (Bonacic et al. 2016).

Betancor et al. (2017) used a molecular approach to evaluated lipid metabolism in first-feeding Atlantic bluefin tuna *Thunnus thynnus* larvae fed enriched rotifers (*Brachionus plicatilis*) and copepod nauplii (*Acartia* sp.). These authors suggest that the absolute docosahexaenoic acid (DHA) level may be important for the survival of Atlantic bluefin tuna larvae but that the DHA:EPA (eicosapentaenoic acid) ratio may be relatively more important for larval growth. An up-regulation in peroxisome proliferator-activated receptor gamma (*ppar*γ), fatty acid binding protein 4 (*fabp4*) and acyl coA oxidase (*aco*) expression in rotifer-fed larvae were associated with a compensatory response to reduced growth, whereas lower expression of lipoprotein lipase (*lpl*) denoted reduced lipid utilization.

Recent studies on the LC-PUFA biosynthesis pathway, based on the expression of fatty acyl desaturases and elongases, suggest that Atlantic bluefin tuna (Morais et al. 2011) and Senegalese sole (Morais et al. 2012) larvae are able to biosynthesise DHA from EPA or, in the case of the later, DHA from α-linolenic acid (Navarro-Guillén et al. 2014). It has also been shown that the dietary DHA:EPA ratio may modulate the expression of LC-PUFA biosynthetic pathway genes, namely delta-6 fatty acyl desaturase (*fads2d6*) (Betancor et al. 2017). However, studies by Pinto et al. (2016) testing microdiets with two lipid and two DHA levels concluded that is not advisable to include low DHA (<5% total fatty acids) and lipid (<7% dry matter) levels in weaning diets for Senegalese sole post-larvae, as a reduction of these levels may compromise post-larval growth performance.

Dietary lipids are a major source of energy and provide essential fatty acids and phospholipids, widely acknowledged as critical success factors for larval fish rearing. The importance of phospholipids in marine larval rearing has been the subject of many recent studies. The results from these studies will not be reviewed in this Chapter, since this subject is addressed in Chap. 5. Concerning the essential fatty acids, arachidonic acid (ARA) has been traditionally the less studied, but its potential to affect growth, survival and stress resistance has been previously acknowledged (Bell and Sargent 2003) and recent studies have provided some knowledge on the molecular mechanisms involved. Marine fish larvae seem to tolerate a wide range of dietary ARA:EPA (up to 3.0) ratios (Alves Martins et al. 2011, 2012). The adaptation to dietary ARA levels in marine fish larvae has been shown to involve the modulation of the expression of genes related to eicosanoid synthesis, lipid metabolism and stress response (Alves Martins et al. 2012; Montero et al. 2015). Nevertheless, despite growth and survival were not affected by the range of dietary ARA:EPA ratios, acute stress coping response seem to be more efficient in Senegalese sole post-larvae fed low (0.7) than high (3.0) ARA:EPA ratios (Alves Martins et al. 2011).

7.2 Larval Muscle Plasticity to Nutrition

Muscle tissue comprises 40–60% of the total body mass in most fish and is the major determinant of fish growth. The axial muscle of most teleost species mainly consists of a deep bulk of fast twitch fibres with glycolytic metabolism and rapidly fatigue, and a superficial thin strip of slow-twitch fibres that are fatigue-resistant and have an oxidative metabolism (Sanger and Stoiber 2001). The skeletal muscle characteristics and the way muscle tissue grows are major factors influencing overall growth capacity in fish, as in mammals and poultry (Johnston 1999; Rehfeldt et al. 2011). In teleost, not only skeletal muscle fibre size change after birth in response to environmental factors and physiological conditions, but new fibres may also continue to be recruited into adulthood determining the maximum size attained by a species (Valente et al. 2013; Vélez et al. 2017). Muscle growth is very plastic in fish and there is increasing evidence that early events imprint an individual physiological memory (Campos et al. 2013b, c, 2014), resulting in long-term effects on postnatal growth and physiological function, irreversibly affecting growth potential and final size attained.

The importance of nutritional status in regulating protein metabolism and muscle growth is widely accepted, and the role of macronutrients is clearly implicated in vertebrates developmental programming (McMillen et al. 2008). In mammal models, studies have highlighted a link between nutritional conditioning during early life stages and survival, growth, learning process, lipid and glucose synthesis in later life, suggesting that developmental programming may have an epigenetic component as epigenetic marks such as DNA methylation or histone tail modifications could provide a persistent memory of earlier nutritional states (Lucas 1998; Sharples et al. 2016; Vickers 2014).

In fact, studies on the epigenetic modulation of DNA methylation are a recent trend in fish and up to now no published literature is available concerning nutritional conditioning of muscle growth, and just a few reports related to early temperature exposure. Campos et al. (2014) have suggested that an epigenetic mechanism could promote differential gene expression and modulate Senegalese sole muscle growth in response to different thermal conditions. Different rearing temperatures during the pelagic phase induced changes in the methylation status of the myogenin putative promoter, its mRNA transcript levels and in the expression of *dnmt1* and *dnmt3b* DNA methyltransferases, which catalyse the methylation of CpG dinucleotides, silencing gene expression (Campos et al. 2013b; Fig. 7.1). These changes resulted in alterations in the fast twitch muscle cellularity of Senegalese sole during metamorphosis climax, and influenced subsequent somatic growth in later stages (Campos et al. 2013a). Campos et al. (2014) suggested the pelagic phase in Senegalese sole as a critical developmental time window prone to epigenetic modifications with long-lasting effects on the regulation of myogenesis and subsequent influence on the potential for growth.

The impact of nutritional factors on genetic pathways regulating muscle fibre determination and growth has been poorly studied in fish and the few studies

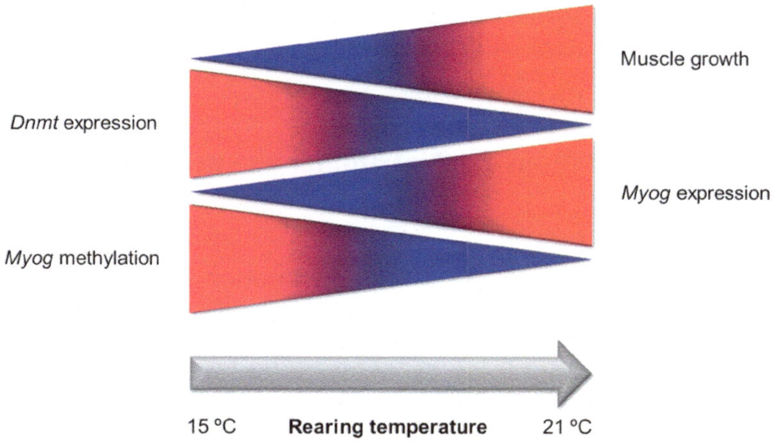

Fig. 7.1 Temperature-induced phenotypic plasticity model of muscle growth in Senegalese sole. A higher rearing temperature lead to lower methylation levels of the myog, which correlated with a decrease in dnmt1 and dnmt3b and an increase in myog expression and muscle growth. Blue and red indicate lower and higher levels of methylation, gene expression and muscle growth, respectively. Original figure in Campos et al. (2013b)

focused on juveniles (Alami-Durante et al. 2010; Campos et al. 2010; Lopes et al. 2017; Valente et al. 2016). In marine fish larvae, Canada (2017) have recently hypothesized that manipulating the formulation of dietary protein might lead to changes in the dietary protein absorption and body retention, and subsequently modify the post-prandial availability of methyl-donors, such methionine, with possible consequences on the DNA methylation and regulation of gene expression in different tissues. In fact, the transcript levels of key genes regulating myogenesis changed between Senegalese sole fed a practical microdiet supplemented with encapsulated crystalline amino acids, during the metamorphosis climax and at the 51 days after hatching (Canada et al. 2016a). The group fed the supplemented diet had lower *dnmt3b* mRNA levels compared to the control group. A lower post-prandial availability of methyl-donors as a result of a lower protein intake might explain the reduced expression of *dnmt's* in those fish. Whether this would also imply an overall DNA hypomethylation in skeletal muscle remains to be clarified. The dietary protein complexity was also shown to induce changes in protein utilization (Canada et al. 2017) and the regulation of myogenesis in Senegalese sole larvae (Canada 2017). Microdiets mostly based on highly hydrolysed protein sources were associated with reduced muscle growth potential and the up-regulation in the transcript levels of genes encoding for de novo DNA methyltransferases. These results support the hypothesis that nutrition may induce changes in post-prandial availability of methyl-donors that might alter the expression of *dnmt's*, with putative consequences on DNA methylation.

In conclusion, the molecular and epigenetic mechanisms underlying skeletal muscle memory in response to early life nutrition are far from being understood in fish. Further research is required to help understanding the 'epi'-memory of skeletal muscle enabling nutritionists to tailor microdiets able to enhance muscle growth.

7.3 New Formulations for Nutritional Modulation in Fish Larvae

Larvae of most marine fish species, and even many freshwater species, rely for a period ranging from one week to one or more months on live prey to grow. Considerable progress was done in microdiet technology in recent years, nevertheless for most marine fish larvae the feeding regime is still based on live prey (rotifers and/or *Artemia* sp.). The progress was delayed until recent years by several factors: low attractiveness of microdiets and consequent low ingestion rates; poor digestibility of microdiets by the larva; high leaching losses of soluble molecules such as FAA, peptides, vitamins and minerals in the microdiets; and difficulties to formulate complete and well balanced microdiets due to lack of knowledge on larval nutritional requirements (Hamre et al. 2013). Larval nutritional requirements are still largely undetermined for marine fish species.

To overcome this limitation on understand of nutritional requirements, larval microdiets are normally formulated by excess of the different essential nutrients, including amino acids, fatty acids, phospholipids, vitamins and minerals. In addition to a formulation that fulfill the nutritional requirements, microdiets for fish larvae need to comply with several structural and functional characteristics: (1) be stable enough to prevent disintegration after the immersion in water, and to avoid major losses of water-soluble nutrients as amino acids, vitamins and minerals—given the small size of the feed particles, with a high surface/volume ratio that reduces the diffusion distance from the core to the surface; (2) have a high floatability so to be accessible to larval fish in the water column for a period as long as possible (3) be attractive to larvae and have an appropriate diameter; and (4) be easily digestible by the larval digestive system. Meeting simultaneously all these requisites is a technological challenge, in particular to balance between stability to prevent excessive leaching of nutrients and high digestibility.

Larval diet formulations rely on protein hydrolysates to improve diet attractiveness, accelerate maturation of the digestive system and improve digestibility. Several diet formulations strategies for marine larvae using protein hydrolysates have been tested with different degrees of success: increasing incorporation rates (de Vareilles Sommières 2013; Savoie et al. 2011; Srichanun et al. 2014); different degree of hydrolysis (Cai et al. 2015; Canada et al. 2017; de Vareilles et al. 2012; Srichanun et al. 2014); different raw material for the protein hydrolysates, from vegetable (Canada et al. 2017), to marine (Cai et al. 2015; de Vareilles et al. 2012; Delcroix et al. 2015; Srichanun et al. 2014), and from animal by-products to yeast (Gisbert et al. 2012; Skalli et al. 2014). In general, as commented in Sect. 7.1.1, moderate inclusion levels of protein hydrolysates give best results, independent of the base raw material used. For instance, the substitution of fish hydrolysates in microdiets for gilthead seabream (Gisbert et al. 2012) and European seabass *Dicentrarchus labrax* (Skalli et al. 2014) by animal by-products or yeast (6–12% in seabream and 9–12% in seabass) showed that is feasible to incorporate alternative sources in larval microdiets at moderate levels. Still, the incorporation success may be species-specific.

The quality of the oils and protein-rich ingredients used in larval microdiets is also of paramount importance. Typically, larval microdiets use the highest quality ingredients available, and nowadays mostly cold-extrusion is used to prepare microdiets so to preserve to the maximum their nutritional value. Marine ingredients (i.e., fish meal, fish oil, squid meal, krill meal, krill oil) are usually the main sources of protein and oils for larval microdiets, but yeast, microalgae, soy lecithin, and vegetable protein concentrates (e.g., wheat gluten, pea protein concentrates, soy isolates/concentrates) are also commonly used.

Larvae of different fish species seem to perform better on different protein sources. When testing a very high quality fish meal, squid meal and a mix of vegetable protein concentrates as main ingredients for gilthead seabream (Conceição et al. 2015), squid meal was the best protein source in what concerns growth performance. However, fish meal seemed to make seabream larvae more robust. This may be due to the presence of some micronutrient(s) in fish meal, and absent in squid meal and the vegetable mixture. Gilthead seabream larvae did not perform well with the mixture of vegetable protein concentrates and both digestive capacity and lipid metabolism were negatively affected, be it due to the presence of an antinutritional factor, and/or to a deficiency in one or more micronutrients present in marine ingredients (Conceição et al. 2015). However, Senegalese sole performed well with a microdiet having the same mixture of vegetable protein concentrates as the main protein source (Pinto et al. 2016). In a study with Ballan wrasse *Labrus bergylta* larvae, a combination of cod muscle meal and shrimp meal performed better than microdiets using fish meal alone, or fish meal in combination with krill hydrolysate or shrimp meal (Kousoulaki et al. 2015).

The larval requirements for *n*-3 HUFAs (highly unsaturated fatty acids) is normally satisfied by DHA-rich fish oils, but it has been shown that at least for gilthead seabream fish oil may be fully replaced by different microalgal products rich in essential fatty acids (Eryalcin et al. 2013). Even though fish meal contains phospholipids, normally these are not sufficient to meet fish larval requirements, even when fish meal is incorporated at high levels. Normally used phospholipids sources are krill oil and soybean lecithin, mainly the latter due to price. When the two were compared in gilthead seabream larvae, krill oil was shown to have a higher effectiveness in promoting survival, growth and skeletal mineralization in comparison with soybean lecithin (Saleh et al. 2015).

Micronutrients as vitamin and minerals are typically supplied as premixes in larval microdiets, which most likely should be different from premixes for juveniles. Studies with gilthead seabream (Atalah et al. 2012), European seabass (Betancor et al. 2013) and meagre *Argyrosomus regius* (El Kertaoui et al. 2017), have pointed out to the importance of having high dietary levels (compared to accepted requirements for juvenile fish) of vitamin E and C to protect DHA and other essential fatty acids from oxidation, in larval microdiets which are typically rich in these highly unsaturated fatty acids. In this regard, the high efficiency of selenium as an antioxidant factor for early weaning diets for gilthead seabream has also been demonstrated (Saleh et al. 2014). Increasing attention has recently been paid to levels and sources of the oxidative stress-related minerals selenium, zinc and manganese that should be supplied in microdiets for marine fish larvae, but if in excess are toxic to larvae. A recent study showed the need to supplement gilthead seabream microdiets based on squid meal and krill oil with one or more of these antioxidant minerals, to promote larval growth and bone mineralization and to prevent skeleton anomalies (Izquierdo et al. 2017). This study also showed that organic minerals are more effective than inorganic forms and nanometals in promoting mineralization and stress resistance.

7.4 Nutritional Programming

Lucas (1998) named the concept programming when during a "critical window" an early stimulus or insult may re-set some physiological pathways with consequences in the long-term. For Aquaculture purposes a "critical window" is usually during egg or larva stage, or during the spawning season (broodstock). The yolk-modification through glucose microinjection in zebrafish *Danio rerio* (Rocha et al. 2014, 2015), prey enrichment with glucose at mouth opening in gilthead seabream (Rocha et al. 2016b), microdiet at early stage formulated to change lipid metabolism in European seabass (Vagner et al. 2007a, b), or protein metabolism in Senegalese sole (Canada et al. 2016a) are different strategies to identify "critical windows" in larvae at early stage of development.

The concept of nutritional programming in fish, is a relative "new" research area in Aquaculture. The first insights in marine fish were in European seabass larvae aiming to improve lipid utilization during juvenile stage (Vagner et al. 2007a, b, 2009). During 39 days seabass larvae were fed with a Low or a High HUFA microdiet, and after 3 months in order to test the concept fish was fed a HUFA experimental deprived diet. The results were auspicious since the relative expression of the delta-6 desaturase was higher in juveniles that during larval stage were fed the Low HUFA diet when compared to juveniles originated from larvae fed the High HUFA diet. Despite this first success experiments regarding nutritional programming in marine fish larvae are still scarce. The data that appear in the following years was acquired using freshwater species as a model (Balasubramanian et al. 2016; Fang et al. 2014; Geurden et al. 2013, 2014; Rocha et al. 2015). Only more recently new data started to be published in gilthead seabream, regarding carbohydrate (Rocha et al. 2016a, b) or lipid utilization (Izquierdo et al. 2015; Turkmen et al. 2017), and in Senegalese sole regarding protein utilization (Canada et al. 2016a).

Nutritional programming as a strategy to enhance nutrient utilization in animal production may lead to higher biological efficiency, meaning a higher production at lower production cost, and minimal environmental impacts. Worldwide aquaculture production is moving to a sustainable production. Meaning that is imperative that marine fish species start to be fed with microdiets containing none or very low amounts of marine ingredients like fish meal and fish oil. Programming carnivorous fish to better utilize dietary carbohydrates would give an advantage to the sector, since the inclusion of carbohydrates in the diet may improve protein and lipid retention by replacing the energy substrate (Kamalam et al. 2017). In gilthead seabream a glucose incorporation during live prey feeding period was able to enhance digestion of carbohydrates and hepatic lipogenesis, suggesting that the early glucose stimuli may alter carbohydrate utilization in seabream juveniles (Rocha et al. 2016a). Enhancing lipid utilization to improve aquaculture sustainability is based on the complete or very low incorporation of fish oil in fish microdiets. For this the nutritional programming should modify the pathways of fatty acid synthesis and the gene expression of key molecular markers such as Δ-6 desaturase. Besides the experiment in European seabass, one experiment was done with gilthead seabream broodstock (Izquierdo et al. 2015). Gilthead seabream broodstock were fed four diets with different replacement levels of fish oil by linseed oil during six months and the progeny was fed microdiets containing low levels of fish meal and fish oil, and high levels of vegetable oil and vegetable meal. A 60% replacement of fish oil by linseed oil in the broodstock diets produced juveniles with a better ability to utilize diet formulated with low fish meal and fish oil and higher growth (Izquierdo et al. 2015).

DNA methylation is mediated by de novo DNA methyltransferases (*dnmt*; *dnmt1*, *dnmt3a* and *dnmt3b*) that are involved in the regulation of gene expression, by repressing transcription. *dnmt3a* and *dnmt3b* are responsible for addition of methyl group de novo, whereas *dnmt1* is responsible for maintenance of DNA methylation patterns during cell replication (Zhang 2015). Epigenetic modifications, such DNA methylation are labile in response to nutritional cues. Very few studies in fish studied how a nutritional cue could change gene regulation during development and juvenile or adult stage (Beaver et al. 2017; Canada 2017; Canada et al. 2016a). Fish larvae have high protein requirements and high amino acid losses for energy production (Conceição et al. 2011), suggesting that dietary IAA levels may be a limiting factor for growth. Senegalese sole larvae were fed with a microdiet where the dietary protein quality was changed by increasing the IAA/DAA ratio, to improve the larvae capacity to retain dietary protein by promoting protein accretion in skeletal muscle. Although *dntm3b* expression was reduced in the larvae fed the IAA:DAA corrected diet, no correlation was established with the regulation of myogenesis (Canada et al. 2016a). Also in Senegalese sole, post-larvae that were fed a diet mostly based on highly hydrolysed protein (high content of 5 kDa peptides) displayed higher transcript levels of *dnmt3a* and *dnmt3b*, which was attributed to increased dietary methionine content (Canada 2017). There are several evidences that nutritional programming is possible in several species, in fish the first results indicate that nutrition change the *dnmt's* expression in the short-term. Future epigenetics studies are needed to identify "critical windows" and the type of stimulus that will re-set the pathways in the long-term.

7.5 Conclusions

Broodstock and early larval nutrition shapes the individual physiological pathways, resulting in long-term effects on postnatal growth and physiological functions. The importance of nutrition in regulating protein metabolism and muscle growth is widely accepted, and the role of macronutrients in vertebrates developmental programming has been clearly demonstrated. Growth is essentially muscle protein deposition and, dependent of the balance between protein synthesis and degradation. Due to the methodological difficulties in determining amino acid requirements in fish larvae, indispensable amino acid profiles of whole-larvae or egg protein have been used as an indicator. Consequently, the use of microdiets designed for a species according to the whole-larvae amino acid profile of the species has been suggested to enhance growth performance. However, during the formulation process one should keep in mind that even when apparently the amino acid requirements for growth are being covered by amino acid balanced microdiets, these may be insufficient to cover metabolic processes other than growth and energy supply. Regarding protein hydrolysates, a moderate rate of inclusion of protein hydrolysates is advisable in larval microdiets, but a higher inclusion may be beneficial only when hydrolysates with a lower degree of hydrolysis is used, and possibly only for the early larval stages. A more complex protein should be incorporated into the larval microdiets as the maturation of the digestive system progresses and a higher proteolytic capacity is acquired by the fish. Most research on larval lipid nutrition has been centred on essential fatty acid and phospholipid requirements and ratios, since its importance to larval growth and quality. However, recent studies have shown that total lipid content, and ratios between phospholipids and triacylglycerols, in microdiets should also be considered to understand larval nutritional requirements. The impact of nutritional factors on genetic pathways regulating muscle fibre determination and growth has been poorly studied in fish, and the few existing studies focused on juveniles. Recent data regarding a possible consequence on the DNA methylation and regulation of gene expression in muscle tissue in marine larvae supports the hypothesis that nutrition may manipulate fish 'epi'-memory. The thorough study of long-term consequences of early nutrition for aquaculture production is an important challenge for the future.

Acknowledgements Sofia Engrola and Cláudia Aragão acknowledge a FCT investigator grant (IF/00482/2014/CP1217/CT0005) and post-doc grant (SFRH/BDP/65578/2009), respectively, funded by the European Social Fund, the Operational Programme Human Potential and the Foundation for Science and Technology of Portugal (FCT—Portugal). This work was funded by Project CCMAR/Multi/04326/2013 (Portugal) from FCT (Portugal).

References

Alami-Durante H, Wrutniak-Cabello C, Kaushik SJ et al (2010) Skeletal muscle cellularity and expression of myogenic regulatory factors and myosin heavy chains in rainbow trout (*Oncorhynchus mykiss*): effects of changes in dietary plant protein sources and amino acid profiles. Comp Biochem Physiol Part A 156(4):561–568. https://doi.org/10.1016/j.cbpa.2010.04.015

Alves Martins D, Engrola S, Morais S et al (2011) Cortisol response to air exposure in *Solea senegalensis* post-larvae is affected by dietary arachidonic acid-to-eicosapentaenoic acid ratio. Fish Physiol Biochem 37(4):733–743. https://doi.org/10.1007/s10695-011-9473-4

Alves Martins D, Rocha F, Martínez-Rodríguez G et al (2012) Teleost fish larvae adapt to dietary arachidonic acid supply through modulation of the expression of lipid metabolism and stress response genes. Br J Nutr 108(5):864–874. https://doi.org/10.1017/s0007114511006143

Aragão C, Conceição LEC, Fyhn HJ et al (2004) Estimated amino acid requirements during early ontogeny in fish with different life styles: gilthead seabream (*Sparus aurata*) and Senegalese sole (*Solea senegalensis*). Aquaculture 242:589–605. https://doi.org/10.1016/j.aquaculture.2004.09.015

Atalah E, Hernandez-Cruz CM, Ganga R et al (2012) Enhancement of gilthead seabream (*Sparus aurata*) larval growth by dietary vitamin E in relation to two different levels of essential fatty acids. Aquac Res 43(12):1816–1827. https://doi.org/10.1111/j.1365-2109.2011.02989.x

Balasubramanian MN, Panserat S, Dupont-Nivet M et al (2016) Molecular pathways associated with the nutritional programming of plant-based diet acceptance in rainbow trout following an early feeding exposure. BMC Genomics 17. http://dx.doi.org/10.1186/s12864-016-2804-1

Beaver LM, Nkrumah-Elie YM, Truong L et al (2017) Adverse effects of parental zinc deficiency on metal homeostasis and embryonic development in a zebrafish model. J Nutr Biochem 43:78–87. https://doi.org/10.1016/j.jnutbio.2017.02.006

Bell JG, Sargent JR (2003) Arachidonic acid in aquaculture feeds: current status and future opportunities. Aquaculture 218(1):491–499. https://doi.org/10.1016/S0044-8486(02)00370-8

Betancor MB, Caballero MJ, Benitez-Santana T et al (2013) Oxidative status and histological changes in sea bass larvae muscle in response to high dietary content of docosahexaenoic acid DHA. J Fish Dis 36(5):453–465. https://doi.org/10.1111/j.1365-2761.2012.01447.x

Betancor MB, Ortega A, de la Gándara F et al (2017) Lipid metabolism-related gene expression pattern of Atlantic bluefin tuna (*Thunnus thynnus* L.) larvae fed on live prey. Fish Physiol Biochem 43(2):493–516. https://doi.org/10.1007/s10695-016-0305-4

Bonacic K, Campoverde C, Sastre M et al (2016) Mechanisms of lipid metabolism and transport underlying superior performance of Senegalese sole (*Solea senegalensis*, Kaup 1858) larvae fed diets containing n-3 polyunsaturated fatty acids. Aquaculture 450:383–396. https://doi.org/10.1016/j.aquaculture.2015.07.009

Cai Z, Li W, Mai K et al (2015) Effects of dietary size-fractionated fish hydrolysates on growth, activities of digestive enzymes and aminotransferases and expression of some protein metabolism related genes in large yellow croaker (*Larimichthys crocea*) larvae. Aquaculture 440:40–47. https://doi.org/10.1016/j.aquaculture.2015.01.026

Campos C, Fernandes JMO, Conceição LEC et al (2013a) Thermal conditions during larval pelagic phase influence subsequent somatic growth of Senegalese sole by modulating gene expression and muscle growth dynamics. Aquaculture 414:46–55. https://doi.org/10.1016/j.aquaculture.2013.07.039

Campos C, Valente LMP, Borges P et al (2010) Dietary lipid levels have a remarkable impact on the expression of growth-related genes in Senegalese sole (*Solea senegalensis* Kaup). J Exp Biol 213(2):200–209. https://doi.org/10.1242/jeb.033126

Campos C, Valente LMP, Conceição LEC et al (2013b) Temperature affects methylation of the myogenin putative promoter, its expression and muscle cellularity in Senegalese sole larvae. Epigenetics 8(4):389–397. https://doi.org/10.4161/epi.24178

Campos C, Valente LMP, Conceição LEC et al (2014) Molecular regulation of muscle development and growth in Senegalese sole larvae exposed to temperature fluctuations. Aquaculture 432:418–425. https://doi.org/10.1016/j.aquaculture.2014.04.035
Campos C, Valente LMP, Conceição LEC et al (2013c) Incubation temperature induces changes in muscle cellularity and gene expression in Senegalese sole (*Solea senegalensis*). Gene 516 (2):209–217. https://doi.org/10.1016/j.gene.2012.12.074
Canada P (2017) Improving growth potential in Senegalese sole (*Solea senegalensis*) through dietary protein: an integrated approach using muscle cellularity, tracer studies and gene expression. Ph.D. thesis, University of Porto, Porto, Portugal
Canada P, Conceição LEC, Mira S et al (2017) Dietary protein complexity modulates growth, protein utilisation and the expression of protein digestion-related genes in Senegalese sole larvae. Aquaculture 479:273–284. https://doi.org/10.1016/j.aquaculture.2017.05.028
Canada P, Engrola S, Mira S et al (2016a) The supplementation of a microdiet with crystalline indispensable amino-acids affects muscle growth and the expression pattern of related genes in Senegalese sole (*Solea senegalensis*) larvae. Aquaculture 458:158–169. https://doi.org/10.1016/j.aquaculture.2016.03.010
Canada P, Engrola S, Richard N et al (2016b) Dietary indispensable amino acids profile affects protein utilization and growth of Senegalese sole larvae. Fish Physiol Biochem 42(6):1493–1508. https://doi.org/10.1007/s10695-016-0235-1
Carter CG, Houlihan DF (2001) Protein synthesis. In: Wright PA, Anderson AJ (eds) Nitrogen excretion. Fish physiology, vol 20. Academic Press, San Diego, pp 31–75
Conceição L, Saleh R, Martos-Sitcha JA et al (2015) Dietary protein source has major impact in growth performance of gilthead seabream larvae. Paper presented at the Aquaculture Europe 2015, 20–23 Oct, Rotterdam, The Netherlands
Conceição LEC, Aragão C, Rønnestad I (2011) Proteins. In: Holt J (ed) Larval fish nutrition. Wiley, UK, pp 83–116
de Vareilles M, Richard N, Gavaia PJ et al (2012) Impact of dietary protein hydrolysates on skeleton quality and proteome in *Diplodus sargus* larvae. J Appl Ichthyol 28(3):477–487. https://doi.org/10.1111/j.1439-0426.2012.01986.x
de Vareilles L, Sommières MDMS (2013) Dietary nitrogen utilisation and protein expression in fish. Ph.D. thesis, University of Algarve, Faro, Portugal
Delcroix J, Gatesoupe F-J, Desbruyères E et al (2015) The effects of dietary marine protein hydrolysates on the development of sea bass larvae, *Dicentrarchus labrax*, and associated microbiota. Aquac Nutr 21:98–104. https://doi.org/10.1111/anu.12139
El Kertaoui N, Hernandez-Cruz CM, Montero D et al (2017) The importance of dietary HUFA for meagre larvae (*Argyrosomus regius*; Asso, 1801) and its relation with antioxidant vitamins E and C. Aquac Res 48(2):419–433. https://doi.org/10.1111/are.12890
Eryalcin KM, Roo J, Saleh R et al (2013) Fish oil replacement by different microalgal products in microdiets for early weaning of gilthead sea bream (*Sparus aurata*, L.). Aquac Res 44(5):819–828. http://dx.doi.org/10.1111/j.1365-2109.2012.03237.x
Fang L, Liang X-F, Zhou Y et al (2014) Programming effects of high-carbohydrate feeding of larvae on adult glucose metabolism in zebrafish, *Danio rerio*. Br J Nutr 111(5):808–818. https://doi.org/10.1017/s0007114513003243
Geurden I, Borchert P, Balasubramanian MN et al (2013) The positive impact of the early-feeding of a plant-based diet on its future acceptance and utilisation in rainbow trout. Plos One 8(12). http://dx.doi.org/10.1371/journal.pone.0083162
Geurden I, Mennigen J, Plagnes-Juan E et al (2014) High or low dietary carbohydrate: protein ratios during first-feeding affect glucose metabolism and intestinal microbiota in juvenile rainbow trout. J Exp Biol 217(19):3396–3406. https://doi.org/10.1242/jeb.106062
Gisbert E, Skalli A, Fernández I et al (2012) Protein hydrolysates from yeast and pig blood as alternative raw materials in microdiets for gilthead sea bream (*Sparus aurata*) larvae. Aquaculture 338:96–104. https://doi.org/10.1016/j.aquaculture.2012.01.007
Goto T, Matsumoto T, Murakami S et al (2003) Conversion of cysteate into taurine in liver of fish. Fish Sci 69:216–218. https://doi.org/10.1046/j.1444-2906.2003.00610.x

Hachero-Cruzado I, Rodríguez-Rua A, Román-Padilla J et al (2014) Characterization of the genomic responses in early Senegalese sole larvae fed diets with different dietary triacylglycerol and total lipids levels. Comp Biochemi Physiol Part D 12:61–73. https://doi.org/10.1016/j.cbd.2014.09.005

Hamre K (2016) Nutrient profiles of rotifers (*Brachionus* sp.) and rotifer diets from four different marine fish hatcheries. Aquaculture 450:136–142. https://doi.org/10.1016/j.aquaculture.2015.07.016

Hamre K, Yúfera M, Rønnestad I et al (2013) Fish larval nutrition and feed formulation: knowledge gaps and bottlenecks for advances in larval rearing. Rev Aquac 5(Suppl. 1):S26–S58. https://doi.org/10.1111/j.1753-5131.2012.01086.x

Hawkyard M, Laurel B, Barr Y et al (2015) Evaluation of liposomes for the enrichment of rotifers (*Brachionus* sp.) with taurine and their subsequent effects on the growth and development of northern rock sole (*Lepidopsetta polyxystra*) larvae. Aquaculture 441:118–125. https://doi.org/10.1016/j.aquaculture.2015.02.012

Hawkyard M, Laurel B, Langdon C (2014) Rotifers enriched with taurine by microparticulate and dissolved enrichment methods influence the growth and metamorphic development of northern rock sole (*Lepidopsetta polyxystra*) larvae. Aquaculture 424–425:151–157. https://doi.org/10.1016/j.aquaculture.2013.12.035

Hawkyard M, Stuart K, Langdon C et al (2016) The enrichment of rotifers (*Brachionus plicatilis*) and *Artemia franciscana* with taurine liposomes and their subsequent effects on the larval development of California yellowtail (*Seriola lalandi*). Aquac Nutr 22(4):911–922. https://doi.org/10.1111/anu.12317

Izquierdo MS, Ghrab W, Roo J et al (2017) Organic, inorganic and nanoparticles of Se, Zn and Mn in early weaning diets for gilthead seabream (*Sparus aurata*; Linnaeus, 1758). Aquac Res 48(6):2852–2867. https://doi.org/10.1111/are.13119

Izquierdo MS, Turkmen S, Montero D et al (2015) Nutritional programming through broodstock diets to improve utilization of very low fishmeal and fish oil diets in gilthead sea bream. Aquaculture 449:18–26. https://doi.org/10.1016/j.aquaculture.2015.03.032

Johnston IA (1999) Muscle development and growth: potential implications for flesh quality in fish. Aquaculture 177(1–4):99–115. https://doi.org/10.1016/s0044-8486(99)00072-1

Kamalam BS, Medale F, Panserat S (2017) Utilisation of dietary carbohydrates in farmed fishes: New insights on influencing factors, biological limitations and future strategies. Aquaculture 467:3–27. https://doi.org/10.1016/j.aquaculture.2016.02.007

Katagiri R, Sasaki T, Diaz A et al (2017) Effect of taurine enrichment in rotifer (*Brachionus* sp.) on growth of larvae of Pacific bluefin tuna *Thunnus orientalis* (Temminck & Schlegel) and yellowfin tuna *T. albacares* (Temminck & Schlegel). Aquac Res 48(6):3013–3031. http://dx.doi.org/10.1111/are.13134

Kim S-K, Matsunari H, Takeuchi T et al (2008) Comparison of taurine biosynthesis ability between juveniles of Japanese flounder and common carp. Amino Acids 35(1):161–168. https://doi.org/10.1007/s00726-007-0600-6

Kim Y-S, Sasaki T, Awa M et al (2016) Effect of dietary taurine enhancement on growth and development in red sea bream *Pagrus major* larvae. Aquac Res 47(4):1168–1179. https://doi.org/10.1111/are.12573

Kousoulaki K, Bogevik AS, Skiftesvik AB et al (2015) Marine raw material choice, quality and weaning performance of Ballan wrasse (*Labrus bergylta*) larvae. Aquac Nutr 21(5):644–654. https://doi.org/10.1111/anu.12186

Kvåle A, Harboe T, Mangor-Jensen A et al (2009) Effects of protein hydrolysate in weaning diets for Atlantic cod (*Gadus morhua* L.) and Atlantic halibut (*Hippoglossus hippoglossus* L.). Aquac Nutr 15:218–227. https://doi.org/10.1111/j.1365-2095.2008.00586.x

Li P, Mai K, Trushenski J et al (2009) New developments in fish amino acid nutrition: towards functional and environmentally oriented aquafeeds. Amino Acids 37(1):43–53. https://doi.org/10.1007/s00726-008-0171-1

Li S, Mai K, Xu W et al (2016) Effects of dietary lipid level on growth, fatty acid composition, digestive enzymes and expression of some lipid metabolism related genes of orange-spotted

grouper larvae (*Epinephelus coioides* H.). Aquac Res 47(8):2481–2495. http://dx.doi.org/10.1111/are.12697
Li W, Ai Q, Mai K et al (2013) Effects of dietary amino acid patterns on growth and protein metabolism of large yellow croaker (*Larimichthys crocea*) larvae. Aquaculture 406:1–8. https://doi.org/10.1016/j.aquaculture.2013.04.029
Liu J, Mai K, Xu W et al (2015) Effects of dietary glutamine on survival, growth performance, activities of digestive enzyme, antioxidant status and hypoxia stress resistance of half-smooth tongue sole (*Cynoglossus semilaevis* Günther) post larvae. Aquaculture 446:48–56. https://doi.org/10.1016/j.aquaculture.2015.04.012
Lopes G, Castro LFC, Valente LMP (2017) Total substitution of dietary fish oil by vegetable oils stimulates muscle hypertrophic growth in Senegalese sole and the upregulation of *fgf6*. Food Funct 8(5):1869–1879. https://doi.org/10.1039/c7fo00340d
Lucas A (1998) Programming by early nutrition: an experimental approach. In: Symposium: the effects of childhood diet on adult health and disease, New Orleans, LA, 8 April 1997 1998. pp 401S–406S
Matsunari H, Hashimoto H, Iwasaki T et al (2013) Effect of feeding rotifers enriched with taurine on the growth and survival of larval amberjack *Seriola dumerili*. Fish Sci 79(5):815–821. https://doi.org/10.1007/s12562-013-0657-y
McMillen IC, MacLaughlin SM, Muhlhausler BS et al (2008) Developmental origins of adult health and disease: the role of periconceptional and foetal nutrition. Basic Clin Pharmacol Toxicol 102(2):82–89. https://doi.org/10.1111/j.1742-7843.2007.00188
Montero D, Terova G, Rimoldi S et al (2015) Modulation of the expression of components of the stress response by dietary arachidonic acid in European sea bass (*Dicentrarchus labrax*) larvae. Lipids 50(10):1029–1041. https://doi.org/10.1007/s11745-015-4057-1
Morais S, Castanheira F, Martinez-Rubio L et al (2012) Long chain polyunsaturated fatty acid synthesis in a marine vertebrate: Ontogenetic and nutritional regulation of a fatty acyl desaturase with Δ4 activity. BBA Mol Cell Biol L 1821(4):660–671. http://dx.doi.org/10.1016/j.bbalip.2011.12.011
Morais S, Mourente G, Ortega A et al (2011) Expression of fatty acyl desaturase and elongase genes, and evolution of DHA:EPA ratio during development of unfed larvae of Atlantic bluefin tuna (*Thunnus thynnus* L.). Aquaculture 313(1):129–139. http://dx.doi.org/10.1016/j.aquaculture.2011.01.031
Navarro-Guillén C, Engrola S, Castanheira F et al (2014) Effect of varying dietary levels of LC-PUFA and vegetable oil sources on performance and fatty acids of Senegalese sole post larvae: puzzling results suggest complete biosynthesis pathway from C18 PUFA to DHA. Comp Biochem Physiol Part B 167:51–58. https://doi.org/10.1016/j.cbpb.2013.10.001
NRC (2011) Nutrient requirements of fish and shrimp. Washington, DC, USA
Pinto W, Engrola S, Santos A et al (2016) Can Senegalese sole post-larvae effectively grow on low dietary DHA and lipid levels during weaning? Aquaculture 463:234–240. https://doi.org/10.1016/j.aquaculture.2016.05.027
Pinto W, Figueira L, Dinis MT et al (2009) How does fish metamorphosis affect aromatic amino acid metabolism? Amino Acids 36:177–183. https://doi.org/10.1007/s00726-008-0045-6
Pinto W, Figueira L, Santos A et al (2013) Is dietary taurine supplementation beneficial for gilthead seabream (*Sparus aurata*) larvae? Aquaculture 384–387:1–5. https://doi.org/10.1016/j.aquaculture.2012.12.011
Pinto W, Ribeiro L, Yúfera M et al (2010a) Dietary taurine supplementation enhances metamorphosis and growth potential of *Solea senegalensis* larvae. Aquaculture 309(1–4):159–164. https://doi.org/10.1016/j.aquaculture.2010.08.031
Pinto W, Rodrigues V, Dinis MT et al (2010b) Can dietary aromatic amino acid supplementation be beneficial during fish metamorphosis? Aquaculture 310:200–205. https://doi.org/10.1016/j.aquaculture.2010.10.026
Rehfeldt C, Te Pas MFW, Wimmers K et al (2011) Advances in research on the prenatal development of skeletal muscle in animals in relation to the quality of muscle-based food.

I. Regulation of myogenesis and environmental impact. Animal 5(5):703–717. https://doi.org/10.1017/S1751731110002089

Richard N, Engrola S, Palma PS et al (2015) Assessment of protein digestive capacity and metabolic utilisation during ontogeny of Senegalese sole larvae: a tracer study using in vivo produced radiolabelled polypeptide fractions. Aquaculture 441:35–44. https://doi.org/10.1016/j.aquaculture.2015.02.003

Rocha F, Dias J, Engrola S et al (2014) Glucose overload in yolk has little effect on the long-term modulation of carbohydrate metabolic genes in zebrafish (*Danio rerio*). J Exp Biol 217 (7):1139–1149. https://doi.org/10.1242/jeb.095463

Rocha F, Dias J, Engrola S et al (2015) Glucose metabolism and gene expression in juvenile zebrafish (*Danio rerio*) challenged with a high carbohydrate diet: effects of an acute glucose stimulus during late embryonic life. Br J Nutr 113(3):403–413. https://doi.org/10.1017/s0007114514003869

Rocha F, Dias J, Geurden I et al (2016a) Dietary glucose stimulus at larval stage modifies the carbohydrate metabolic pathway in gilthead seabream (*Sparus aurata*) juveniles: an in vivo approach using 14C-starch. Comp Biochem Physiol A 201:189–199. https://doi.org/10.1016/j.cbpa.2016.07.016

Rocha F, Dias J, Geurden I et al (2016b) High-glucose feeding of gilthead seabream (*Sparus aurata*) larvae: effects on molecular and metabolic pathways. Aquaculture 451:241–253. https://doi.org/10.1016/j.aquaculture.2015.09.015

Román-Padilla J, Rodríguez-Rúa A, Ponce M et al (2017) Effects of dietary lipid profile on larval performance and lipid management in Senegalese sole. Aquaculture 468 Part 1:80–93. http://doi.org/10.1016/j.aquaculture.2016.10.005

Rønnestad I, Conceição LEC, Aragão C et al (2000) Free amino acids are absorbed faster and assimilated more efficiently than protein in postlarval Senegal sole (*Solea senegalensis*). J Nutr 130:2809–2812

Rotman F, Stuart K, Drawbridge M (2017) Effects of taurine supplementation in live feeds on larval rearing performance of California yellowtail *Seriola lalandi* and white seabass *Atractoscion nobilis*. Aquac Res 48(3):1232–1239. https://doi.org/10.1111/are.12964

Saavedra M, Candeias-Mendes A, Castanho S et al (2015) Amino acid profiles of meagre (*Argyrosomus regius*) larvae: towards the formulation of an amino acid balanced diet. Aquaculture 448:315–320. https://doi.org/10.1016/j.aquaculture.2015.06.003

Saavedra M, Conceição LEC, Barr Y et al (2010) Tyrosine and phenylalanine supplementation on *Diplodus sargus* larvae: effect on growth and quality. Aquac Res 41(10):1523–1532. https://doi.org/10.1111/j.1365-2109.2009.02446.x

Saavedra M, Conceição LEC, Pousão-Ferreira P et al (2006) Amino acid profiles of *Diplodus sargus* (L., 1758) larvae: implications for feed formulation. Aquaculture 261:587–593. https://doi.org/10.1016/j.aquaculture.2006.08.016

Saleh R, Betancor MB, Roo J et al (2015) Effect of krill phospholipids versus soybean lecithin in microdiets for gilthead seabream (*Sparus aurata*) larvae on molecular markers of antioxidative metabolism and bone development. Aquac Nutr 21(4):474–488. https://doi.org/10.1111/anu.12177

Saleh R, Betancor MB, Roo J et al (2014) Selenium levels in early weaning diets for gilthead seabream larvae. Aquaculture 426:256–263. https://doi.org/10.1016/j.aquaculture.2014.02.011

Salze G, Craig SR, Smith BH et al (2011) Morphological development of larval cobia *Rachycentron canadum* and the influence of dietary taurine supplementation. J Fish Biol 78 (5):1470–1491. https://doi.org/10.1111/j.1095-8649.2011.02954.x

Salze G, McLean E, Craig SR (2012) Dietary taurine enhances growth and digestive enzyme activities in larval cobia. Aquaculture 362–363:44–49. https://doi.org/10.1016/j.aquaculture.2012.07.021

Sanger AM, Stoiber W (2001) Muscle fiber diversity and plasticity. In: Johnston IA (ed) Muscle development and growth, vol 18. Academic Press, London, pp 187–250

Savoie A, Le Francois NR, Lamarre SG et al (2011) Dietary protein hydrolysate and trypsin inhibitor effects on digestive capacities and performances during early-stages of spotted wolffish: suggested mechanisms. Comp Biochem Physiol A 158(4):525–530. https://doi.org/10.1016/j.cbpa.2010.12.017

Sharples AP, Stewart CE, Seaborne RA (2016) Does skeletal muscle have an 'epi'-memory? The role of epigenetics in nutritional programming, metabolic disease, aging and exercise. Aging Cell 15(4):603–616. https://doi.org/10.1111/acel.12486

Skalli A, Zambonino-Infante JL, Kotzamanis Y et al (2014) Peptide molecular weight, distribution of soluble protein fraction affects growth performance and quality in European sea bass (*Dicentracrchus labrax*) larvae. Aquac Nutr 20(2):118–131. https://doi.org/10.1111/anu.12058

Srichanun M, Tantikitti C, Kortner TM et al (2014) Effects of different protein hydrolysate products and levels on growth, survival rate and digestive capacity in Asian seabass (*Lates calcarifer* Bloch) larvae. Aquaculture 428:195–202. https://doi.org/10.1016/j.aquaculture.2014.03.004

Turkmen S, Castro PL, Caballero MJ et al (2017) Nutritional stimuli of gilthead seabream (*Sparus aurata*) larvae by dietary fatty acids: effects on larval performance, gene expression and neurogenesis. Aquac Res 48(1):202–213. https://doi.org/10.1111/are.12874

Vagner M, Robin JH, Zambonino-Infante JL et al (2009) Ontogenic effects of early feeding of sea bass (*Dicentrarchus labrax*) larvae with a range of dietary n-3 highly unsaturated fatty acid levels on the functioning of polyunsaturated fatty acid desaturation pathways. Br J Nutr 101 (10):1452–1462. https://doi.org/10.1017/s0007114508088053

Vagner M, Robin JH, Zambonino Infante JL et al (2007a) Combined effects of dietary HUFA level and temperature on sea bass (*Dicentrarchus labrax*) larvae development. Aquaculture 266:179–190. https://doi.org/10.1016/j.aquaculture.2007.02.040

Vagner M, Zambonino Infante JL, Robin JH et al (2007b) Is it possible to influence European sea bass (*Dicentrarchus labrax*) juvenile metabolism by a nutritional conditioning during larval stage? Aquaculture 267:165–174. https://doi.org/10.1016/j.aquaculture.2007.01.031

Valente LMP, Cabral EM, Sousa V et al (2016) Plant protein blends in diets for Senegalese sole affect skeletal muscle growth, flesh texture and the expression of related genes. Aquaculture 453:77–85. https://doi.org/10.1016/j.aquaculture.2015.11.034

Valente LMP, Moutou KA, Conceição LEC et al (2013) What determines growth potential and juvenile quality of farmed fish species? Rev Aquac 5(Suppl. 1):S168–S193. https://doi.org/10.1111/raq.12020

Vélez EJ, Lutfi E, Azizi S et al (2017) Understanding fish muscle growth regulation to optimize aquaculture production. Aquaculture 467:28–40. https://doi.org/10.1016/j.aquaculture.2016.07.004

Vickers MH (2014) Early life nutrition, epigenetics and programming of later life disease. Nutrients 6(6):2165–2178. https://doi.org/10.3390/nu6062165

Yokoyama M, Takeuchi T, Park GS et al (2001) Hepatic cysteinesulphinate decarboxylase activity in fish. Aquac Res 32(s1):216–220. https://doi.org/10.1046/j.1355-557x.2001.00017.x

Yuan Y, Li S, Zhang L et al (2017) Influence of dietary lipid on growth performance and some lipogenesis-related gene expression of tongue sole (*Cynoglossus semilaevis*) larvae. Aquac Res 48(3):767–779. https://doi.org/10.1111/are.12921

Zhang N (2015) Epigenetic modulation of DNA methylation by nutrition and its mechanisms in animals. Anim Nutr 1(3):144–151. https://doi.org/10.1016/j.aninu.2015.09.002

Chapter 8
Fish Pigmentation. A Key Issue for the Sustainable Development of Fish Farming

Laura Cal, Paula Suarez-Bregua, Paloma Moran, José Miguel Cerdá-Reverter and Josep Rotllant

Abstract Pigment pattern defects are one of the leading causes of performance loss in aquaculture. Thus, despite the improved methodologies for farming of marine and fresh water fish species, there is still a high incidence of larvae with pigmentation abnormalities. The reason for the relatively high incidence of color abnormalities is unknown but is probably the result of complex interactions between genetic and environmental factors. Nevertheless, and in spite of the importance of the problem, there is scanty information about the pigmentation pattern development in aquaculture fish species, which makes the control of this problem difficult. In mammals, the pigment pattern is mainly dependent on the type of melanin that is synthesized and on the shape of melanocyte. However, in fish the pigment patterns result by the combination and patterned distribution of different pigment cell types, each expressing chemically distinct pigment. Thus showing that the cellular basis for pigment pattern formation is different in the two groups. Therefore, better understanding of the molecular regulation of pigment pattern development and pigment abnormalities in fish will lead to new approaches to increase the efficiency of fish farming by reducing pigment anomalies in aquaculture. This chapter will summarize the current state of knowledge of the processes through which fish coloration is produced and controlled. We will summarize the available data, guiding the reader through the cellular and molecular determinants of fish pigmentation. We will also explore the current research of pigment abnormalities observed in fish species and the possible reasons of their high prevalence in commercial fish farming.

Keywords Chromatophores · Pigment pattern · Fish · Malpigmentation

L. Cal · P. Suarez-Bregua · J. Rotllant (✉)
Instituto de Investigaciones Marinas (IIM-CSIC), Vigo, Spain
e-mail: rotllant@iim.csic.es

P. Moran
Departamento genética, Universidad de Vigo, Vigo, Spain

J. M. Cerdá-Reverter
Instituto de Acuicultura de Torre la Sal (IATS-CSIC), Castellón, Spain

M. Yúfera (ed.), *Emerging Issues in Fish Larvae Research*,
https://doi.org/10.1007/978-3-319-73244-2_8

8.1 Pigmentation

Color patterns in animal species are involved in a wide range of functions. Although the significance of animal coloration is often poorly understood, pigment patterns have been associated with camouflage (Belk and Smith 1996; Dice and Blossom 1937; Protas and Patel 2008; Rudh and Qvarnström 2013), foraging success (Tso et al. 2002), thermoregulation (Ellis 1980; Rudh and Qvarnström 2013), photo-protection (Brenner and Hearing 2008; Rudh and Qvarnström 2013) or mate selection (Bajer et al. 2011; Kodric-Brown and Nicoletto 2001; Maan and Sefc 2013; Protas and Patel 2008). Coloration is composed of structural and pigmentary colors. Structural colors are produced by the interaction between the light and tissue nanostructures (Parker and Martini 2006; Roberts et al. 2012). Examples of structural colors are light reflection in iridophores (fish, amphibians and reptiles) (Kawaguti 1965; Olsson et al. 2013; Rudh and Qvarnström 2013), light scattering in bird barbules (Roulin and Ducrest 2013), or diffraction gratings in antenna hairs of some crustaceans (Parker and Martini 2006). Pigmentary colors are based on chemical pigments (Olsson et al. 2013; Prum 2006; Roulin and Ducrest 2013), deposited in specific organelles (chromatosomes) contained within pigment cells (chromatophores) (Fujii 2000). Pigmentary colors are frequently associated with structural colors since pigment cells are distributed in layers in the integument structures (Olsson et al. 2013; Prum 2006; Roulin and Ducrest 2013). Pigment patterns are primarily generated by the organized distribution of neural crest-derived pigment cells or chromatophores. In mammals, the color pattern is determined by the distribution and production of two pigments, pheomelanin (red/yellow) and eumelanin (brown/black) by a single dedicated cell type, the melanocyte. In fish and specially teleosts, the diversity and complexity of pigment patterns are remarkably more complex, which have up to six different types of chromatophores with distinctive pigments. Teleost genomic duplication (TGD) and the functional divergence of pigmentation ohnologue genes have been suggested as the possible causes of this diversity (Braasch et al. 2008). The coloration of an adult fish is determined by the superposition of two different pigmentation patterns. For example, it has been shown that zebrafish *Danio rerio* have two distinct adult pigment patterning mechanisms—an ancient dorsal-ventral patterning mechanism based on dorsal-ventral differential *asip1* gene expression, and a more recent striping mechanism based on cell-cell interactions (Ceinos et al. 2015). However, these genetically shaped pigment-patterning mechanisms can be changed to some extent in response to different environmental stimuli and represent one of the most captivating features in fish pigmentation.

Additionally, fish can exhibit diverse patterns of pigmentation depending on their state of development. The larval color phenotype, for example, is commonly different from the adult phenotype (Kelsh et al. 2009; Parichy et al. 2011). Many pigmentary defects may result from alterations during metamorphosis from the larval phenotype to the adult phenotype because an apparently normal body color pattern in fish larvae occasionally results in abnormal adult pigmentation (Bolker and Hill 2000; Darias et al. 2013b; Ceinos et al. 2015).

8.1.1 Pigment Cells

Pigment cells derive from neural crest cells, a type of multipotent stem cell which produce intermediate precursors of specific cell types (Le Douarin and Dupin 2003). Those intermediate precursors migrate to their final position, where they develop to pigment cells, glia cells and neurons of the peripheral system, among others (Le Douarin and Dupin 2003; Le Douarin and Kalcheim 1999; Raible and Eisen 1994).

Pigment pattern formation in vertebrates has been widely studied. Mammals and birds only have one pigment cell type in the skin, the melanocyte, which produces two types of melanin: dark/brown eumelanin or yellow/red pheomelanin (Barsh 1996; Hubbard et al. 2010; Lin and Fisher 2007). In contrast, amphibians, reptiles and fish present several pigment cell types (chromatophores), each one synthesizing a chemically distinct pigment: melanophores (eumelanin), xanthophores and erythrophores (pteridines and/or carotenoids), and iridophores (guanine light-reflecting platelets) (Bagnara and Matsumoto 2006). Interestingly, contrary to birds and mammals, pheomelanin is not produced in fish or reptiles (Kottler et al. 2015; Olsson et al. 2013). Additionally, some fish species also show leucophores (guanine light-reflecting platelets) (Fujii 1993) and cyanophores (unknown blue pigment) (Bagnara and Matsumoto 2006; Goda and Fujii 1995; Kelsh 2004).

The pigment pattern in mammals is mainly dependent on the type of melanin that is synthesized in melanocytes and also on the shape of melanocytes (Ito and Wakamatsu 2011; Roulin and Ducrest 2013; Slominski 2004) (Fig. 8.1a). Fish pigment patterns result from the combination and patterned distribution of different pigment cell types (Fig. 8.1b) (Fujii 2000; Kelsh 2004; Kelsh et al. 2009; Olsson et al. 2013).

The dark blue coloration of stripes in zebrafish, for example, is formed by four layers of pigment cells. The most external layer is composed of stellate and pale yellow xanthophores, which is followed by a thin layer of "loose" bluish type S iridophores (uniform size and high number of platelets). A melanophore layer appears beneath of iridophores type S, and the deepest layer is an iridophores Type L (large size and low number of platelets) layer (Frohnhöfer et al. 2013; Hirata et al. 2003; Mahalwar et al. 2014; Watanabe and Kondo 2015). Conversely, the iridescent/golden coloration of interstripes in zebrafish is achieved by two layers: a layer of dense xanthophores covering the deepest layer of dense iridophores type S (Fig. 8.2) (Frohnhöfer et al. 2013; Hirata et al. 2003; Mahalwar et al. 2014; Watanabe and Kondo 2015). A similar organization of pigment cells has been found in Turbot *Scophthalmus maximus*. Turbot melanophores and xanthophores are found in the most superficial pigment layer and the iridophores in the deepest layer in dorsal skin; only a single layer of iridophores exists in light ventral

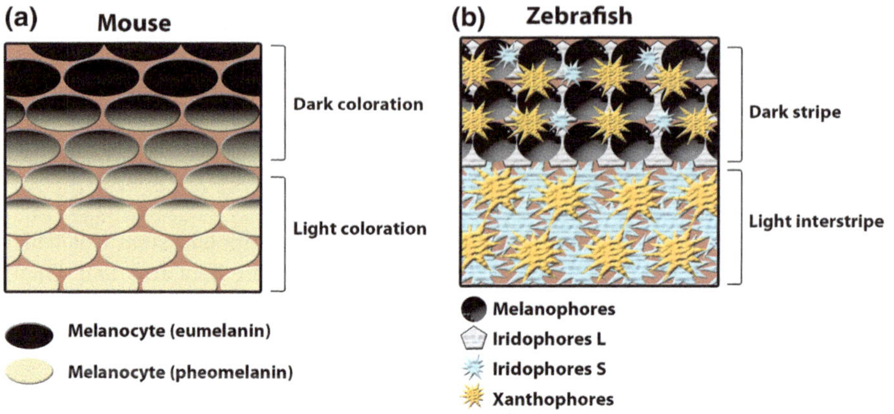

Fig. 8.1 Comparisons of the cellular composition of adult color pattern in mammals and fish. **a** In mouse, melanocytes contain melanosomes that can produce eumelanin (dark pigment) or pheomelanin (light pigment), different ratios resulting in darker or lighter coloration. **b** In zebrafish, the adult pigment pattern is based on dark stripes and light interstripes. Different fish chromatophores produce chemically different pigments, so the fish pigment pattern is obtained by a specific distribution of different pigment cells types. Dark stripes are formed by four layers of chromatophores: type S iridophores (high number and uniform size of platelets), xanthophores, melanophores and type L iridophores (low number and large size of platelets), while light interstripes are composed of two layers: a xanthophore layer and a type S iridophore layer. Adapted from Kelsh (2004) and Irion et al. (2016)

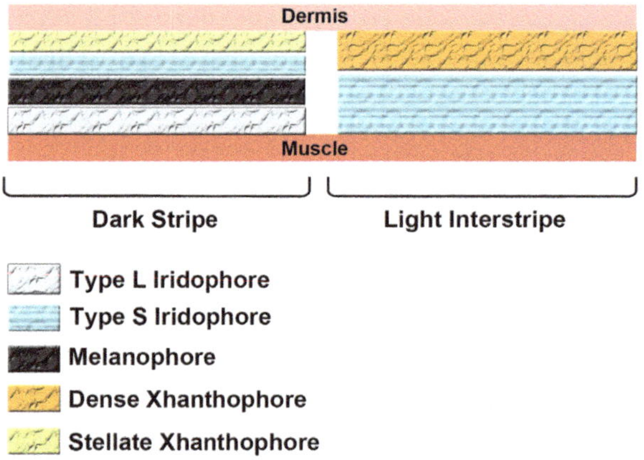

Fig. 8.2 Scheme of layered organization of pigment cells in the striped pattern of zebrafish. Stripes are formed by the following layers from top to bottom: a thin layer of stellate xanthophores, a thin layer of "loose" S-iridophores, a layer of melanophores and a final layer of L-iridophores. Interstripes are organized in two layers: a more superficial dense xanthophore layer and a deeper dense S-iridophore layer

skin (Faílde et al. 2014). These different colorations of dorsal skin in turbot are due to different quantities of pigment cell types. Dark-brown turbot show a higher number of melanophores, yellowish-brown turbot a higher number of xanthophores and a greyish-brown turbot a similar number of melanophores and xanthophores (Faílde et al. 2014). The specific organization and amounts of the different pigment cell types are fundamental for the differences in pigment patterns and colorations.

8.1.2 The Cell-Cell Interaction Mechanism (Turing-Mechanism)

The cellular and molecular mechanisms underlying pigment pattern formation have been widely studied in the last decade. In fish, most studies on pigment patterning have focused on the stripe formation of zebrafish, in which a core striping mechanism dependent upon interactions between different pigment cell types has been identified (Frohnhöfer et al. 2013; Maderspacher and Nüsslein-Volhard 2003; Singh et al. 2014; Takahashi and Kondo 2008). Each pigment cell type tends to completely cover the skin; however, the presence of different pigment cell types forces them to interact with each other, producing the stripe pattern (Frohnhöfer et al. 2013; Irion et al. 2016; Mahalwar et al. 2016). Thus, zebrafish iridophores from the first interstripe are responsible for setting the pattern of adult stripe formation at the beginning of adulthood (Frohnhöfer et al. 2013), and the complete stripe pattern results from homotypic and heterotypic interactions between every pigment cell (Frohnhöfer et al. 2013; Mahalwar et al. 2016; Walderich et al. 2016). Iridophores and melanophores present short-range repulsion forces, which cause melanophore aggregation; moreover, iridophores and xanthophores exhibit mutual attraction, while xanthophores and melanophores repel each other (Frohnhöfer et al. 2013). Iridophores and xanthophores have a positive long-range effect on melanophore aggregation and survival (Frohnhöfer et al. 2013; Irion et al. 2016). Furthermore, feedback mechanisms also interact between pigment cells. Thus, iridophores are required for xanthophore organization, while xanthophores and melanophores are responsible for the localization of iridophores (Frohnhöfer et al. 2013; Irion et al. 2014, 2016). Additionally, the specific shape-change of each chromatophore type (dense S-iridophores and "loose" S-iridophores, dense xanthophores and stellate xanthophores) is essential for the appropriate stripe pattern formation (Fadeev et al. 2015; Irion et al. 2016; Mahalwar et al. 2014; Watanabe and Kondo 2015). Similar to the zebrafish model, Senegal sole *Solea senegalensis* chromatophores show similar interactions. Senegal sole iridophores are restricted to melanophore-free areas, and xanthophores seems to repel melanophores (Darias et al. 2013a). Therefore, although some of the cellular and molecular features of pattern formation still remain to be elucidated, present evidence strongly suggests that the underlying mechanism is mathematically equivalent to the mechanism postulated by Adan Turing more than half a century ago (Watanabe and Kondo 2015).

8.1.3 The Countershading Pattern Mechanism

Most of vertebrate pigment patterns include a light ventrum and dark dorsum. Although the most thoroughly studied mechanism in fish is the striped pattern of zebrafish, it has only recently been demonstrated that zebrafish use two distinct adult pigment patterning mechanisms: a more recent striping mechanism based on cell-cell interactions, and an ancient dorsal-ventral countershading pattern mechanism that is present in all vertebrates (Ceinos et al. 2015). The agouti-signaling protein has been identified as the paracrine factor that drives this common dorsal-ventral countershading pattern in mammals and also in birds, (Chandramohan et al. 2013; Fontanesi et al. 2012; Le Pape et al. 2008; Lu et al. 1994; Siracusa 1994; Norris and Whan 2008; Oribe et al. 2012). A mammalian agouti orthologue has also been identified in fish *(asip1)* (Cerdá-Reverter et al. 2005; Kurokawa et al. 2006). In amphibians, the melanization inhibiting factor (MIF) (Fukuzawa and Ide 1988; Fukuzawa et al. 1995) has been proposed as the potential mammalian agouti orthologue (Cerdá-Reverter et al. 2005). All this suggests that the melanocortin system is the key system responsible for the dorsal-ventral countershading pattern in vertebrates.

The melanocortin system acts through a type of G-protein couplet receptor known as melanocortin receptors (MCRs). Melanocortin 1 receptor (MC1R) is usually expressed in the skin and contributes to pigmentation in mammals and fish (Cone 2006; Cooray and Clark 2011; Gross et al. 2009; Sánchez et al. 2010). MC1R is typically identified as the α-MSH receptor. MC1R is activated by α-MSH, which induces cAMP production via adenylyl cyclase (AC) activation in the melanocyte (García-Borrón et al. 2005; Sánchez et al. 2010). In mammals, it is well known that this increase in cAMP activates protein kinase A (PKA), promoting microphthalmia-associated transcription factor (*mitf*) expression, which results in the activation of the melanin synthesis pathway and de novo dark/brown eumelanin production (Buscà and Ballotti 2000; García-Borrón et al. 2005) (Fig. 8.3). Thus, several polymorphisms in *Mc1r* gene are related to different pigment phenotypes in mammals (Dun et al. 2007; Newton et al. 2000, 2007; Robbins et al. 1993; Valverde et al. 1995).

As mentioned above, Mc1r is also involved in fish pigmentation (Cerdá-Reverter et al. 2011; Fujii 2000) and one polymorphism has been detected which appears to produce hypopigmentation in fish (Gross et al. 2009). The α-Msh also stimulates melanophore proliferation and melanin production in fish (Fujii 2000; Yamanome et al. 2007). Mc1r is also stimulated by α-Msh, increasing cytosolic levels of cAMP, which activates Pka (Sánchez et al. 2010; Sugimoto et al. 1997) and, as in mammals, *mitf* gene induces melanization in fish (Lister et al. 1999). However, the best known function of α-Msh is the rapid dispersion of melanosome in fish

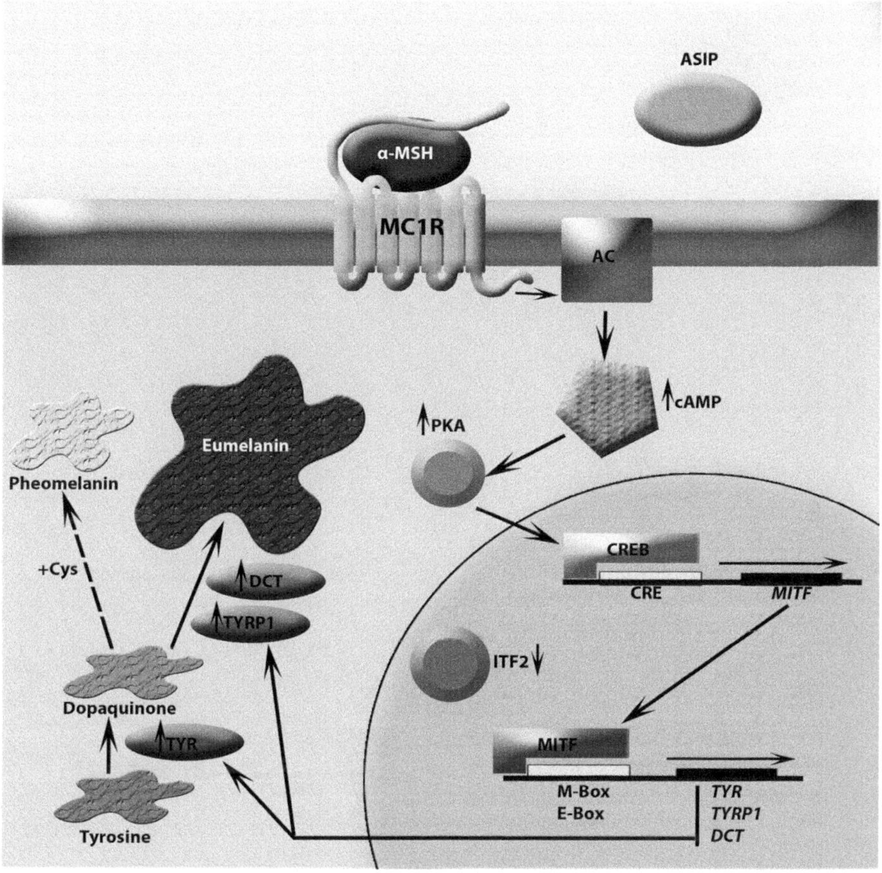

Fig. 8.3 Schematic representation of the melanogenic effects of α-melanocyte-stimulating hormone (α-MSH)/melanocortin-1 receptor MC1R signaling pathway in mammalian melanocytes. Abbreviations: α-MSH, α-Melanocyte Stimulating Hormone; AC, Adenylyl Cyclase; ASIP, Agouti Signaling Protein; cAMP, Cyclic Adenosine Monophosphate; CRE, cAMP Responsive Element; CREB, cAMP Responsive Element Binding Protein; Cys, Cysteine; DCT, Dopachrome Tautomerase; ITF2, Initiation Transcription Factor 2; MITF, Microphthalmia-associated Transcription Factor; MC1R, Melanocortin 1 Receptor; PKA, Protein Kinase A; TYR, Tyrosinase; TYRP1, Tyrosinase Related Protein 1. Modified from García-Borrón et al. (2005)

melanophores (Fujii and Miyashita 1982; Kobayashi et al. 2009; Logan et al. 2006; Mizusawa et al. 2013; Sugimoto et al. 1997).

In mammals, the α-MSH effect is antagonized by ASIP, which binds to MC1R (Lu et al. 1994; Ollmann et al. 1998) and promotes the change from eumelanin to pheomelanin synthesis (Miller et al. 1993). *Asip* is expressed in mammalian dermal papillae of hair follicles (Millar et al. 1995) and binds to MC1R, which results in a reduction in cytosolic cAMP and down-regulation of the melanogenic pathway (Aberdam et al. 1998; Furumura et al. 1998; Suzuki et al. 1997; Yang et al. 1997)

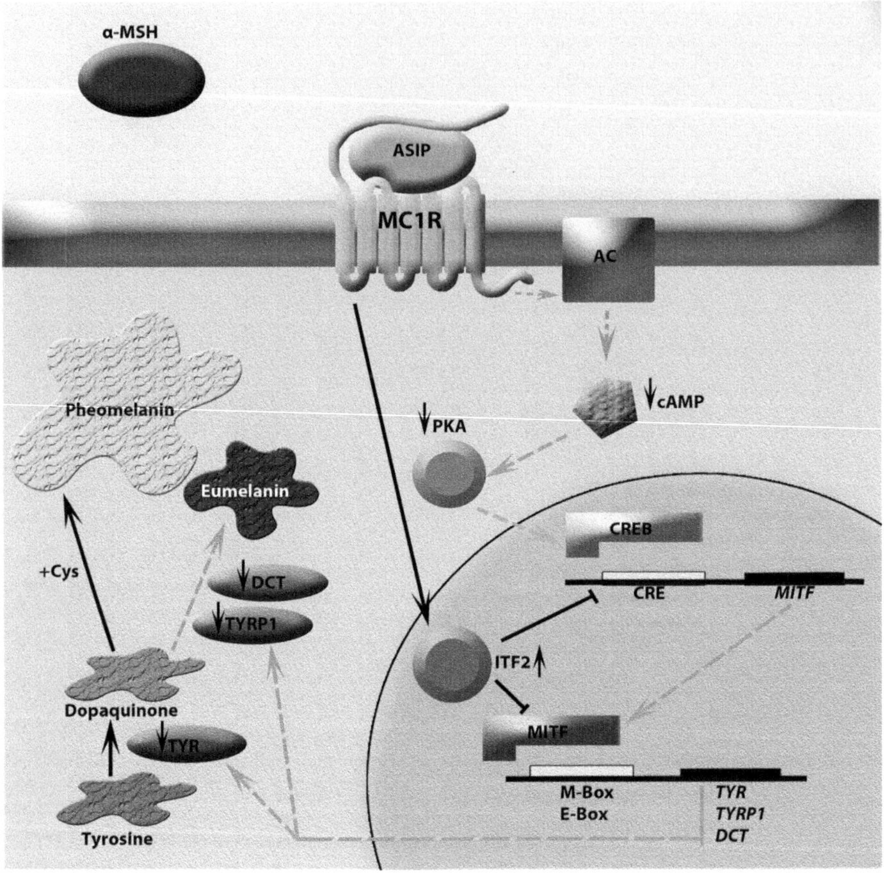

Fig. 8.4 Melanogenesis. ASIP-MC1R signaling axis in mammalian melanocyte. Abbreviations: α-MSH, α-Melanocyte Stimulating Hormone; AC, Adenylyl Cyclase; ASIP, Agouti Signaling Protein; cAMP, Cyclic Adenosine Monophosphate; CRE, cAMP Responsive Element; CREB, cAMP Responsive Element Binding Protein; Cys, Cysteine; DCT, Dopachrome Tautomerase; ITF2, Initiation Transcription Factor 2; MITF, Microphthalmia Transcription Factor; MC1R, Melanocortin 1 Receptor; PKA, Protein Kinase A; TYR, Tyrosinase; TYRP1, Tyrosinase Related Protein 1

(Fig. 8.4). Thus, the dorsal-ventral countershading pattern in mammals is achieved by high *Asip* expression levels in ventral skin (Lightner 2009; Manceau et al. 2011; Vrieling et al. 1994).

Unlike mammals, fish do not produce pheomelanin (Kottler et al. 2015), and the dorsal-ventral countershading pattern is established through the differential distribution of different pigment cell types (Bagnara and Matsumoto 2006). However, fish also show high levels of *asip1* (orthologue of mammalian *Asip*) in the ventral skin area (Agulleiro et al. 2014; Ceinos et al. 2015; Cerdá-Reverter et al. 2005; Guillot et al. 2012; Kurokawa et al. 2006), which inhibits α-Msh-induced melanosome movement (Cerdá-Reverter et al. 2005). A similar pigment role has been suggested for fish Asip1 (Agulleiro et al. 2014; Ceinos et al. 2015; Cerdá-Reverter et al. 2005; Guillot et al. 2012; Kurokawa et al. 2006) although its role is mediated by different mechanisms from those acting in mammals (Ceinos et al. 2015).

In fish, Asip1 acts as an inhibitor of pigment dispersion (Cerdá-Reverter et al. 2005; Guillot et al. 2012) and reduces the number of differentiated melanophores (Ceinos et al. 2015). Apparently, the Asip1 function is restricted to the adult pigment pattern. Ceinos et al. (2015) analyzed the *asip1* expression pattern during zebrafish larval development and reported that *asip1* starts to be highly and differentially expressed during metamorphosis. Comparable results were found in spotted gar *Lepisosteus oculatus* (Cal et al. 2017). Additionally, ectopic overexpression of *asip1* in zebrafish and turbot only affect the adult pigment pattern, resulting in dorsal skin paling (Ceinos et al. 2015; Guillot et al. 2012). Therefore showing that dorsal-ventral pigment pattern is commonly stablished after metamorphosis. Early larval stages pigment pattern in all fish species examined did not display countershading. In zebrafish, larval pigment pattern is formed by melanophores aggregated in stripes in dorsal and ventral regions of the animal (Parichy et al. 2011); and Senegalese sole display plentiful melanophores and xanthophores along the larval body, including abdominal regions (Darias et al. 2013a). However, factors affecting larvae, such as nutritional deficiencies, can produce deregulation of adult pigment pathways and, therefore, can result in a malpigmented dorsal-ventral phenotype (Darias et al. 2013b).

8.1.4 Physiological and Morphological Color Changes

Apart from the different genetically shaped pigment-patterning mechanisms that determine the final color of an organism, some organisms have evolved body coloration plasticity, so called color change, in order to cope with distinct backgrounds. Fish color changes are produced by two different mechanisms: physiological and morphological. Physiological color change is caused by short-term stimuli and is based on the aggregation or dispersion of pigment granules (chromatosomes) in skin chromatophores (Sköld et al. 2016) or changes in the distance and angle between light-reflecting platelets in motile iridophores (Fujii 2000). The pigment-containing organelles (chromatosomes) display bidirectional movement. A network of microtubules in the chromatophores is responsible for the aggregation of chromatosomes in a central mass of spherical shape or their dispersal through the cytoplasm to the periphery (Sköld et al. 2016). By contrast, iridophores do not have pigment organelles, but they do have stacks of crystals in the cytoplasm (Fujii 2000). However, some types of iridophores present cellular motility. Motile iridophores shift the position and distance of their crystals, which results in changes in the wavelengths of reflected light (Fujii 2000; Mäthger et al. 2003). It has been suggested that this motile capability is mediated also by microtubular structures (Oshima and Fujii 1987).

Physiological color change is controlled by both sympathetic and endocrine systems. In the former case, noradrenalin has been shown to induce chromatosome aggregation (Aspengren et al. 2003; Biswas et al. 2014; Fujii 2000), while in the endocrine system, several hormones are involved in pigmentation. Two melanin concentrating hormones (Mch1 and Mch2) play an important role in pigment granule aggregation (Berman et al. 2009; Fujii 2000; Kang and Kim 2013; Mizusawa et al. 2013, 2015). The α-melanocyte stimulating hormone (α-Msh) induces pigment granule dispersion (Fujii 2000; Fujii and Miyashita 1982; Yamanome et al. 2007). Although the adrenocorticotropic hormone (Acth) is commonly related to the stress response (Wendelaar-Bonga et al. 1994), its ability to disperse chromatosomes has also been reported (Fujii 2000). Indeed, the existence of melanocortin 2 receptor (*mc2r*) expression, which is the most well-known receptor for ACTH (Schiöth et al. 2005), has been detected in European sea bass *Dicentrarchus labrax* skin (Agulleiro et al. 2013). Other hormones involved in color change are melatonin, which has been reported to act as chromatosome aggregator (Aspengren et al. 2003; Sköld et al. 2008), and prolactin, which has a pigment-dispersing effect in xanthophores and erytrophores (Fujii 2000; Sköld et al. 2008). During physiological color change, noradrenalin and Mch are commonly released on light backgrounds (Fig. 8.5a). By contrast, α-Msh plasma levels are increased on dark backgrounds (Fig. 8.5b) (Kawauchi et al. 1983; Logan et al. 2006; Sugimoto 2002; Mizusawa et al. 2013).

Interestingly, the dispersion ability of α-Msh can be regulated by the degree of acetylation at the N-terminal domain. Pigment dispersion by Monoacetyl-α-Msh has been reported in tilapia *Oreochromis mossambicus*, goldfish *Carassius auratus*, barfin flounder *Verasper moseri* and Japanese flounder *Paralichthys olivaceus* (Kobayashi et al. 2010, 2012a, b; van der Salm et al. 2005). The diacetyl-α-Msh also shows high capacity of pigment dispersion in goldfish xanthophores (Kobayashi et al. 2012a). The dispersion capability of desacetyl-α-Msh has also

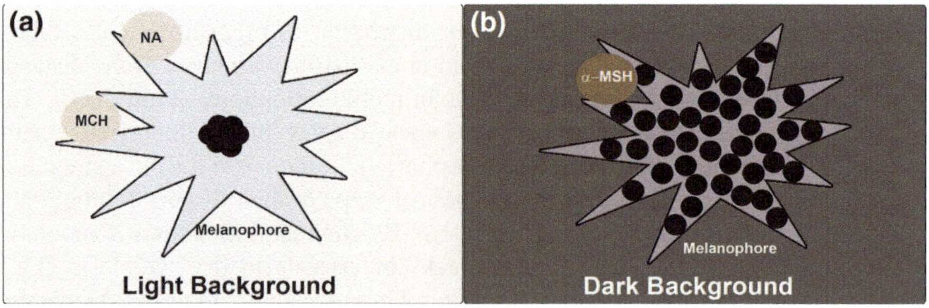

Fig. 8.5 Schematic diagram of sympathetic and endocrine control of skin pigmentation on **a** light and **b** dark backgrounds. α-Msh, α-Melanocyte stimulating hormone; Mch, Melanin concentrating hormone; Na, Noradrenalin; black circles represent melanosomes

been detected in both barfin and Japanese flounder (Kobayashi et al. 2010, 2012). Kobayashi et al. (2016) reported the ligand-dependent phenomenon of pigment dispersion in flounder. Barfin and Japanese flounder xanthophores express only one type of receptor for α-Msh, the melanocortin 5 receptor (Mc5r), and their xanthosomes are dispersed by both desacetyl-α-Msh and monoacetyl-α-Msh; however, barfin and Japanese flounder melanophores expresses two different α-Msh receptors, melanocortin 1 receptor (Mc1r) and Mc5r, and their melanosomes are dispersed only by desacetyl-α-Msh (Kobayashi et al. 2010, 2012). Kobayashi et al. (2016) suggested the heterodimerization of Mc1r/Mc5r as an explanation for the differential reaction to the acetylation degree of α-Msh in chromatophores.

Morphological color change is caused by long-term stimuli and is mediated by apoptosis or the proliferation of skin chromatophores as well as changes in their morphology and amounts of pigment (Sköld et al. 2016). Several studies have demonstrated that fish change their color by decreasing or increasing the number, morphology and size of melanophores under long-term adaptation to light or dark background, respectively; the opposite response is also observed in the number of iridophores (Sugimoto et al. 2000, 2005; van der Salm et al. 2005). Also, melanophores proliferate or die under dark or light background adaptation, respectively (Sugimoto et al. 2005; van der Salm et al. 2005). Both physiological and morphological color changes appear to be controlled by similar mechanisms. As mentioned above, Mch has an aggregating effect over fish chromatophores and is usually released during white background adaptation (Suzuki et al. 1995; Takahashi et al. 2004). Mch does not only aggregate chromatosomes, but also inhibits α-Msh dispersing activity (Mizusawa et al. 2011) and even the release of α-Msh (Baker et al. 1986; Barber et al. 1987), which prevents melanosome dispersion. However, Mch does not seem to have any effect on melanophore death since melanophore sensitivity to Mch decreases during long-term stimulation (Sugimoto 2002). As regard melanophore proliferation during black background adaptation, α-Msh has been identified as the main factor involved not only in melanosome dispersal, but also in melanophore development (Sugimoto 2002). In vivo long-term treatment with α-Msh promotes an increase in melanophore density in some fish species like tilapia (van Eys and Peters 1981) or barfin flounder (Yamanome et al. 2007), and high plasmatic levels of α-Msh have also been reported during black background adaptation in trout (Rodrigues and Sumpter 1984). Interestingly, the dynamics of plasmatic levels of α-Msh levels during background adaptation is not the same in all species that have been studied (Baker et al. 1984; Gilham and Baker 1984; Rotllant et al. 2003; Mizusawa et al. 2013; van Eys and Peters 1981; van der Salm et al. 2005). It has been suggested that this might be due to the existence of different α-Msh isoforms with different bioactivity intensities (Leclercq et al. 2009); or perhaps there exists a multiple endocrine control of background in fish, or the role of α-Msh in background adaptation may be species-specific.

8.2 Fish Pigmentation and Aquaculture

Malpigmentation is a common problem in commercial fish production and a common cause of losses in aquaculture. Abnormal pigmentation has been reported in several fish species, but particularly in flatfish (Akyol and Şen 2012; Purchase et al. 2002; Macieira et al. 2006). Normal pigmentation in flatfish is characterized with a high number of melanophores on the ocular side and low number on the blind side, which results in a dark ocular side and light blind side (Bolker and Hill 2000; Venizelos and Benetti 1999). The most common pigment defects in flatfish are:

- Albinism, pseudo-albinism or hypomelanosis: from complete absence of pigmentation to white areas on the ocular side (Bolker and Hill 2000; Venizelos and Benetti 1999).
- Hypermelanosis or ambicoloration: from complete pigmentation to pigmented areas on the blind side (Bolker and Hill 2000; Venizelos and Benetti 1999).

It has been shown that pigment defects are not restricted to cultured populations. Several cases of wild malpigmented adults have been reported in various flatfish species like common sole *Solea solea* (Akyol and Şen 2012; Cerím et al. 2016; Paris and Quignard 1968), plainfin sole *Achirus declivis* (Macieira et al. 2006), lined sole *Achirus lineatus* (Macieira et al. 2006), hogchoker *Trinectes maculatus* (Moore and Posey 1974), stone flounder *Kareius bicoloratus* (Fujita 1980), Patagonian flounder *Paralichthys patagonicus* (Díaz de Astarloa et al. 2006), among others. However, flatfish hatcheries have the highest reported rates of pigment abnormalities (Bolker and Hill 2000).

8.2.1 Factors Associated with Pigmentation Defects

Numerous studies have suggested different putative factors that may be associated with pigmentation defects. In flatfish, the developmental stage preceding metamorphosis was identified as the crucial period when environmental factors can strongly disturb the standard pigment pattern formation (Næss and Lie 1998). This period is also known as the "pigmentation window" (Izquierdo and Koven 2011).

One of the main studied factors implicated in flatfish malpigmentation is feeding, and the diet composition is often considered an essential aspect for improving pigmentation in cultured flatfish species. The correct ratio and absolute amount of docosahexaenoic acid (DHA), eicosapentaenoic acid (EPA) and arachidonic acid (ARA) has been considered essential for achieving the standard skin pigmentation (Izquierdo and Koven 2011; Sargent et al. 1999). The precise moment for feeding with the optimum ratio of specific dietary compounds is also a key aspect to take into account: the "pigmentation window" (Boglino et al. 2014; Izquierdo and Koven 2011; Næss and Lie 1998). In several flatfish species, such as Senegalense sole, common sole or yellowtail flounder *Limanda ferruginea*; the "pigmentation window" takes place in a short period of time during pre-metamorphic stages (Boglino et al. 2014; Copeman et al. 2002; Lund et al. 2008). Vitamin A, retinoic acid, has also been found important for the proper development of pigments in flatfish. Optimal levels of vitamin A prevents abnormal pigmentation but an excess produces developmental problems, *e.g.* skeletal malformation (Miki et al. 1990; Takeuchi et al. 1995). Vitamin A is involved in multiple biological functions, so it is difficult to establish the optimal level for this dietary nutrient (Bolker and Hill 2000).

Other aspects identified as important factors in fish pigment development are stress conditions and endocrine disorders. Ambicoloration in flatfish has been associated with high levels of cortisol, which indicates that stress during rearing conditions is a possible factor for pigment abnormalities (Yamada et al. 2011). This could also be related with the malpigmentation linked to substratum type. It has been suggested that flatfish that have the possibility to bury themselves in a sandy substratum show lower pigment defects than others without such an opportunity (Ottesen and Strand 1996; Stickney and White 1975). Endocrine disorders play a role in pigmentation through thyroid hormone, which produces pigment defects when it is over- or under-expressed (McMenamin et al. 2014; Yoo et al. 2000). Thyroid hormone was identified as a factor involved in adult pigment pattern development through its inhibition of melanophore number and the stimulation of xanthophore differentiation in zebrafish (Guillot et al. 2016; McMenamin et al. 2014). Additionally, in flatfish, an excess of thyroid hormone results in a high incidence of albinism, which suggests that thyroid hormone also inhibits melanophore proliferation in Japanese flounder *Paralichthys olivaceus* (Yoo et al. 2000). Other endocrine disorders could be the deregulation of α-Msh production, which has been suggested as a factor of malpigmentation in summer flounder *Paralichthys dentatus* after metamorphosis (Bolker and Hill 2000; Itoh et al. 2012). Itoh et al. (2012) describe that some of the environmental factors which result in a high incidence of pigment defects are related with the light-brain-pituitary axis and seems to be able to disrupt normal α-Msh synthesis and secretion in flounder. Moreover, Mch1 and Mch2, which are also endocrine hypothalamo-pituitary peptides, may also be involved in flatfish pigmentary defects. Low *mch* gene expression was suggested as the potential cause of malpigmentation in starry flounder *Platichthys stellatus* (Kang and Kim 2013). It is clear that further studies into α-Msh and Mchs regulation and pigment abnormalities are needed.

8.2.2 Causes of Pigment Abnormalities: Hypothesis

The complexity of pigment pattern establishment mechanisms in fish makes it difficult to reduce the likelihood of pigment defects occurring in aquaculture fish species. Two hypotheses have been put forward to explain the pigment abnormalities in flatfish.

The first suggests that nutritional deficits during fish development result in visual defects, which interfere with the endocrine signaling pathway necessary for melanophore differentiation. Kanazawa (1993) suggested that deficient ratios of DHA, EPA and Vitamin A produce rhodopsin malformation in flatfish eye. Subsequent studies on the lipid composition of neural tissues in abnormally pigmented fish showed that they have different contents of DHA, EPA and ARA from normal pigmented fish (Estevez and Kanazawa 1996). Additionally, neural alterations and eye degeneration were reported in flatfish as a result of the deficient intake of amino and fatty acids (Estevez et al. 1997). Taken together, these data support the first hypothesis: the signal pathway from the retina to the central nervous system could be disturbed by nutritional insufficiencies. This disturbance might be the cause of the inappropriate production of α-Msh in the pituitary, which could produce pigment abnormalities in flatfish (Fig. 8.6) (Bolker and Hill 2000).

The second hypothesis proposes that pigment defects in flatfish are the consequence of the deregulation of the mechanisms responsible for the establishment of the ocular- and blind-side skin structures during larvae metamorphosis (Seikai and Matsumoto 1994). Bolker et al. (2005) histologically analyzed normal pigmented and malpigmented skin and also proposed that pigment abnormalities may be

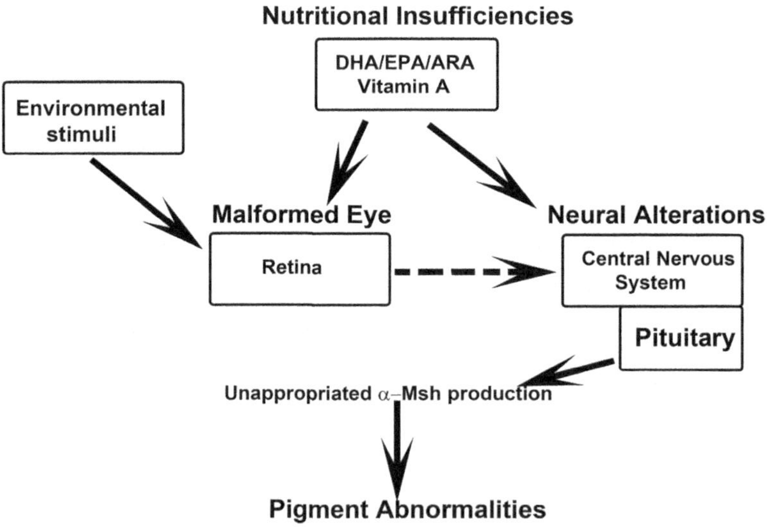

Fig. 8.6 Schematic representation of the malpigmentation hypothesis based on visual defects

caused by the normal regulatory pathway on the incorrect side. Recently, our studies showed that the establishment of both the asymmetrical body conformation and the adult pigment pattern is synchronized (Darias et al. 2013a), and that several deregulated genes are responsible for the pigmentary defects in adult pigment pattern in flatfish (Darias et al. 2013b; Guillot et al. 2012). Among others, *asip1*, an endogenous melanocortin inverse agonist that is commonly more expressed in blind-side skin, is deregulated in malpigmented fish. In adult Senegal sole and turbot, ocular-side non-pigmented spots show abnormally high levels of *asip1* mRNA (Guillot et al. 2012). Moreover, Guillot et al. (2012) noted that higher levels in ocular-side light spots are similar to blind-side *asip1* levels. Interestingly, the malpigmentation caused by nutritional defects in sole larvae is accompanied by *asip1* mRNA up-regulation in adult sole compared to normally pigmented fish (Darias et al. 2013b). Therefore, *asip1* deregulation has been proposed as a potential cause of pseudoalbinism (Darias et al. 2013b; Guillot et al. 2012). Whatever the case, *asip1* regulation seems to have a crucial role in pigment abnormalities.

Additionally, other genes have been seen to be deregulated in malpigmented flatfish. For example, Darias et al. (2013b) suggested that, surprisingly, in pseudo-albino sole melanophore differentiation is stimulated by upregulation of *somatolactin* (*sl*) gene, which is implicated in melanophore differentiation (Fukamachi et al. 2009). Other genes include *paired box protein 3* (*pax3*) gene, which participates in melanocyte development (Kubic et al. 2008); *tyrosinase* (*tyr*) gene, which is an important element of melanogenesis (García-Borrón et al. 2005); and *mitf* gene, which is considered a master regulator of melanogenesis (García-Borrón et al. 2005). Furthermore, the *mc1r* gene was also upregulated. However, melanogenesis and the final differentiation of melanophores seem to be disrupted. It was suggested that such melanogenic disruption is the result of (i) the downregulation of *mast/stem cell growth factor receptor Kit* (*cKit*) gene, which is responsible for melanocyte survival and melanogenesis (Alexeev and Yoon 2006); (ii) the downregulation of *sodium/potassium/calcium exchanger 5* (*slc24a5*) gene,

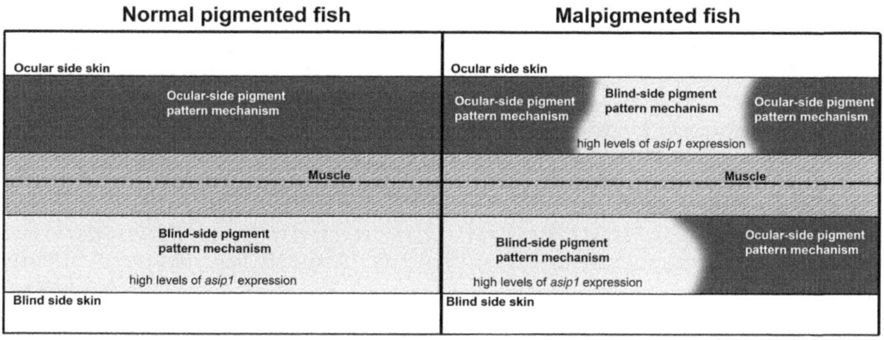

Fig. 8.7 Schematic representation of the malpigmentation hypothesis based on normal pigmentation pathway in the wrong side

which is involved in mammalian melanogenesis (Ginger et al. 2008); (iii) the upregulation of *asip1* and also *mitf*, whose overexpression has been suggested to result in blocking final melanocyte differentiation (Darias et al. 2013b). This gene expression profile seems to inhibit *tyrosinase-related protein 1 (trp1)* gene expression and the melanogenic pathway. Thus, Darias et al. (2013b) suggested that nutritional deficiencies affect the ordinary gene expression on the ocular side of sole. Therefore, it has also hypothesized by Guillot et al. (2012) and Darias et al. (2013b) that pigmentary defects could be the result of the establishment of pigment pattern mechanisms on the wrong side of flatfish (Fig. 8.7).

These two hypotheses are not mutually exclusive (Bolker and Hill 2000). Indeed, both hypotheses suggest an effect on different mechanisms, among them the regulation of different melanogenic genes and different melanocortin system participants (α-Msh, Mc1r, Asip1), which are known to be involved in dorsal-ventral countershading pattern in mammals.

References

Aberdam E, Bertolotto C, Sviderskaya EV et al (1998) Involvement of microphthalmia in the inhibition of melanocyte lineage differentiation and of melanogenesis by agouti signal protein. J Biol Chem 273:19560–19565

Agulleiro MJ, Cortés R, Fernández-Durán B et al (2013) Melanocortin 4 receptor becomes an ACTH teceptor by coexpression of melanocortin receptor accessory protein 2. Mol Endocrinol 27(11):1934–1945. https://doi.org/10.1210/me.2013-1099

Agulleiro MJ, Cortés R, Leal E et al (2014) Characterization, tissue distribution and regulation by fasting of the agouti family of peptides in the sea bass (*Dicentrarchus labrax*). Gen Comp Endocrinol 205:251–259. https://doi.org/10.1016/j.ygcen.2014.02.009

Akyol O, Şen H (2012) First record of abnormal pigmentation in a wild common sole, *Solea solea* L., from the Aegean Sea. Turkish J Vet Anim Sci 36:727–729. https://doi.org/10.3906/vet-1110-7

Alexeev V, Yoon K (2006) Distinctive role of the cKit receptor tyrosine kinase signaling in mammalian melanocytes. J Invest Dermatol 126:1102–1110. https://doi.org/10.1038/sj.jid.5700125

Aspengren S, Sköld HN, Quiroga G et al (2003) Noradrenaline- and melatonin-mediated regulation of pigment aggregation in fish melanophores. Pigment Cell Res 16:59–64 (003 [pii])

Díaz de Astarloa JM, Rico R, Acha M (2006) First report of a totally ambicoloured Patagonian flounder *Paralichthys patagonicus* (Paralichthyidae) with dorsal fin anomalies. Cybium 30:73–76

Bagnara JT, Matsumoto J (2006) Comparative anatomy and physiology of pigment cells in nonmammalian tissues. In: Nordlund JJ, Boissy RE, Hearing VJ, King RA, Oetting WS, J-PO (eds) The pigmentary system: physiology and pathophysiology, 2nd edn. Blackwell Publishing, Oxford, pp 11–59

Bajer K, Molnár O, Török J et al (2011) Ultraviolet nuptial colour determines fight success in male European green lizards (*Lacerta viridis*). Biol Lett 7:866–868. https://doi.org/10.1098/rsbl.2011.0520

Baker B, Bird D, Buckingham J (1986) Effects of chronic administration of melanin-concentrating hormone on corticotrophin, melanotrophin and pigmentation in the trout. Gen Comp Endocrinol 63:62–69

Baker BI, Wilson JF, Bowley TJ (1984) Changes in pituitary and plasma levels of MSH in teleosts during physiological colour change. Gen Comp Endocrinol 55:142–149. https://doi.org/10.1016/0016-6480(84)90138-2

Barber LD, Baker BI, Penny JC et al (1987) Melanin concentrating hormone inhibits the release of αMSH from teleost pituitary glands. Gen Comp Endocrinol 65:79–86. https://doi.org/10.1016/0016-6480(87)90225-5

Barsh GS (1996) The genetics of pigmentation: from fancy genes to complex traits. Trends Genet 12:299–305

Belk MC, Smith MH (1996) Pelage coloration in oldfield mice (*Peromyscus polionotus*): antipredator adaptation? J Mammal 77:882–890

Berman JR, Skariah G, Maro GS et al (2009) Characterization of two melanin-concentrating hormone genes in zebrafish reveals evolutionary and physiological links with the mammalian MCH system. J Comp Neurol 517:695–710. https://doi.org/10.1002/cne.22171

Biswas SP, Jadhao AG, Palande NV (2014) Role of catecholamines and nitric oxide on pigment displacement of the chromatophores of freshwater snakehead teleost fish, *Channa punctatus*. Fish Physiol Biochem 40:457–467. https://doi.org/10.1007/s10695-013-9857-8

Boglino A, Wishkerman A, Darias MJ et al (2014) Senegalese sole (*Solea senegalensis*) metamorphic larvae are more sensitive to pseudo-albinism induced by high dietary arachidonic acid levels than post-metamorphic larvae. Aquaculture 433:276–287. https://doi.org/10.1016/j.aquaculture.2014.06.012

Bolker JA, Hill CR (2000) Pigmentation development in hatchery-reared flatfishes. J Fish Biol 56:1029–1052. https://doi.org/10.1006/jfbi.2000.1260

Bolker JA, Hakala TF, Quist JE (2005) Pigmentation development, defects, and patterning in summer flounder (*Paralichthys dentatus*). Zoology 108:183–193. https://doi.org/10.1016/j.zool.2005.05.001

Braasch I, Volff J-N, Schartl M (2008) The evolution of teleost pigmentation and the fish-specific genome duplication. J Fish Biol 73:1891–1918. https://doi.org/10.1111/j.1095-8649.2008.02011.x

Brenner M, Hearing VJ (2008) The protective role of melanin against UV damage in human skin. Photochem Photobiol 84:539–549. https://doi.org/10.1111/j.1751-1097.2007.00226.x

Buscà R, Ballotti R (2000) Cyclic AMP a key messenger in the regulation of skin pigmentation. Pigment Cell Res 13:60–69

Cal L, Megías M, Cerdá-Reverter JM et al (2017) BAC recombineering of the agouti loci from spotted gar and zebrafish reveals the evolutionary ancestry of dorsal-ventral pigment asymmetry in fish. J Exp Zool B Mol Dev Evol. https://doi.org/10.1002/jez.b.22748, 24 May 2017

Ceinos RM, Guillot R, Kelsh RN et al (2015) Pigment patterns in adult fish result from superimposition of two largely independent pigmentation mechanisms. Pigment Cell Melanoma Res 28:196–209. https://doi.org/10.1111/pcmr.12335

Cerdá-Reverter JM, Agulleiro MJ, Guillot R et al (2011) Fish melanocortin system. Eur J Pharmacol 660:53–60. https://doi.org/10.1016/j.ejphar.2010.10.108

Cerdá-Reverter JM, Haitina T, Schiöth HB et al (2005) Gene structure of the goldfish agouti-signaling protein: a putative role in the dorsal-ventral pigment pattern of fish. Endocrinology 146:1597–1610. https://doi.org/10.1210/en.2004-1346

Cerím H, Celík M, Yapici S (2016) Occurrence of colour abnormalities and morphological aberration in common sole *Solea solea* (L., 1758) captured from the Aegean Sea. Cah Biol Mar 57:85–87

Chandramohan B, Renieri C, La Manna V et al (2013) The alpaca agouti gene: genomic locus, transcripts and causative mutations of eumelanic and pheomelanic coat color. Gene 521:303–310. https://doi.org/10.1016/j.gene.2013.03.060

Cone RD (2006) Studies on the physiological functions of the melanocortin system. Endocr Rev 27:736–749. https://doi.org/10.1210/er.2006-0034

Cooray SN, Clark AJL (2011) Melanocortin receptors and their accessory proteins. Mol Cell Endocrinol 331:215–221. https://doi.org/10.1016/j.mce.2010.07.015

Copeman LA, Parrish CC, Brown JA et al (2002) Effects of docosahexaenoic, eicosapentaenoic, and arachidonic acids on the early growth, survival, lipid composition and pigmentation of yellowtail flounder (*Limanda ferruginea*): a live food enrichment experiment. Aquaculture 210:285–304. https://doi.org/10.1016/S0044-8486(01)00849-3

Darias MJ, Andree KB, Boglino A et al (2013a) Coordinated regulation of chromatophore differentiation and melanogenesis during the ontogeny of skin pigmentation of *Solea senegalensis* (Kaup, 1858). PLoS ONE 8:e63005. https://doi.org/10.1371/journal.pone.0063005

Darias MJ, Andree KB, Boglino A et al (2013b) Morphological and molecular characterization of dietary-Induced pseudo-albinism during post-embryonic development of *Solea senegalensis* (Kaup, 1858). PLoS ONE 8:e68844. https://doi.org/10.1371/journal.pone.0068844

Dice LR, Blossom PM (1937) Studies of mammalian ecology in southwestern North America, with special attention to the colors of desert mammals. Carnegie Inst Wash 485:1–129

Dun G, Li X, Cao H et al (2007) Variations of melanocortin receptor 1 (MC1R) gene in three pig breeds. J Genet Genomics 34:777–782. https://doi.org/10.1016/S1673-8527(07)60088-5

Ellis HI (1980) Metabolism and solar radiation in dark and white herons in hot climates. Physiol Zool 53:358–372

Estevez A, Kanazawa A (1996) Fatty acid composition of neural tissues of normally pigmented and unpigmented juveniles of Japanese flounder using rotifer and *Artemia* enriched in n-3 HUFA. Fish Sci 62:88–93. https://doi.org/10.2331/fishsci.62.88

Estevez A, Sameshima M, Ishikawa M et al (1997) Effect of diets containing low levels of methionine and oxidized oil on body composition, retina structure and pigmentation success of Japanese flounder. Aquac Nutr 3:201–216. https://doi.org/10.1046/j.1365-2095.1997.00092.x

Fadeev A, Krauss J, Frohnhöfer HG et al (2015) Tight junction protein 1a regulates cell organisation during zebrafish colour patterning. Elife. https://doi.org/10.7554/eLife.06545

Faílde LD, Bermúdez R, Vigliano F et al (2014) Morphological, immunohistochemical and ultrastructural characterization of the skin of turbot (*Psetta maxima* L.). Tissue Cell 46:334–342. https://doi.org/10.1016/j.tice.2014.06.004

Fontanesi L, Rustempašić A, Brka M et al (2012) Analysis of polymorphisms in the agouti signalling protein (ASIP) and melanocortin 1 receptor (MC1R) genes and association with coat colours in two Pramenka sheep types. Small Rumin Res 105:89–96. https://doi.org/10.1016/j.smallrumres.2012.02.008

Frohnhöfer HG, Krauss J, Maischein H-M et al (2013) Iridophores and their interactions with other chromatophores are required for stripe formation in zebrafish. Development 140:2997–3007. https://doi.org/10.1242/dev.096719

Fujii R (1993) Coloration and chromatophores. In: Evans D (ed) The physiology of fishes. FL CRC Press, Boca Raton, pp 535–562

Fujii R (2000) The regulation of motile activity in fish chromatophores. Pigment Cell Res 13:300–319. https://doi.org/10.1034/j.1600-0749.2000.130502.x

Fujii R, Miyashita Y (1982) Receptor mechanisms in fish chromatophore—V. MSH disperses melanosomes in both dermal and epidermal melanophores of a catfish (*Parasilurus asotus*). Comp Biochem Physiol 71C:1–6

Fujita K (1980) Reversed ambicolorate flounder, *Kareius bicoloratus*, caught from Tokyo Bay. Jpn J Ichthyol 27:175–178

Fukamachi S, Yada T, Meyer A et al (2009) Effects of constitutive expression of somatolactin alpha on skin pigmentation in medaka. Gene 442:81–87. https://doi.org/10.1016/j.gene.2009.04.010

Fukuzawa T, Ide H (1988) A ventrally localized inhibitor of melanization in *Xenopus laevis* skin. Dev Biol 129:25–36

Fukuzawa T, Samaraweera P, Mangano FT et al (1995) Evidence that MIF plays a role in the development of pigmentation patterns in the frog. Dev Biol 167:148–158

Furumura M, Sakai C, Potterf SB et al (1998) Characterization of genes modulated during pheomelanogenesis using differential display. Proc Natl Acad Sci U S A 95:7374–7378. https://doi.org/10.1073/pnas.95.13.7374

García-Borrón JC, Sánchez-Laorden BL, Jiménez-Cervantes C (2005) Melanocortin-1 receptor structure and functional regulation. Pigment Cell Res 18:393–410. https://doi.org/10.1111/j.1600-0749.2005.00278.x

Gilham ID, Baker BI (1984) Evidence for the participation of a melanin-concentrating hormone in physiological colour change in the eel. J Endocrinol 102:237–243

Ginger RS, Askew SE, Ogborne RM et al (2008) SLC24A5 encodes a trans-Golgi network protein with potassium-dependent sodium-calcium exchange activity that regulates human epidermal melanogenesis. J Biol Chem 283:5486–5495. https://doi.org/10.1074/jbc.M707521200

Goda M, Fujii R (1995) Blue chromatophores in two species of callionymid fish. Zoolog Sci 12:811–813. https://doi.org/10.2108/zsj.12.811

Gross JB, Borowsky R, Tabin CJ (2009) A novel role for Mc1r in the parallel evolution of depigmentation in independent populations of the cavefish *Astyanax mexicanus*. PLoS Genet 5: e1000326. https://doi.org/10.1371/journal.pgen.1000326

Guillot R, Ceinos RM, Cal R et al (2012) Transient ectopic overexpression of agouti-signalling protein 1 (Asip1) induces pigment anomalies in flatfish. PLoS ONE 7:e48526. https://doi.org/10.1371/journal.pone.0048526

Guillot R, Muriach B, Rocha A et al (2016) Thyroid hormones regulate zebrafish melanogenesis in a gender-specific manner. PLoS ONE 11(11):e0166152. https://doi.org/10.1371/journal.pone.0166152

Hirata M, Nakamura K, Kanemaru T et al (2003) Pigment cell organization in the hypodermis of zebrafish. Dev Dyn 227:497–503. https://doi.org/10.1002/dvdy.10334

Hubbard JK, Uy JAC, Hauber ME et al (2010) Vertebrate pigmentation: from underlying genes to adaptive function. Trends Genet 26:231–239. https://doi.org/10.1016/j.tig.2010.02.002

Irion U, Frohnhöfer HG, Krauss J et al (2014) Gap junctions composed of connexins 41.8 and 39.4 are essential for colour pattern formation in zebrafish. Elife 3:e05125. https://doi.org/10.7554/eLife.05125

Irion U, Singh AP, Nüsslein-Volhard C (2016) The developmental genetics of vertebrate color pattern formation: lessons from zebrafish. Curr Top Dev Biol 117:141–169. https://doi.org/10.1016/bs.ctdb.2015.12.012

Ito S, Wakamatsu K (2011) Human hair melanins: what we have learned and have not learned from mouse coat color pigmentation. Pigment Cell Melanoma Res 24:63–74. https://doi.org/10.1111/j.1755-148X.2010.00755.x

Itoh K, Washio Y, Fujinami Y et al (2012) Continuous illumination through larval development suppresses dopamine synthesis in the suprachiasmatic nucleus, causing activation of α-MSH synthesis in the pituitary and abnormal metamorphic skin pigmentation in flounder. Gen Comp Endocrinol 176:215–221. https://doi.org/10.1016/j.ygcen.2012.01.017

Izquierdo M, Koven W (2011) Lipids. In: Holt GJ (ed) Larval fish nutrition, 1st edn. Wiley, New York, pp 47–81

Kanazawa A (1993) Nutritional mechanisms involved in the occurrence of abnormal pigmentation in hatchery-reared flatfish. J World Aquac Soc 24:162–166

Kang DY, Kim HC (2013) Functional characterization of two melanin-concentrating hormone genes in the color camouflage, hypermelanosis, and appetite of starry flounder. Gen Comp Endocrinol 189:74–83. https://doi.org/10.1016/j.ygcen.2013.04.025

Kawaguti S (1965) Electron microscopy on iridophores in the scale of the blue wrasse. Proc Jpn Acad 41:610–613

Kawauchi H, Kawazoe I, Tsubokawa M et al (1983) Characterization of melanin-concentrating hormone in chum salmon pituitaries. Nature 305:321–323

Kelsh RN (2004) Genetics and evolution of pigment patterns in fish. Pigment Cell Res 17:326–336. https://doi.org/10.1111/j.1600-0749.2004.00174.x

Kelsh RN, Harris ML, Colanesi S et al (2009) Stripes and belly-spots—a review of pigment cell morphogenesis in vertebrates. Semin Cell Dev Biol 20:90–104. https://doi.org/10.1016/j.semcdb.2008.10.001

Kobayashi Y, Tsuchiya K, Yamanome T et al (2010) Differential expressions of melanocortin receptor subtypes in melanophores and xanthophores of barfin flounder. Gen Comp Endocrinol 168:133–142. http://doi.org/10.1016/j.ygcen.2010.04.017

Kobayashi Y, Hamamoto A, Takahashi A et al (2016) Dimerization of melanocortin receptor 1 (MC1R) and MC5R creates a ligand-dependent signal modulation: potential participation in physiological color change in the flounder. Gen Comp Endocrinol 230–231:103–109. https://doi.org/10.1016/j.ygcen.2016.04.008

Kobayashi Y, Mizusawa K, Chiba H et al (2012a) Further evidence on acetylation-induced inhibition of the pigment-dispersing activity of α-melanocyte-stimulating hormone. Gen Comp Endocrinol 176:9–17. http://doi.org/10.1016/j.ygcen.2011.12.001

Kobayashi Y, Mizusawa K, Saito Y et al (2012b) Melanocortin systems on pigment dispersion in fish chromatophores. Front Endocrinol (Lausanne) 3:9. https://doi.org/10.3389/fendo.2012.00009

Kobayashi Y, Mizusawa K, Yamanome T et al (2009) Possible paracrine function of alpha-melanocyte-stimulating hormone and inhibition of its melanin-dispersing activity by N-terminal acetylation in the skin of the barfin flounder, Verasper moseri. Gen Comp Endocrinol 161:419–424. https://doi.org/10.1016/j.ygcen.2009.02.009

Kodric-Brown A, Nicoletto P (2001) Female choice in the guppy (*Poecilia reticulata*): the interaction between male color and display. Behav Ecol Sociobiol 50:346–351. https://doi.org/10.1007/s002650100374

Kottler VA, Künstner A, Schartl M (2015) Pheomelanin in fish? Pigment Cell Melanoma Res 28:355–356. https://doi.org/10.1111/pcmr.12359

Kubic JD, Young KP, Plummer RS et al (2008) Pigmentation PAX-ways: the role of Pax3 in melanogenesis, melanocyte stem cell maintenance, and disease. Pigment Cell Melanoma Res 21:627–645. https://doi.org/10.1111/j.1755-148X.2008.00514.x

Kurokawa T, Murashita K, Uji S (2006) Characterization and tissue distribution of multiple agouti-family genes in pufferfish, *Takifugu rubripes*. Peptides 27:3165–3175. https://doi.org/10.1016/j.peptides.2006.09.013

Le Douarin N, Kalcheim C (1999) The neural crest. Developmental and cell biology series, 2nd edn. Cambridge University Press, Cambridge

Le Douarin NM, Dupin E (2003) Multipotentiality of the neural crest. Curr Opin Genet Dev 13:529–536. https://doi.org/10.1016/j.gde.2003.08.002

Le Pape E, Wakamatsu K, Ito S et al (2008) Regulation of eumelanin/pheomelanin synthesis and visible pigmentation in melanocytes by ligands of the melanocortin 1 receptor. Pigment Cell Melanoma Res 21:477–486. https://doi.org/10.1111/j.1755-148X.2008.00479.x

Leclercq E, Taylor JF, Migaud H (2009) Morphological skin colour changes in teleosts. Fish Fish 11:159–193. https://doi.org/10.1111/j.1467-2979.2009.00346.x

Lightner JK (2009) Genetics of coat color II: the agouti signaling protein (ASIP) gene. Answers Res J 2:79–84

Lin JY, Fisher DE (2007) Melanocyte biology and skin pigmentation. Nature 445:843–850. https://doi.org/10.1038/nature05660

Lister JA, Robertson CP, Lepage T et al (1999) Nacre encodes a zebrafish microphthalmia-related protein that regulates neural-crest-derived pigment cell fate. Development 126:3757–3767

Logan DW, Burn SF, Jackson IJ (2006) Regulation of pigmentation in zebrafish melanophores. Pigment Cell Res 19:206–213. https://doi.org/10.1111/j.1600-0749.2006.00307.x

Lu D, Willard D, Patel IR et al (1994) Agouti protein is an antagonist of the melanocyte-stimulating-hormone receptor. Nature 371:799–802. https://doi.org/10.1038/371799a0

Lund I, Steenfeldt SJ, Banta G et al (2008) The influence of dietary concentrations of arachidonic acid and eicosapentaenoic acid at various stages of larval ontogeny on eye migration, pigmentation and prostaglandin content of common sole larvae (*Solea solea* L.). Aquaculture 276:143–153. https://doi.org/10.1016/j.aquaculture.2008.01.004

Maan ME, Sefc KM (2013) Colour variation in cichlid fish: developmental mechanisms, selective pressures and evolutionary consequences. Semin Cell Dev Biol 24:516–528. https://doi.org/10.1016/j.semcdb.2013.05.003

Macieira RM, Joyeux JC, Chagas LP (2006) Ambicoloration and morphological aberration in the sole Achirus declivis (Pleuronectiformes: Achiridae) and two other cases of color abnormalities in achirid soles from southeastern Brazil. Neotrop Ichthyol 4:287–290. https://doi.org/10.1590/S1679-62252006000200016

Maderspacher F, Nüsslein-Volhard C (2003) Formation of the adult pigment pattern in zebrafish requires leopard and obelix dependent cell interactions. Development 130:3447–3457. https://doi.org/10.1242/dev.00519

Mahalwar P, Singh AP, Fadeev A et al (2016) Heterotypic interactions regulate cell shape and density during color pattern formation in zebrafish. Biol Open. https://doi.org/10.1242/bio.022251

Mahalwar P, Walderich B, Singh AP et al (2014) Local reorganization of xanthophores fine-tunes and colors the striped pattern of zebrafish. Science 345:1362–1364. https://doi.org/10.1126/science.1254837

Manceau M, Domingues VS, Mallarino R et al (2011) The developmental role of Agouti in color pattern evolution. Science 331:1062–1065. https://doi.org/10.1126/science.1200684

Mäthger LM, Land MF, Siebeck UE et al (2003) Rapid colour changes in multilayer reflecting stripes in the paradise whiptail, *Pentapodus paradiseus*. J Exp Biol 206:3607–3613. https://doi.org/10.1242/jeb.00599

McMenamin SK, Bain EJ, McCann AE et al (2014) Thyroid hormone-dependent adult pigment cell lineage and pattern in zebrafish. Science 345:1358–1361. https://doi.org/10.1126/science.1256251

Miki N, Taniguchi T, Hamakawa H (1990) Reduction of albinism in hatchery-reared flounder "hirame", *Paralichthys olivaceus* by feeding on rotifer enriched with vitamin-A. Suisanzoshoku 38:147–155

Millar SE, Miller MW, Stevens ME et al (1995) Expression and transgenic studies of the mouse agouti gene provide insight into the mechanisms by which mammalian coat color patterns are generated. Development 121:3223–3232

Miller MW, Duhl DM, Vrieling H et al (1993) Cloning of the mouse agouti gene predicts a secreted protein ubiquitously expressed in mice carrying the lethal yellow mutation. Genes Dev 7:454–467. https://doi.org/10.1101/gad.7.3.454

Mizusawa K, Kawashima Y, Sunuma T et al (2015) Involvement of melanin-concentrating hormone 2 in background color adaptation of barfin flounder *Verasper moseri*. Gen Comp Endocrinol 214:140–148. https://doi.org/10.1016/j.ygcen.2014.07.008

Mizusawa K, Kobayashi Y, Sunuma T et al (2011) Inhibiting roles of melanin-concentrating hormone for skin pigment dispersion in barfin flounder, *Verasper moseri*. Gen Comp Endocrinol 171:75–81. https://doi.org/10.1016/j.ygcen.2010.12.008

Mizusawa K, Kobayashi Y, Yamanome T et al (2013) Interrelation between melanocyte-stimulating hormone and melanin-concentrating hormone in physiological body color change: roles emerging from barfin flounder *Verasper moseri*. Gen Comp Endocrinol 181:229–234. https://doi.org/10.1016/j.ygcen.2012.09.026

Moore CJ, Posey CR (1974) Pigmentation and morphological abnormalities in the hogchoker, *Trinectes maculatus* (Pisces, Soleidae). Copeia 3:660–670

Næss T, Lie Ø (1998) A sensitive period during first feeding for the determination of pigmentation pattern in Atlantic halibut, *Hippoglossus hippoglossus* L., juveniles: The role of diet. Aquac Res 29:925–934

Newton J, Wilkie A, He L et al (2000) Melanocortin 1 receptor variation in the domestic dog. Mamm Genome 11:24–30

Newton RA, Roberts DW, Leonard JH et al (2007) Human melanocytes expressing MC1R variant alleles show impaired activation of multiple signaling pathways. Peptides 28:2387–2396. https://doi.org/10.1016/j.peptides.2007.10.003

Norris BJ, Whan VA (2008) A gene duplication affecting expression of the ovine ASIP gene is responsible for white and black sheep. Genome Res 18:1282–1293. https://doi.org/10.1101/gr.072090.107

Ollmann MM, Lamoreux ML, Wilson BD et al (1998) Interaction of Agouti protein with the melanocortin 1 receptor in vitro and in vivo. Genes Dev 12:316–330

Olsson M, Stuart-Fox D, Ballen C (2013) Genetics and evolution of colour patterns in reptiles. Semin Cell Dev Biol 24:529–541. https://doi.org/10.1016/j.semcdb.2013.04.001

Oribe E, Fukao A, Yoshihara C et al (2012) Conserved distal promoter of the agouti signaling protein (ASIP) gene controls sexual dichromatism in chickens. Gen Comp Endocrinol 177:231–237. https://doi.org/10.1016/j.ygcen.2012.04.016

Oshima N, Fujii R (1987) Motile mechanism of blue damselfish (*Chrysiptera cyanea*) iridophores. Cell Motil Cytoskeleton 8:85–90. https://doi.org/10.1002/cm.970080112

Ottesen OH, Strand HK (1996) Growth, development, and skin abnormalities of halibut (*Hippoglossus hippoglossus* L.) juveniles kept on different bottom substrates. Aquaculture 146:17–25. https://doi.org/10.1016/S0044-8486(96)01359-2

Parichy DM, Elizondo MR, Mills MG et al (2011) Normal table of post-embrionic zebrafish development: staging by externally visible anatomy of the living fish. Dev Dyn 238:2975–3015. https://doi.org/10.1002/dvdy.22113.Normal

Paris J, Quignard J (1968) Quelques cas d'ambicoloration et d'albinisme chez *Solea solea vulgaris* Quensel. Rev Trav Inst Pech Marit 32:507–510

Parker AR, Martini N (2006) Structural colour in animals—simple to complex optics. Opt Laser Technol 38:315–322. https://doi.org/10.1016/j.optlastec.2005.06.037

Protas ME, Patel NH (2008) Evolution of coloration patterns. Annu Rev Cell Dev Biol 24:425–446. https://doi.org/10.1146/annurev.cellbio.24.110707.175302

Prum RO (2006) Anatomy, physics, and evolution of structural colors. In: Hill GE, McGraw KJ (eds) Bird coloration: mechanisms and measurements. Harvard University Press

Purchase CF, Boyce DL, Brown JA (2002) Occurrence of hypomelanization in cultured yellowtail flounder *Limanda ferruginea*. Aquac Res 33:1191–1193. https://doi.org/10.1046/j.1365-2109.2002.00764.x

Raible DW, Eisen JS (1994) Restriction of neural crest cell fate in the trunk of the embryonic zebrafish. Development 120:495–503

Robbins LS, Nadeau JH, Johnson KR et al (1993) Pigmentation phenotypes of variant extension locus alleles result from point mutation that alter MSH receptor function. Cell 72:827–834

Roberts NW, Marshall NJ, Cronin TW (2012) High levels of reflectivity and pointillist structural color in fish, cephalopods, and beetles. Proc Natl Acad Sci U S A 109:E3387; author reply E3388. https://doi.org/10.1073/pnas.1216282109

Rodrigues K, Sumpter J (1984) Effects of background adaptation on the pituitary and plasma concentrations of some proopiomelanocortin-related peptides in the rainbow trout (*Salmo gairdneri*). J Endocrinol 101:277–284

Rotllant J, Tort L, Montero D et al (2003) Background colour influence on the stress response in cultured red porgy *Pagrus pagrus*. Aquaculture 223:129–139. https://doi.org/10.1016/S0044-8486(03)00157-1

Roulin A, Ducrest A-L (2013) Genetics of colouration in birds. Semin Cell Dev Biol 24:594–608. https://doi.org/10.1016/j.semcdb.2013.05.005

Rudh A, Qvarnström A (2013) Adaptive colouration in amphibians. Semin Cell Dev Biol 24:553–561. https://doi.org/10.1016/j.semcdb.2013.05.004

Sánchez E, Rubio VC, Cerdá-Reverter JM (2010) Molecular and pharmacological characterization of the melanocortin type 1 receptor in the sea bass. Gen Comp Endocrinol 165:163–169. https://doi.org/10.1016/j.ygcen.2009.06.008

Sargent J, McEvoy L, Estevez A et al (1999) Lipid nutrition of marine fish during early development: current status and future directions. Aquaculture 179:217–229. https://doi.org/10.1016/S0044-8486(99)00191-X

Schiöth HB, Haitina T, Ling MK et al (2005) Evolutionary conservation of the structural, pharmacological, and genomic characteristics of the melanocortin receptor subtypes. Peptides 26:1886–1900. https://doi.org/10.1016/j.peptides.2004.11.034

Seikai T, Matsumoto J (1994) Mechanism of pseudoalbinism in flatfish: an association between pigment cell and skin differentiation. J World Aquac Soc 25:78–85

Singh AP, Schach U, Nüsslein-Volhard C (2014) Proliferation, dispersal and patterned aggregation of iridophores in the skin prefigure striped colouration of zebrafish. Nat Cell Biol 16:607–614. https://doi.org/10.1038/ncb2955

Siracusa LD (1994) The agouti gene: turned on to yellow. Trends Genet 10:423–428

Sköld HN, Amundsen T, Svensson PA et al (2008) Hormonal regulation of female nuptial coloration in a fish. Horm Behav 54:549–556. https://doi.org/10.1016/j.yhbeh.2008.05.018

Sköld HN, Aspengren S, Cheney KL, Wallin M (2016) Fish chromatophores-from molecular motors to animal behavior. Elsevier Inc.

Slominski A (2004) Melanin pigmentation in mammalian skin and its hormonal regulation. Physiol Rev 84:1155–1228. https://doi.org/10.1152/physrev.00044.2003

Stickney RR, White DB (1975) Ambicoloration in tank cultured flounder, *Paralichthys dentatus*. Trans Am Fish Soc 104:158–160

Sugimoto M (2002) Morphological color changes in fish: regulation of pigment cell density and morphology. Microsc Res Tech 58:496–503. https://doi.org/10.1002/jemt.10168

Sugimoto M, Nagamori H, Yasui H et al (1997) Regulation of melanophore responsiveness in the background-adapted medaka, *Oryzias latipes*: change in the intracellular signaling system. Comp Biochem Physiol C: Pharmacol Toxicol Endocrinol 117:259–265

Sugimoto M, Uchida N, Hatayama M (2000) Apoptosis in skin pigment cells of the medaka, *Oryzias latipes* (Teleostei), during long-term chromatic adaptation: the role of sympathetic innervation. Cell Tissue Res 301:205–216. https://doi.org/10.1007/s004410000226

Sugimoto M, Yuki M, Miyakoshi T et al (2005) The influence of long-term chromatic adaptation on pigment cells and striped pigment patterns in the skin of the zebrafish, *Danio rerio*. J Exp Zool 303A:430–440. https://doi.org/10.1002/jez.a.177.LONG-TERM

Suzuki I, Tada A, Ollmann MM et al (1997) Agouti signaling protein inhibits melanogenesis and the response of human melanocytes to alpha-melanotropin. J Invest Dermatol 108:838–842

Suzuki M, Narnaware Y, Baker B et al (1995) Influence of environmental colour and diurnal phase on MCH gene expression in the trout. J Neuroendocrinol 7:319–328

Takahashi A, Tsuchiya K, Yamanome T et al (2004) Possible involvement of melanin-concentrating hormone in food intake in a teleost fish, barfin flounder. Peptides 25:1613–1622. https://doi.org/10.1016/j.peptides.2004.02.022

Takahashi G, Kondo S (2008) Melanophores in the stripes of adult zebrafish do not have the nature to gather, but disperse when they have the space to move. Pigment Cell Melanoma Res 21:677–686. https://doi.org/10.1111/j.1755-148X.2008.00504.x

Takeuchi T, Dedi J, Ebisawa C et al (1995) The effect of beta-carotene and vitamin A enriched *Artemia* nauplii on the malformation and color abnormality of larval Japanese flounder. Fish Sci 61:141–148

Tso I-M, Tai P-L, Ku T-H et al (2002) Colour-associated foraging success and population genetic structure in a sit-and-wait predator *Nephila maculata* (Araneae: Tetragnathidae). Anim Behav 63:175–182. https://doi.org/10.1006/anbe.2001.1878

Valverde P, Healy E, Jackson I et al (1995) Variants of the melanocyte-stimulating hormone receptor gene are associated with red hair and fair skin in humans. Nat Genet 11:328–330

van der Salm AL, Metz JR, Bonga SEW et al (2005) Alpha-MSH, the melanocortin-1 receptor and background adaptation in the Mozambique tilapia, *Oreochromis mossambicus*. Gen Comp Endocrinol 144:140–149. https://doi.org/10.1016/j.ygcen.2005.05.009

van Eys GJ, Peters PT (1981) Evidence for a direct role of alpha-MSH in morphological background adaptation of the skin in Sarotherodon mossambicus. Cell Tissue Res 217:361–372

Venizelos A, Benetti DD (1999) Pigment abnormalities in flatfish. Aquaculture 176:181–188. https://doi.org/10.1016/S0044-8486(99)00060-5

Vrieling H, Duhl DM, Millar SE et al (1994) Differences in dorsal and ventral pigmentation result from regional expression of the mouse agouti gene. Proc Natl Acad Sci U S A 91:5667–5671

Walderich B, Singh AP, Mahalwar P et al (2016) Homotypic cell competition regulates proliferation and tiling of zebrafish pigment cells during colour pattern formation. Nat Commun 7:11462. https://doi.org/10.1038/ncomms11462

Watanabe M, Kondo S (2015) Comment on "Local reorganization of xanthophores fine-tunes and colors the striped pattern of zebrafish". Science 348:297. https://doi.org/10.1126/science.1254837

Wendelaar-Bonga SE, Balm PHM, Lamers AE (1994) The involvement of acth and msh in the stress response in teleost fish. Neth J Zool 45:103–106. https://doi.org/10.1163/156854295X00672

Yamada T, Donai H, Okauchi M et al (2011) Induction of ambicoloration by exogenous cortisol during metamorphosis of spotted halibut *Verasper variegatus*. Comp Biochem Physiol - B Biochem Mol Biol 160:174–180. https://doi.org/10.1016/j.cbpb.2011.08.004

Yamanome T, Chiba H, Takahashi A (2007) Melanocyte-stimulating hormone facilitates hypermelanosis on the non-eyed side of the barfin flounder, a pleuronectiform fish. Aquaculture 270:505–511. https://doi.org/10.1016/j.aquaculture.2007.05.037

Yang YK, Ollmann MM, Wilson BD et al (1997) Effects of recombinant agouti-signaling protein on melanocortin action. Mol Endocrinol 11:274–280. https://doi.org/10.1210/mend.11.3.9898

Yoo JH, Takeuchi T, Tagawa M et al (2000) Effect of thyroid hormones on the stage-specific pigmentation of the Japanese flounder *Paralichthys olivaceus*. Zoolog Sci 17:1101–1106. https://doi.org/10.2108/zsj.17.1101

Chapter 9
Novel Aspects of Phosphate Endocrine Control: A Key Element for the Long-Term Sustainability of Finfish Aquaculture

Paula Suarez-Bregua, Laura Cal, Pedro M. Guerreiro and Josep Rotllant

Abstract Phosphorus, in the form of inorganic phosphate (Pi), is one of the most important macronutrients for all organisms, including fish. Among other functions, it is indispensable for the formation and development of hard tissues such as bones and scales. However, its deficiency has implications not only for those hard tissues, where it is responsible for defective mineralization, leading to skeletal malformation, but also for disturbances of intermediary metabolism, especially energy metabolism, leading to impairment of growth. Nevertheless, the endocrine mechanisms for regulation of Pi balance in fish have largely been overlooked. Currently, in commercial fish culture systems Pi enriched diets are generally used to avoid skeletal malformations ensure health and increase growth but, however, excess levels can have harmful effects on fish. Additionally, the excess of unused/excreted Pi in the effluents from these culture systems turns intensive fish farming in a major source of eutrophication in the aquatic systems. Therefore, the improvement of foods used and the reduction of the outputs of these dissolved wastes it will be a key element for the long-term sustainability of aquaculture. One of the fundamental points to achieve this goal is to understand the mechanisms that regulate Pi homeostasis in fish, which ultimately will contribute to the equilibrium between the requirements for optimal physiology, fast growth and reduced environmental impact. Therefore, in this chapter, we attempt to describe the current state of knowledge regarding the recently characterized endocrine and non-endocrine factors involved in the regulation of phosphate homeostasis in fish.

Keywords Mineral homeostasis · Phosphate · Fish farming · Skeletal disorders
Bone mineralization and remodeling

P. Suarez-Bregua · L. Cal · J. Rotllant (✉)
Institute of Marine Research, Spanish National Research Council (IIM–CSIC), Vigo, Spain
e-mail: rotllant@iim.csic.es

P. M. Guerreiro
Center for Marine Sciences (CCMAR), University of Algarve, Faro, Portugal

M. Yúfera (ed.), *Emerging Issues in Fish Larvae Research*,
https://doi.org/10.1007/978-3-319-73244-2_9

9.1 Phosphate Homeostasis

Phosphorus is an essential element in vertebrates since plays a vital role in numerous biological functions. Widely distributed, phosphorus is present in the body as stable inorganic phosphate (Pi) and organic phosphate compounds (e.g., phospholipids, phosphoproteins, nucleotides and some carbohydrates) (Bansal 1990; Dean et al. 2015). Thus, it is involved in the maintenance of the cell membrane integrity (phospholipids cellular bilayer), in the structure of nucleic acids (DNA and RNA) and in multiple biochemical processes of the cellular metabolism such as the generation, store and release of energy (ATP), acid-base homeostasis or skeletal mineralization (Favus et al. 2006; Penido and Alon 2012). Because of the importance of this mineral in the whole-organism physiology, optimal phosphate balance is critical for fish health and welfare.

Phosphate homeostasis is determined by the modulation of intracellular and extracellular fluxes among three main target organs: intestine (absorption site), kidney (excretion and reabsorption site) and skeleton (storage site) (Lall 2002; Witten and Huysseune 2009). The ionized Pi in the serum represents the metabolically active fraction that can be freely interchangeable among extracellular fluid (ECF), intracellular fluid (ICF) and bone (Favus et al. 2006; Witten and Huysseune 2009) in order to maintain the serum Pi levels within a narrow physiological range. Additionally, fish have continuous access to ions from surrounding waters and they are able to exchange them via gills, skin and oral epithelium (Abbink et al. 2004; Flik and Verbost 1993). Branchial epithelium is the major contact site between fish and surrounding water as gills contain a high number of ionocytes (i.e., ion-transporting cells) or also named chloride cells (Flik and Verbost 1993). Circulations Pi levels are closely associated to those of calcium. In freshwater and seawater, the calcium levels are unlimited so that fish, as other aquatic vertebrates, are able to uptake calcium from surrounding waters. However, the dissolved phosphorus concentration in aquatic environments is too low and it is not considered a enough external source for body functions (Lall 2002; Lall and Lewis-McCrea 2007), and no phosphate transporters are known in the branchial epithelium of teleosts. Interestingly, a recent study (Schultz et al. 2014) showed Pi transporters in the Pacific hagfish *Eptatretus stoutii* skin, which could be important for epithelial absorption while scavenging in whale carcasses where the medium Pi concentration may be elevated. Nonetheless, phosphate homeostasis relies essentially on dietary phosphorus as the main external mineral source required for skeletal metabolism and growth in fish (Lall 2002; Witten and Huysseune 2009).

Fish skeleton is the main calcium and phosphate reservoir in the body. About 99 and 85% of the whole-body calcium and phosphate, respectively, are stored in the skeleton as hydroxyapatite (HA) crystals, a rigid complex of calcium phosphate associated with the bone matrix (Lovell 1998). In vertebrates, the skeleton takes part in the regulation of calcium and phosphate homeostasis via bone remodeling. Calcium and phosphate are provided from bone through bone resorption to compensate the inadequate availability of minerals and maintain relatively constant the ionic levels in blood. Conversely, if the serum calcium and phosphate levels are high, both minerals are deposited in the skeleton for storage (Kini and Nandeesh 2012). Bone remodeling is a lifelong process that is carried out by the coordinated action of osteoblasts (i.e., bone forming cells), osteoclasts (i.e., bone resorbing cells) as well as osteocytes (i.e., entrapped cells inside the bone matrix) (Hadjidakis and Androulakis 2006).

In mammals, bone remodeling is triggered by osteocytes as these cells acts as local sensors and respond to surrounding mechanical signals and factors regulating calcium and phosphate homeostasis through coordination of bone surface cells (Seeman 2009). Bone remodeling starts with the osteoclasts formation and activation, osteoclast-mediated bone resorption, a reversal period and a long period of bone matrix formation mediated by osteoblasts that is followed by the mineralization of new bone matrix formed (Sims and Gooi 2008) (Fig. 9.1).

Similar to mammals, most of the basal teleosts (such as cyprinids and salmonids) have a dynamic cellular bone with osteocytes, osteoblasts and a prevalence of multinucleated osteoclasts. However, advanced teleosts (such as perciforms) have a lacking-osteocytes bone (i.e., acellular bone) and a predominance of mononucleated osteoclasts (Witten et al. 2001; Witten and Huysseune 2009). In absence of osteocytes, the acellular fish bone is also capable to remodel (Atkins et al. 2014), so

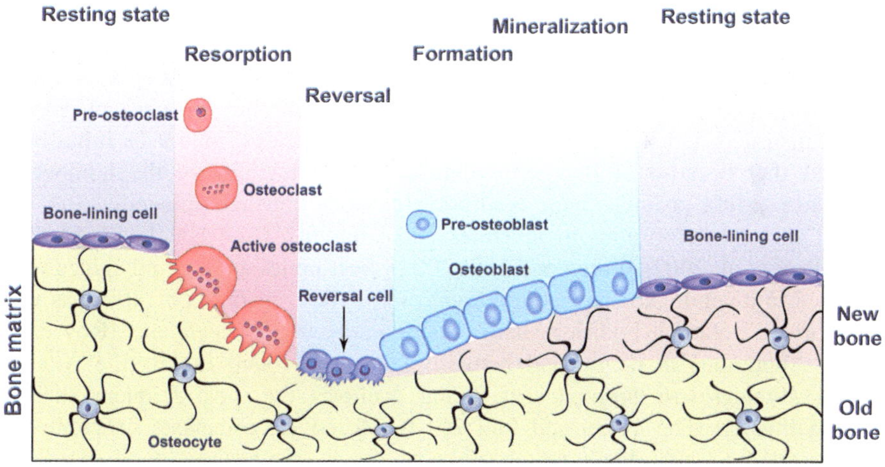

Fig. 9.1 Schematic representation of bone remodeling process (read sketch from left to right). In response to specific bone remodeling signals, the lining cells cover the bone surface in the "resting state" in order to prepare the bone matrix for osteoclast attachment. The differentiation and activation of osteoclasts starts and active osteoclasts degrade the old mineralized bone matrix ("resorption"). After osteoclasts leave the bone surface, specific mononucleated phagocytic cells remove demineralized matrix from bone surface in a phase named "reversal", and the bone formation process is stimulated. Then, bone "formation" and "mineralization" are mediated by mature osteoblasts to lead a newly deposited and mineralized bone matrix. Finally, bone surface returns to a "resting state" where the osteoblasts differentiate into osteocytes or lining cells

it has been hypothesized that other cells (e.g., osteoblasts) could trigger remodeling via alternative pathways. Hence, osteocytes seem not to be the only regulators of bone resorption and formation in teleosts. The evolution of acellular bone from cellular bone and the dominance of lacking-osteocytes bone in advanced teleosts have been supposed to bring some functional advantage, although it remains to be proven (Shahar and Dean 2013).

On the other hand, teleosts with mononucleated osteoclasts have also been shown to reabsorb bone as they have typical marker enzymes from multinucleated osteoclasts, such as tartrate resistant acid phosphatase (TRAP) or cathepsin and therefore, a bone resorption process different from their mammalian counterparts has been accepted (Nemoto et al. 2007).

In fish, the phosphate bioavailability appears to be more critical than that of calcium, so that hypophosphatemic and hyperphosphatemic conditions are more likely to trigger skeletal remodeling. However, calcium shortage conditions have been supposed to cause the resorption of postcranial dermal skeleton (i.e., scales) more probably than the endoskeleton resorption (Rotllant et al. 2005; Witten and Huysseune 2009). Scale regeneration is a common event both in freshwater and seawater fish (De Vrieze et al. 2011; Guerreiro et al. 2013; Persson et al. 1999), showing similar cell types (scleroblasts and scleroclasts) and the processes involved in bone remodeling, and may well be a preferential trigger for ion-resorption in fish, which may in a first instance, protect the endoskeleton. Interestingly, the Ca:Pi ratio varies slightly accordingly to the species and type of environment, with freshwater species generally showing lower Pi in relation to total calcium in scale (Guerreiro et al. 2013; Herrmann-Erlee and Flik 1989).

Diverse studies have shown that phosphate deficiencies, caused by shortage of dietary phosphorus or starvation, produce an increase in the frequency of physiological disorders and skeletal malformations in fish species (Table 9.1). It has been reported that because of phosphate deficiency, the increase in the number of osteoclasts could promote bone resorption causing impaired mineralization and several skeletal deformities in juveniles and adult fish (Lall and Lewis-McCrea 2007; Roy et al. 2002; Witten and Hall 2003). Soft bones and curved spines were observed in juvenile Atlantic salmon *Salmo salar* (Baeverfjord et al. 1998) as well as deformed vertebral bodies in haddock *Melanogrammus aeglefinus* (Roy et al. 2002). Alterations in neural and hemal spines were found in juvenile Atlantic halibut *Hippoglossus hippoglossus*, being scoliosis the most frequent skeletal abnormality (Lewis-McCrea and Lall 2010). Moreover, phosphate deficient diet was reported to alter the mineralization of newly synthesized bone matrix in juvenile Atlantic salmon leading to uncoupling between bone formation and mineralization (Witten et al. 2016). Interestedly, an increase of bone resorption as well as anomalies were not observed (Witten et al. 2016), unlike other investigations where the lack of mineralization is due to bone resorption and appears to be

Table 9.1 Physiological disorders and skeletal malformations reported in several fish species under phosphate deficiency

Species	Common disorders
Atlantic halibut (*Hippoglossus hippoglossus*)	Skeletal deformity (scoliosis, hemal and neural spines abnormalities)
Atlantic Salmon (*Salmo salar*)	Reduced bone mineralization (soft bones), skeletal deformity (twisted spines), reduced growth and poor feed conversion
Channel catfish (*Ictalurus punctatus*)	Reduced bone mineralization, reduced growth and poor feed conversion
Chinook salmon (*Oncorhynchus tshawytscha*)	Poor feed conversion
Common carp (*Cyprinus carpio*)	Reduced bone mineralization, skeletal deformity, cranial deformity, reduced growth and increased visceral fat
Japanese eel (*Anguilla japonica*)	Reduced growth and anorexia
Red sea bream (*Chrysophrys major*)	Reduced bone mineralization, skeletal deformity (curved and enlarged spongy vertebrae), reduced growth and poor feed conversion
Haddock (*Melanogrammus aeglefinus*)	Reduced bone mineralization, skeletal deformity (curvature vertebrae), increased number of osteoclasts and reduced growth
Rainbow trout (*Oncorhynchus mykiss*)	Reduced bone mineralization, skeletal deformity, reduced growth and poor feed conversion

(From Lall 2002; Lewis-McCrea and Lall 2010; Roy et al. 2002)

responsible for development of multiples skeletal deformities in phosphate shortage conditions. Because fish can store phosphorus in skeletal structures to compensate for high plasma levels, excess of phosphate ions over the immediate needs in haddock contribute to increase the number of osteoblasts and enhance mineralization of bone matrix (Roy et al. 2002; Roy and Lall 2003).

Additionally, phosphate insufficiency produce metabolic disturbances that may lead to impairment of growth performance and reduced body mass (Sugiura et al. 2004; Witten et al. 2016). On the other hand, hyperphosphatemic conditions have been reported to affect the synthesis of protein as well as zinc absorption inducing a decrease of growth rate in rainbow trout *Oncorhynchus mykiss* (Satoh et al. 1996).

Therefore, adequate absorption of dietary phosphorus is essential for phosphate homeostasis in fish and it must be precisely balanced to prevent a variety of phosphate metabolism related-disorders such as reduced growth, decreased feed efficiency or skeletal malformations among others (Lall 2002; Lall and Lewis-McCrea 2007). In this context, dietary phosphorus intake is closely related to the regulation of phosphate metabolism, which is under the control of endocrine and non-endocrine factors.

9.2 Control of Phosphate Metabolism

An endogenous molecular control is responsible for mineral metabolism in vertebrates. Similar to calcium homeostasis, the maintenance of adequate phosphate levels involves a coordinated interrelationship of numerous systemic and local factors acting in intestine, kidney and skeleton (Favus et al. 2006). Thus, this highly controlled process is carried out by endocrine factors, cell-signaling products acting in a paracrine/autocrine fashion and delicate feedback loops (Fig. 9.2). Historically, parathyroid hormone (PTH), vitamin D (1,25(OH)2D) and calcitonin (CT) have been considerate the most important hormones for phosphate and calcium homeostasis in mammals (Bergwitz and Jüppner 2010; Mundy and Guise 1999; Potts 2005). Over the past decade, genetic analysis information from human disorders of mineral metabolism have led to discovery of novel hormones involved in phosphate balance, far less understood than that of calcium. Particularly, the identification and characterization of fibroblast growth factor 23 (FGF23) and other local bone-derived local factors have revealed novel regulatory axis of the mineral metabolism (Bergwitz and Jüppner 2010; Sapir-Koren and Livshits 2011). In addition to this complex network of endocrine, autocrine, and paracrine signals, recent studies have pointed out the importance of the vertebrate brain as a central regulator of bone remodeling, which directly influences into the calcium and phosphate homeostasis (Quiros-Gonzalez and Yadav 2014; Takeda 2008). Thus, three main pathways have been defined in the neural control of the bone mass. The first pathway involves neuroendocrine factors from hypothalamic neurons, which are processed through pituitary gland (e.g., gonadotrophin hormone releasing hormone (GnRH), thyrotropin releasing hormone (TRH), growth hormone releasing hormone (GHRH) and oxytocin). The second pathway entails hypothalamic neuropeptides directly released into the bloodstream (e.g., pro-opiomelanocortin-derived peptides (POMC) and cocaine-and amphetamine-regulated transcript (CART), neuropeptide Y (NPY) and agouti-related peptide (AgRP) (Driessler and Baldock 2010; Quiros-Gonzalez and Yadav 2014). The third pathway consists of efferent neural signals that control the bone mass through the sympathetic (SNS) and parasympathetic nervous systems (PSNS) (Bajayo et al. 2012; De Vernejoul 2013).

Despite importance of phosphate for numerous biological processes, little is known about the molecular regulation of phosphate homeostasis in fish. Similar to mammals, the mineral metabolism in fish has traditionally been linked to the endocrine regulation of calcium (Guerreiro and Fuentes 2007), and only secondarily, of phosphate. However, newly identified factors in mammals but also in fish have been investigated in different teleost species in order to explore conserved or alternative regulatory mechanisms. Here, we describe the current knowledge about molecular factors involved in phosphate metabolism in fish. Due to the inevitable link between phosphate and calcium for bone mineral homeostasis we also outline the involvement of these factors in the calcium balance (Fig. 9.2).

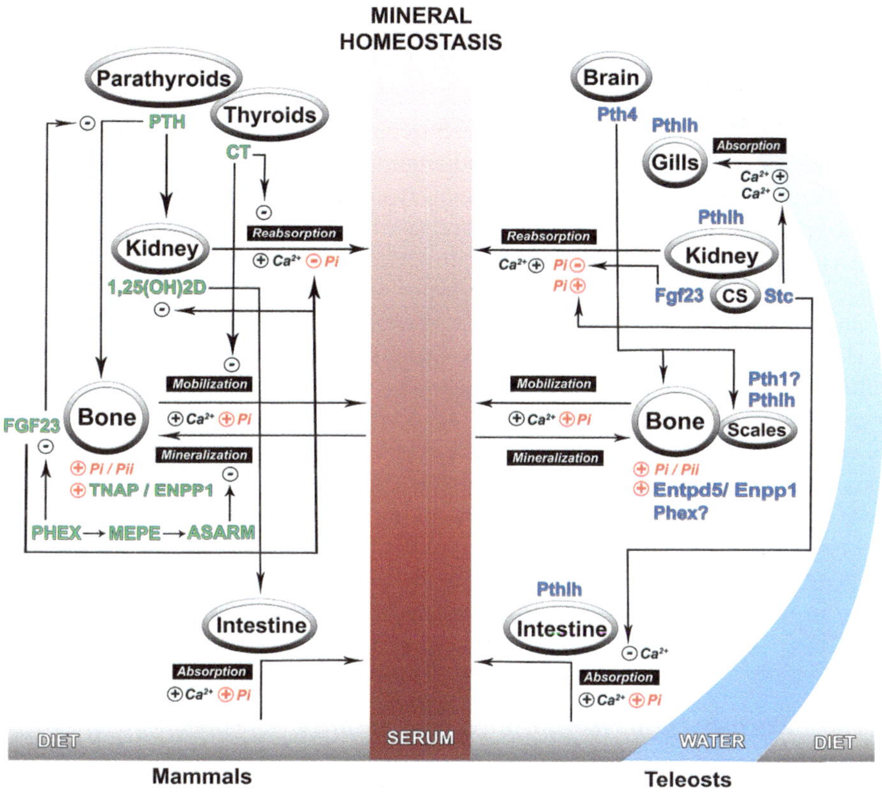

Fig. 9.2 Complex network of systemic and local factors involved in the mineral homeostasis in mammals (left side) and teleosts (right side). Normal physiological levels of serum phosphate and calcium are maintained by the interplay of intestinal absorption from diet, branchial absorption from surrounding water (in teleosts), kidney reabsorption and mobilization/mineralization in skeleton. Similar endocrine factors, local signals acting in paracrine/autocrine pathways and negative feedback mechanisms preserve the mineral homeostasis in both mammals and teleosts

9.2.1 Stanniocalcin

Stanniocalcin (Stc) was originally identified in the fish-specific endocrine glands known as corpuscles of Stannius (CS) (Wagner et al. 1986). CS are small spherical bodies generally located on the ventral surface of the kidney in bony fish (Yeung et al. 2012). Stc is the predominant hormone that regulates calcium and phosphate metabolism in fish. An increase in the calcium plasma levels produce the secretion of Stc into the blood, which acts in target organs (such as gills, intestine and kidney) in order to restore the normocalcemia (Lall 2002).

Teleost fish, like all vertebrates, have two isoforms of stanniocalcin (*i.e. stc1* and *stc2*), as result of two rounds of genome duplication that occurred in early vertebrate evolution (1R/2R WGD) (Cañestro 2012). Besides, teleost fish specifically underwent a third round of genome duplication (3R or teleost specific genome duplication) so that two genes for each isoform (i.e., *stc1a/stc1b* and *stc2a/stc2b*) were also identified in the teleost genome (Braasch and Postlethwait 2012; Schein et al. 2012).

Stc genes are not exclusively produced in the CS and studies demonstrated that these factors are expressed in numerous tissues suggesting a possible mechanism of local regulation besides an endocrine pathway (Mccudden et al. 2001). In response to a rise of serum calcium levels, Stc is released into the bloodstream and acts as a hypocalemic/hyperphosphatemic factor in order to control the calcium and phosphate homeostasis. Thus, this endocrine factor acts in gills and intestine to reduce or inhibit the calcium transport via apical epithelial calcium channel (ECaC) (Sundell et al. 1992; Tseng et al. 2009), reduces calcium absorption by increasing calcium efflux and precipitation in intestinal fluid (Fuentes et al. 2010) and promotes the phosphate reabsorption in kidney (Lu et al. 1994). Salmon Stc was reported to stimulate the net phosphate reabsorption by a cAMP-dependent pathway in renal proximal tubule culture of winter flounder *Pseudopleuronectes americanus* (Lu et al. 1994). Unlike the other *stc* genes, *stc1a* gene expression was strongly up-regulated in the CS of spotted green pufferfish *Tetraodon nigroviridis* after an increase in the water calcium concentration, which suggested the involvement of *stc1a* in the extracellular calcium balance (Schein et al. 2012).

9.2.2 Parathyroid Hormone (PTH) Family

The PTH family is composed by a group of peptides which play important roles in calcium-phosphate homeostasis, bone remodeling and in multiples developmental processes in vertebrates (Broadus and Stewart 1994; Guerreiro et al. 2007; Kronenberg 2006; Potts 2005; Pinheiro et al. 2010; Suarez-Bregua et al. 2017). The peptides in the PTH family share a highly conserved PTH-specific domain and mediate their actions through three G-protein coupled receptors (GPCR) (Venkatakrishnan et al. 2013): parathyroid hormone receptor 1 (PTH1R renamed as Pth1ra in teleost fish), PTH2R and PTH3R (renamed as Pth1rb in teleost fish) (Gardella and Jüppner 2001; Hoare et al. 2000; Pinheiro et al. 2012; Rubin et al. 1999; Suarez-Bregua et al. 2017). Unlike mammals, where parathyroid hormone (PTH1 (PTH)) is the master endocrine factor regulating bone mineral homeostasis, in fish parathyroid hormone like hormone (Pth3 (Pthlh)) and parathyroid hormone 4 (Pth4) appears to have a more prominent role in such functions (Guerreiro et al. 2007; Potts 2005; Suarez-Bregua et al. 2017).

Parathyroid Hormone (PTH1) Human parathyroid hormone (hPTH) is secreted from parathyroid glands in response to low blood calcium and high phosphate levels. As an endocrine factor, PTH acts in the kidney to increase the synthesis and activation of 1,25-dihydroxyvitamin D (1,25(OH)2D) that, indirectly, increases the intestinal calcium absorption. Additionally, PTH increases the distal tubular calcium reabsorption and prevents the phosphate reabsorption in kidney (Potts 2005). Eventually, PTH mobilizes calcium and phosphate via PTH1R signaling in bone which directly contributes to maintenance of the serum mineral levels (Carter and Schipani 2006; Gensure et al. 2005; Potts 2005). Although in fish two different forms of PTH (i.e., Pth1a and Pth1b) have been identified and analyzed, neither of them appears to have a decisive role in bone mineral metabolism, and contradictory results have been published in the literature. While on the one hand it has been reported that Pth1a is involved in the maintenance of calcium homeostasis through scales resorption in goldfish *Carassius auratus* (Suzuki et al. 2011) and by promoting the differentiation of ionocytes in low water calcium (Kwong and Perry 2015). On the other, this hypercalcemic effect was not found in gilthead seabream *Sparus aurata* scales and instead, Pth1b appeared to cause the ion mobilization from postcranial dermal skeleton (Canario et al. 2006).

Parathyroid Hormone Like Hormone (PTH3) PTH3 (formerly PTHLH) is involved in calcium homeostasis but also in the embryonic development of the skeleton and other tissues in mammals. It is produced by multitude of organs and mediates its actions via autocrine, paracrine and endocrine pathways across the body (Gardella et al. 2006; Kronenberg 2006; VanHouten et al. 2004). An hypercalcemic factor firstly found in Atlantic cod *Gadus morhua* and eel *Anguilla anguilla* pituitaries by (Parsons et al. 1978) was most likely the Pthlh (i.e., Pth3), subsequently described in the gilthead seabream using immunological methods (Danks et al. 1993; Devlin et al. 1996) and has the first gene from the parathyroid hormone family characterized in fish (gilthead seabream and Japanese pufferfish *Takifugu rubripes*) (Flanagan et al. 2000; Power et al. 2000). Although Pth3 was discovered in the pituitary gland, later Pth3 protein was found in plasma and in several different tissues such as gills, operculum, kidney, pituitary gland, brain, saccus vasculosus, muscle, skin, spleen, liver and intestine (Abbink and Flik 2007; Guerreiro et al. 2007). The widespread distribution of Pth3 across the fish tissues involves a wide range of functions through both endocrine and local pathways. Fish have two co-orthologs (i.e., Pthlha (Pth3a) and Pthlhb (Pth3b)) which seem to have a dual role in bone formation and mineralization but also in mineral homeostasis (Yan et al. 2012). Besides its known hypercalcemic function acting in the branchial and intestinal epithelium in larval and post-larval fish (Guerreiro et al. 2001) and in juveniles and adults (Abbink et al. 2006; Fuentes and Figueiredo 2006; Guerreiro et al. 2001), this peptide has been reported to play a role in the phosphate homeostasis by acting in the kidney of winter flounder. Pth3 increases the net secretion of Pi while promotes calcium uptake in primary culture of kidney proximal tubule cells (Guerreiro et al. 2010). Moreover, it has been postulated that the receptors Ph1ra and Pth1rb could mediate the Pth3 actions since both receptors have been found in renal tissue (Guerreiro et al. 2010). On the other hand, piscine Pth3 showed specific actions in scales where produced an increase of osteoclast activity and, consequently, the calcium mobilization (Rotllant et al. 2005). Due to calcium and phosphate mobilization from skeleton are necessarily linked, it is possible that this peptide also exerts a hyperphosphatemic action via scale resorption.

Parathyroid Hormone 4 (PTH4) Pth4 is a recently identified PTH family member unique to non-placental mammals that plays important roles in bone mineral homeostasis (Suarez-Bregua et al. 2017). Pth4 (formerly Pth-l) was initially isolated in Japanese pufferfish *Takifugu rubripes* and it was reported to stimulate the calcium uptake from water and ion mobilization from gilthead seabream scales (Canario et al. 2006). The hypercalcemic activity was also studied in other tetrapods such as *Xenopus* and chicken, where the Pth4 effect in calcium absorption was similar to that found in fish (Pinheiro et al. 2010). In zebrafish *Danio rerio*, Pth4 is expressed by two clusters of neurons in the lateral hypothalamus and it has been shown to participate in a novel brain to bone pathway regulating bone mass accrual (Suarez-Bregua et al. 2017). Functional studies showed a decrease of bone mineral density in adult fish as well as impaired mineralization during larvae development. Additionally, zebrafish with bone mineral defects exhibited changes in the phosphate homeostasis gene markers expressed in bone but also in kidney and intestine, suggesting that this neuropeptide directly and/or indirectly participates in the phosphate metabolism across the fish body. Particularly, Pth4 was lost during the vertebrate evolution so that it has been hypothesized to be a central regulator of bone mineral metabolism in fish similarly to PTH in mammals, with the exception of that Pth4 is a brain-derived peptide and PTH is an endocrine factor secreted by the parathyroid glands (Gardella et al. 2006; Suarez-Bregua et al. 2017).

9.2.3 Fibroblast Growth Factor 23

Fibroblast growth factor (FGF23) is a novel member from fibroblast growth factor family (FGF) and it has been identified in humans as the causative gene mutated in autosomal-dominant hypophosphatemic rickets (ADHR) (Econs et al. 2000) and overexpressed in tumor-induced osteomalcia (TIO) (Shimada et al. 2001). In mammals, this peptide acts as a phosphatemic endocrine hormone in the bone-kidney regulatory axis. It is secreted by osteoblasts/osteocytes in bone and targets kidney in order to control the phosphate balance (Lu and Feng 2012). The effects of FGF23 in kidney are mediated by the binding to fibroblast growth factor receptor (FGFR) and to an essential cofactor named αKlotho (Urakawa et al. 2006).

FGF23 reduces the physiological reabsorption of phosphate in the renal proximal tubule by down-regulating the expression of type 2a and type 2c sodium-phosphate cotransporters (i.e., NPT2a and NPT2c) (Shimada et al. 2003). Moreover, this hormone controls the vitamin D metabolism by reducing the serum levels of 1,25 (OH)2D through inhibition of the vitamin D metabolizing enzymes in the kidney. Thus, FGF23 indirectly promotes the decrease of the intestinal phosphate absorption, via type 2b sodium-phosphate cotransporter (NPT2b), as well as the PTH synthesis and secretion (Hori et al. 2011; Sapir-Koren and Livshits 2011; Shimada et al. 2003). On the other hand, PTH could play a direct role in the FGF23 regulation but, although the reciprocal regulation between both factors have been studied, the results are still controversial (Lu and Feng 2012).

Of note, αKlotho cofactor, independently of FGF23, has been reported to induce hypophosphatemia through inactivation of NPT2a and NPT2c in the proximal tubule in kidney (Hu et al. 2010).

In zebrafish, *fgf23* gene expression was restricted to CS, which are closely located to kidney and are involved in calcium-phosphate homeostasis regulation in fish. In turn, the gene expression of *αklotho* was predominantly detected in the kidney, but also in brain, pancreas and liver (Mangos et al. 2012). Overexpression of *fgf23* in the CS was found in transient receptor potential melastatin 7 (*trpm7*) zebrafish mutants, which exhibit a dysregulation in the calcium and phosphate homeostasis. Loss of function studies using morpholino showed that *fgf23* expression contributes to kidney stone formation due to an increase in the precipitation of calcium phosphate (Elizondo et al. 2010). Therefore, a conserved FGF23 signaling pathway has been suggested in fish, although its underlying mechanism remains unknown.

9.2.4 Phosphate-Regulating Gene with Homology to Endopeptidase on the X Chromosome

PHEX (phosphate-regulating gene with homology to endopeptidase on the X chromosome) has been identified as the mutated gene responsible for the X-chromosome linked hypophosphatemia (XLH) in humans (Francis et al. 1995). This X-linked inherited syndrome causes a severe disorder of phosphate homeostasis and it is considered the most common of rickets in humans (Tenenhouse 1999). PHEX is predominantly expressed in teeth and bone, and it is located in the membrane of odontoblasts/osteoblasts where exerts its proteolytic activity (Ruchon et al. 1998; Liu et al. 2006). Loss of PHEX causes defective mineralization of skeletal structures, impaired renal phosphate handling and aberrant vitamin D metabolism (Quarles and Drezner 2001). Although PHEX is considered a bone-derived local factor that acts through autocrine/paracrine pathways, it has been reported to indirectly regulate the mineralization and renal phosphate balance by controlling other bioactive peptides in bone (Rowe 2004; Sapir-Koren and Livshits 2011). It has been suggested that PHEX is directly or indirectly involved in FGF23 regulation since deficiency of PHEX produces high levels of *FGF23* gene expression (Ichikawa et al. 2012; Kiela and Ghishan 2009; Yuan et al. 2008). However, the underlying molecular mechanism remains unknown. On the other hand, the interaction between PHEX and other bioactive peptides has also been described. Some studies have shown that PHEX specifically hydrolyzes MEPE, a bone-renal extracellular matrix protein, and release the ASARM peptide (acidic serine aspartate-rich MEPE-associated peptide) (Rowe 2004, 2012). ASARM released as phosphorylated peptide inhibits the bone mineralization (Martin et al. 2008).

In fish, the role of Phex in mineralization and phosphate balance remains unclear. In not bone (*nob)* mutants zebrafish, which exhibit a non-mineralized bone, the *phex* expression levels remained unchanged (Huitema et al. 2012). The same results were found in dragonfish (*dgf)* mutants, characterized by displaying an increased and ectopic mineralization, where levels of *phex* expression were not disturbed (Apschner et al. 2014). However, *phex* gene expression was decreased in zebrafish larvae with impaired craniofacial mineralization. The down-regulation of *phex* expression was correlated with the decrease of *pth4* transcript levels, as consequence of the Pth4 neuronal ablation in larvae, which caused an alteration in the bone and phosphate homeostasis (Suarez-Bregua et al. 2017).

9.2.5 Ectonucleoside Triphosphate/Diphosphohydrolase 5 (entpd5) and Ectonucleotide Pyrophosphatase Phosphodiesterase 1(enpp1)

The phosphate homeostasis and bone mineralization are carried out by the counterbalance of phosphate and calcium efflux/influx among bone, kidney and intestine. In addition, the tight balance of phosphate/pyrophosphate (Pi/PPi) levels in the bone environment has been reported to be crucial for the proper bone mineralization (Sapir-Koren and Livshits 2011). PPi is a mineralization inhibitor since it prevents the formation of hydroxyapatite crystals on the extracellular bone matrix (Terkeltaub 2001). At molecular level, the Pi/PPi ratio in mammals is mainly maintained by the balance between tissue-nonspecific alkaline phosphatase (TNAP) and ENPP1. TNAP hydrolyzes the PPi or ATP to generate Pi, changing ratio towards mineralization (Ciancaglini et al. 2009). ENPP1 generates intracellular PPi from nucleoside triphosphates to inhibit the ability of Pi to crystallize along with calcium (Terkeltaub 2001).

In fish, two important factors also appear to be crucial in the control of Pi/Pii normal physiological levels: Entpd5 and Enpp1 (Apschner et al. 2014; Huitema et al. 2012). Entpd5 is a bone-derived local factor produced by osteoblasts and it has been undiscovered as the causative gene responsible for the *nob* mutant phenotype in zebrafish. Non-mineralized phenotype in *entpd5* mutant fish was rescued by exposition to excess of inorganic phosphate dissolved in water, which demonstrated that Entpd5 regulates the phosphate balance for bone mineralization (Huitema et al. 2012). On the other hand, *enpp1* have been identified as the gene mutated underlying the *dgf* zebrafish mutant phenotype (Huitema et al. 2012). Similar to mammals, Enpp1 generates PPi that is necessary to maintain the Pi/Pii balance for the correct mineralization. Unlike *entpd5*, which is only expressed by osteoblasts in tissues associated to skeletal mineralization, *enpp1* present a ubiquitous expression pattern as in mammals, although a higher level of expression in bone structures was found (Apschner et al. 2014; Murshed et al. 2005). Mutation of *enpp1* gene gives rise to an insufficient proportion of PPi to compensate the Pi levels in the extracellular bone matrix or in other soft tissues such as skin, cartilage structures, notochord sheet and heart. As a result a generalized and ectopic mineralization in the body is produced (Apschner et al. 2014; Huitema et al. 2012; Terkeltaub 2001). Therefore, Entpd5 and Enpp1 are crucial proteins involved in the Pi/Pii balance, which links the phosphate homeostasis and mineralization.

9.3 Dietary Phosphorus: Implications for Sustainable Fish Farming

Aquaculture is the fastest growing animal production sector in the world. Intensive fish farming has expanded in the recent years using high-cost nutrients from commercial standard diets (Cho 1993). In fact, nutrition and feeding represents from 40 to 50% of the total production costs in most fish farming companies (Craig and Helfrich 2009). However, the nutritional composition of the commercial diets and the feeding strategy used are often not appropriate for particular fish requirements which results in a low growth performance as well as an increase of waste outputs in aquatic systems (Cho and Bureau 2001; Hixson 2014). Because of that, long-term sustainable fish farming needs to develop new and nutritionally balanced diets as well as to minimize their environmental impact.

In fish culture, phosphorus must be supplied by the diet as intestinal absorption is the main source of Pi intake (Lall 2002). The bioavailability of phosphorus in fish diets is highly variable and depends on the type of feed ingredients, the chemical form of phosphorus, the digestibility and stability of the diets or the interaction with several nutrients, among others (Cho and Bureau 2001; Lall 2002). Traditionally, commercial fish diets have been composed of a high proportion of fish meal or meat and bone meal which contribute from 1.5 to 3.2% and from 3.5 to 5.5% of phosphorus, respectively (Lall 2002). The relatively high levels of phosphorus in these feed ingredients come from skeletal structures which mostly contains Pi readily available for fish absorption (Hardy and Gatlin 2002). Thus, fish meal or meat and bone meal have been considered adequate sources of phosphorus in diets for fish culture. However, the content of phosphorus in animal byproducts-based commercial diets is substantially higher than the basal requirements of phosphorus in most finfish species. In fact, phosphorus requirements have been reported for several finfish species where the requirements vary between 0.3 and 0.9%, being an average of 0.6% for rainbow trout, Atlantic salmon and chum salmon *Oncorhynchus keta* (reviewed in (Lall 2002)). Therefore, fishmeal or meat and bone meal-based commercial diets contain an excess of phosphorus respect to the basic requirements in fish. In the last two decades there has been a tendency to replace fish meal-based diets by plant protein-based diets but, however, they still have several nutritional drawbacks compared to fish meal, including the presence of indigestible phosphorus types, particularly for carnivorous fish species (Hixson 2014; Olsen and Hasan 2012). Because of that, phosphate enriched diets from animal sources continue to be commonly used in fish farming to ensure growth and prevent skeletal disorders. The most common skeletal anomalies in fish hatcheries include soft bones, curved spines, twisted pleural ribs and several vertebral disorders such as compressed vertebral bodies that can result in scoliosis and cephalic or neck anomalies (Boglione et al. 2013; Lall and Lewis-McCrea 2007). The cellular processes which give rise to low mineralized skeleton could be due to a under-mineralization of the new bone matrix formed by osteoblasts (i.e., osteomalcia or soft bones) or an increase of bone resorption by osteoclasts and/or osteocytes (i.e., osteoporotic bone) (To et al. 2012; Witten and Huysseune 2009). Also, osteoclast-independent resorption could be driven by the halastatic mineral loss that results from a depletion of the mineral content with no degradation of the organic bone matrix (Kacem and Meunier 2003). Similar to humans, low mineralized bone has been related to prevalence of skeletal deformities, which is an important issue for larvae growth and fish production (Sugiura et al. 2004).

Some studies have assessed the appearance of skeletal disorders in reared larval and juvenile fish fed on different concentrations of phosphorus. Juvenile haddock fed on low phosphorus diet (0.42%) exhibited a decrease bone mineral content and deformed vertebrae (Roy et al. 2002). Hyperphosphatemic haddock (1.42% dietary phosphorus) showed reduced vertebral ash content as well as an increase in the plasma phosphate and a high amount of urinary phosphate excretion (Roy et al. 2002). Atlantic salmon post-smolts fed on phosphate deficient diet (0.47% dietary phosphorus) during ten weeks showed osteomalcia (soft bones) (Witten et al. 2016); but, bone deformities were not found. However, in Atlantic salmon juveniles, the occurrence of spinal anomalies has been related with dietary phosphorus restriction (Sullivan et al. 2007). In juvenile rainbow trout, vertebrae and scales were analyzed after feeding low (0.50% total phosphorus) versus standard phosphorus diet (0.92% total phosphorus) (Le Luyer et al. 2014). Although dietary phosphorus restriction caused a decrease in the mineral content of scales, significant mineral loss in vertebrae was not found, which suggests that initial mineral mobilization from exoskeleton takes place under these particular conditions. However, dietary phosphorus restriction in rainbow trout has been associated with increased prevalence of bone anomalies such as biconcave vertebrae, homogeneous compressions, small and widely spaced vertebrae (Le Luyer et al. 2014).

The excess of dietary phosphorus absorbed but not retained by fish is excreted as urine into the water. On the other hand, the ingested excess or non-digestible dietary phosphorus excreted as feces and the additional non-ingested food are deposited in the sediment (in semi-intensive fish farming) or accumulated in the bottom of the tanks (in intensive fish farming) (Hardy and Gatlin 2002; Lall 2002). Thus, phosphate waste outputs from fish culture can either be soluble or particulate. Soluble phosphate wastes are readily available as nutrients in natural waters while particulate forms accumulated in the sediment have a major impact in the ecosystem since the degradation of excessive organic matter leads anoxia and environmental toxicity (Lall 2002; Piedecausa et al. 2012). Phosphate enriched effluents from fish farming will affect the quality of the water and will cause eutrophication of the aquatic ecosystems. In these circumstances, one of the current major concerns of finfish aquaculture is to minimize the phosphorus discharges in natural waters by reducing the level and/or increasing the bioavailability of phosphorus in key feed ingredients.

Therefore, physiological problems will arise not only from phosphate depletion but also from excessive phosphate availability. Increased absorption and hyperphosphatemia create important Ca:Pi imbalance, create a permanent load on hormonal regulatory systems, and ultimately lead to the appearance of diseases and malformations, reduced growth or even death. Many cultured fish show abundant renal calculi, a sign of impaired kidney function probably due to inadequate mineral diet. Ectopic calcification leads to skeletal malformations, with both biological and economic implications, and precocious skeletal mineralization and improper bone remodeling can impair fish growth right from its initial stages.

Although dietary phosphorus plays an important role in bone mineral metabolism of cultured fish larvae and adults, skeletal disorders are probably the result of complex interactions between nutritional, environmental and genetic factors (Lall and Lewis-McCrea 2007).

A better understanding of the molecular mechanisms involved in the phosphate homeostasis and bone metabolism in fish is essential. New molecular markers will help us to broad our knowledge about the factors that are behind the phenotypic changes such as skeletal malformations and impaired growth performance. As for many other aspects of the fish life cycle, it is also likely that phosphate requirements change for each life stage—larvae, juveniles, and adults. Thus, changes in phosphate metabolism markers could be associated to particular nutritional requirements in order to improve the formulation of standard diets for fish farming. As consequence, a balance between the feed costs and optimal growth and healthy fish will be achieved. Additionally, the phosphorus wastes will be reduced minimizing the environmental impact, which contributes to a sustainable aquaculture for the future.

References

Abbink W, Flik G (2007) Parathyroid hormone-related protein in teleost fish. Gen Comp Endocrinol 152:243–251. https://doi.org/10.1016/j.ygcen.2006.11.010

Abbink W, Bevelander GS, Rotllant J et al (2004) Calcium handling in *Sparus auratus*: effects of water and dietary calcium levels on mineral composition, cortisol and PTHrP levels. J Exp Biol 207:4077–4084. https://doi.org/10.1242/jeb.01254

Abbink W, Bevelander GS, Hang X et al (2006) PTHrP regulation and calcium balance in sea bream (*Sparus auratus* L.) under calcium constraint. J Exp Biol 209:3550–3557. https://doi.org/10.1242/jeb.02399

Apschner A, Huitema LF, Ponsioen B et al (2014) Zebrafish enpp1 mutants exhibit pathological mineralization, mimicking features of generalized arterial calcification of infancy (GACI) and pseudoxanthoma elasticum (PXE). Dis Model Mech 7:811–822. https://doi.org/10.1242/dmm.015693

Atkins A, Dean MN, Habegger ML et al (2014) Remodeling in bone without osteocytes: billfish challenge bone structure-function paradigms. Proc Natl Acad Sci U S A 111:16047–16052. https://doi.org/10.1073/pnas.1412372111

Baeverfjord G, Asgard T, Shearer KD (1998) Development and detection of phosphorus deficiency in Atlantic salmon, *Salmo salar* L., parr and post-smolts. Aquac Nutr 4:1–11. https://doi.org/10.1046/j.1365-2095.1998.00095.x

Bajayo A, Bar A, Denes A et al (2012) Skeletal parasympathetic innervation communicates central IL-1 signals regulating bone mass accrual. Proc Natl Acad Sci U S A 109:15455–15460. https://doi.org/10.1073/pnas.1206061109

Bansal VK (1990) Serum Inorganic Phosphorus. In: Walker HK, Hall WD, Hurst JW (ed) Clinical methods: the history, physical, and laboratory examinations. Butterworths, Boston, pp 895–899

Bergwitz C, Jüppner H (2010) Regulation of phosphate homeostasis by PTH, vitamin D, and FGF23. Annu Rev Med 61:91–104. https://doi.org/10.1146/annurev.med.051308.111339

Boglione C, Gavaia P, Koumoundouros G et al (2013) Skeletal anomalies in reared European fish larvae and juveniles. Part 1: normal and anomalous skeletogenic processes. Rev Aquac 5 (Suppl. 1):S99–S120. https://doi.org/10.1111/raq.12015

Braasch I, Postlethwait JH (2012) Polyploidy in Fish and the Teleost Genome Duplication. In: Soltis PS, Soltis DE (eds) Polyploidy and Genome Evolution. Springer, Berlin, Heidelberg, pp 341–383

Broadus A, Stewart A (1994) Parathyroid hormone-related protein: structure, processing, and physiological actions. In: Bilezikian J (ed) The parathyroids. Basic and clinical concepts, Raven Press pp 259–294

Canario AVM, Rotllant J, Fuentes J et al (2006) Novel bioactive parathyroid hormone and related peptides in teleost fish. FEBS Lett 580:291–299. https://doi.org/10.1016/j.febslet.2005.12.023

Cañestro C (2012) Two Rounds of whole-genome duplication: evidence and impact on the evolution of vertebrate innovations. In: Soltis P, Soltis DE (eds) Polyploidy and genome evolution pp 309–339

Carter PH, Schipani E (2006) The roles of parathyroid hormone and calcitonin in bone remodeling: prospects for novel therapeutics. Endocr Metab Immune Disord Drug Targets 6:59–76

Cho CY (1993) Digestibility of feedstuffs as a major factor in aquaculture waste management. In: Kaushik SJ, Luquet P (eds) Fish nutrition in practice, KAUSHIK, S. INRA ed. pp 365–374

Cho CY, Bureau DP (2001) A review of diet formulation strategies and feeding systems to reduce excretory and feed wastes in aquaculture. Aquac Res 32:349–360. https://doi.org/10.1046/j.1355-557x.2001.00027.x

Ciancaglini P, Yadav MC, Sper Simão AM et al (2009) Kinetic analysis of substrate utilization by native and TNAP-, NPP1- or PHOSPHO1-deficient matrix vesicles. J Bone Miner Res 25 (091029140456050):37. https://doi.org/10.1359/jbmr.091023

Craig S, Helfrich LA (2009) Understanding fish nutrition, feeds, and feeding. Virginia Cooperative extension pp 420–256

Danks JA, Devlin AJ, Ho PM et al (1993) Parathyroid hormone-related protein is a factor in normal fish pituitary. Gen Comp Endocrinol 92:201–212. https://doi.org/10.1006/gcen.1993.1156

De Vernejoul MC (2013) More on bone and nerves. IBMS Bonekey 10:1–2. https://doi.org/10.1038/bonekey.2013.47

De Vrieze E, Sharif F, Metz JR et al (2011) Matrix metalloproteinases in osteoclasts of ontogenetic and regenerating zebrafish scales. Bone 48:704–712. https://doi.org/10.1016/j.bone.2010.12.017

Dean MN, Ekstrom L, Monsonego-Ornan E et al (2015) Mineral homeostasis and regulation of mineralization processes in the skeletons of sharks, rays and relatives (Elasmobranchii). Cell Dev Biol 46:51–67. https://doi.org/10.1016/j.semcdb.2015.10.022

Devlin AJ, Danks JA, Faulkner MK et al (1996) Immunochemical detection of parathyroid hormone-related protein in the saccus vasculosus of a teleost fish. Gen Comp Endocrinol 101:83–90. https://doi.org/10.1006/gcen.1996.0010

Driessler F, Baldock P (2010) Hypothalamic regulation of bone. J Mol Endocrinol 45:175–181. https://doi.org/10.1677/JME-10-0015

Econs MJ, Strom TM, White KE et al (2000) Autosomal dominant hypophosphataemic rickets is associated with mutations in FGF23. Nat Genet 26:345–348. https://doi.org/10.1038/81664

Elizondo MR, Budi EH, Parichy DM (2010) trpm7 regulation of in vivo cation homeostasis and kidney function involves stanniocalcin 1 and fgf23. Endocrinology 151:5700–5709. https://doi.org/10.1210/en.2010-0853

Favus MJ, Bushinsky DA, Lemann J (2006) Regulation of calcium, magnesium, and phosphate metabolism. In: American society for bone and mineral research pp 76–117

Flik G, Verbost PM (1993) Calcium transport in fish gills and intestine. J Exp Biol 184:17–29

Francis F, Hennig S, Korn B et al (1995) A gene (PEX) with homologies to endopeptidases is mutated in patients with X-linked hypophosphatemic rickets. The HYP consortium. Nat Genet 11:130–136. https://doi.org/10.1038/ng1095-130

Flanagan JA, Power DM, Bendell LA et al (2000) Cloning of the cDNA for sea bream (*Sparus aurata*) parathyroid hormone-related protein. Gen Comp Endocrinol 118:373–382

Fuentes J, Figueiredo J (2006) Parathyroid hormone-related protein regulates intestinal calcium transport in sea bream (*Sparus auratus*). Am J Physiol Regul Integr Comp Physiol 291:1499–1506. https://doi.org/10.1152/ajpregu.00892.2005

Fuentes J, Power DM, Canario AV (2010) Parathyroid hormone-related protein-stanniocalcin antagonism in regulation of bicarbonate secretion and calcium precipitation in a marine fish intestine. Am J Physiol Regul Integr Comp Physiol 299:R150–R158. https://doi.org/10.1152/ajpregu.00378.2009

Gardella TJ, Jüppner H (2001) Molecular properties of the PTH/PTHrP receptor. Trends Endocrinol Metab 12:210–217

Gardella TJ, Jüppner H, Brown EM et al (2006) Parathyroid hormone and parathyroid hormone–related peptide in the regulation of calcium homeostasis and bone development. In: Jameson JL, De Groot LJ, Kretser D et al. (eds) Endocrinology: adult and pediatric pp 1377–1417

Gensure RC, Gardella TJ, Jüppner H (2005) Parathyroid hormone and parathyroid hormone-related peptide, and their receptors. Biochem Biophys Res Commun 328:666–678. https://doi.org/10.1016/j.bbrc.2004.11.069

Guerreiro PM, Fuentes J (2007) Control of calcium balance in fish. In: Baldisserotto B, Mancera JM, Kapoor BG (eds) Fish osmoregulation. Science Publisher, USA

Guerreiro PM, Fuentes J, Power DM et al (2001) Parathyroid hormone-related protein: a calcium regulatory factor in sea bream (*Sparus aurata* L.) larvae. Am J Physiol Regul Integr Comp Physiol 281:R855–R860

Guerreiro PM, Renfro JL, Power DM et al (2007) The parathyroid hormone family of peptides: structure, tissue distribution, regulation, and potential functional roles in calcium and phosphate balance in fish. Am J Physiol Regul Integr Comp Physiol 292:R679–R696. https://doi.org/10.1152/ajpregu.00480.2006

Guerreiro PM, Canario AVM, Power DM et al (2010) Piscine PTHrP regulation of calcium and phosphate transport in winter flounder renal proximal tubule primary cultures. Am J Physiol Regul Integr Comp Physiol 299:603–611. https://doi.org/10.1152/ajpregu.00509.2009

Guerreiro PM, Costa R, Power DM (2013) Dynamics of scale regeneration in seawater—and brackish water-acclimated sea bass, *Dicentrarchus labrax*. Fish Physiol Biochem 39:917–930. https://doi.org/10.1007/s10695-012-9751-9

Hadjidakis DJ, Androulakis II (2006) Bone remodeling. Ann N Y Acad Sci 1092:385–396. https://doi.org/10.1196/annals.1365.035

Hardy R, Gatlin D (2002) Nutritional strategies to reduce nutrient losses in intensive aquaculture. In: Cruz-Suárez LE., Ricque-Marie D, Tapia- Salazar M, Gaxiola-Cortés MG, Simoes N (eds) Avances en Nutrición Acuícola VI. México, 2002

Herrmann-Erlee M, Flik G (1989) Bone: comparative studies on endocrine involvement in bone metabolism. In: Pang P, Schreibman M (eds) Vertebrate endocrinology: fundamentals and biomedical implications - regulation of calcium and phosphate. Academic Press, San Diego, pp 211–242

Hixson SM (2014) Fish nutrition and current issues in aquaculture: the balance in providing safe and nutritious seafood, in an environmentally sustainable manner. J Aquac Res Dev 5:1–10. https://doi.org/10.4172/2155-9546.1000234
Hoare S, Rubin D, Jüppner H et al (2000) Evaluating the ligand specificity of zebrafish parathyroid hormone (PTH) receptors: comparison of PTH, PTH-related protein, and tuberoinfundibular peptide of 39. Endocrinology 141:3080–3086
Hori M, Shimizu Y, Fukumoto S (2011) Minireview: fibroblast growth factor 23 in phosphate homeostasis and bone metabolism. Endocrinology 152:4–10. https://doi.org/10.1210/en.2010-0800
Hu MC, Shi M, Zhang J et al (2010) Klotho: a novel phosphaturic substance acting as an autocrine enzyme in the renal proximal tubule. FASEB J 24:3438–3450. https://doi.org/10.1096/fj.10-154765
Huitema LF, Apschner A, Logister I et al (2012) Entpd5 is essential for skeletal mineralization and regulates phosphate homeostasis in zebrafish. Proc Natl Acad Sci U S A 109:21372–21377. https://doi.org/10.1073/pnas.1214231110
Ichikawa S, Austin AM, Gray AK et al (2012) A Phex mutation in a murine model of X-linked hypophosphatemia alters phosphate responsiveness of bone cells. J Bone Miner Res 27:453–460. https://doi.org/10.1002/jbmr.544
Kacem A, Meunier FJ (2003) Halastatic demineralization in the vertebrae of Atlantic salmon, during their spawning migration. J Fish Biol 63:1122–1130. https://doi.org/10.1046/j.1095-8649.2003.00229.x
Kiela PR, Ghishan FK (2009) Recent advances in the renal-skeletal-gut axis that controls phosphate homeostasis. Lab Invest 89:7–14. https://doi.org/10.1038/labinvest.2008.114
Kini U, Nandeesh BN (2012) Physiology of bone formation, remodeling, and metabolism. In: Fogelman I, Gnanasegaran G, Wall H (eds) Radionuclide and Hybrid Bone Imaging. Springer, Berlin, Heidelberg, pp 29–57
Kronenberg HM (2006) PTHrP and skeletal development. Ann N Y Acad Sci 1068:1–13. https://doi.org/10.1196/annals.1346.002
Kwong RWM, Perry SF (2015) An Essential Role for Parathyroid Hormone in Gill Formation and Differentiation of Ion-Transporting Cells in Developing Zebrafish. Endocrinology 156:2384–2394. https://doi.org/10.1210/en.2014-1968
Lall SP (2002) The minerals. In: Halver JE, Hardy RW (eds) Fish nutrition. Academic Press, pp 259–308
Lall SP, Lewis-McCrea LM (2007) Role of nutrients in skeletal metabolism and pathology in fish — An overview. Aquaculture 267:3–19. https://doi.org/10.1016/j.aquaculture.2007.02.053
Le Luyer J, Deschamps MH, Proulx E et al (2014) Responses of different body compartments to acute dietary phosphorus deficiency in juvenile triploid rainbow trout (*Oncorhynchus mykiss*, Walbaum). J Appl Ichthyol 30:825–832. https://doi.org/10.1111/jai.12519
Lewis-McCrea LM, Lall SP (2010) Effects of phosphorus and vitamin C deficiency, vitamin A toxicity, and lipid peroxidation on skeletal abnormalities in Atlantic halibut (*Hippoglossus hippoglossus*). J Appl Ichthyol 26:334–343. https://doi.org/10.1111/j.1439-0426.2010.01431.x
Liu S, Zhou J, Tang W et al (2006) Pathogenic role of Fgf23 in Hyp mice. AJP Endocrinol Metab 291:E38–E49. https://doi.org/10.1152/ajpendo.00008.2006
Lovell T (ed) (1998) Nutrition and feeding of fish. Kluwer Academy Publishers
Lu Y, Feng JQ (2012) FGF23 in skeletal modeling and remodeling. Curr Osteoporos Rep 9:103–108. https://doi.org/10.1007/s11914-011-0053-4.FGF23
Lu M, Wagner GF, Renfro JL (1994) Stanniocalcin stimulates phosphate reabsorption by flounder renal proximal tubule in primary culture. Am J Physiol 267:R1356–R1362
Mangos S, Amaral AP, Faul C et al (2012) Expression of fgf23 and αklotho in developing embryonic tissues and adult kidney of the zebrafish, *Danio rerio*. Nephrol Dial Transplant 27:4314–4322. https://doi.org/10.1093/ndt/gfs335
Martin A, David V, Laurence JS et al (2008) Degradation of MEPE, DMP1, and release of SIBLING ASARM-peptides (minhibins): ASARM-peptide(s) are directly responsible for

defective mineralization in HYP. Endocrinology 149:1757–1772. https://doi.org/10.1210/en.2007-1205

Mccudden CR, Kogon MR, Dimattia GE et al (2001) Novel expression of the stanniocalcin gene in fish. J Endocrinol 171:33–44

Mundy GR, Guise T (1999) Hormonal control of calcium homeostasis. Clin Chem 45:1347–1352

Murshed M, Harmey D, Millán JL et al (2005) Unique coexpression in osteoblasts of broadly expressed genes accounts for the spatial restriction of ECM mineralization to bone. Genes Dev 19:1093–1104. https://doi.org/10.1101/gad.1276205

Nemoto Y, Higuchi K, Baba O et al (2007) Multinucleate osteoclasts in medaka as evidence of active bone remodeling. Bone 40:399–408. https://doi.org/10.1016/j.bone.2006.08.019

Olsen RL, Hasan MR (2012) A limited supply of fishmeal: impact on future increases in global aquaculture production. Trends Food Sci Technol 27:120–128

Parsons JA, Gray D, Rafferty B et al (1978) Evidence for a hypercalcaemic factor in the fish pituitary immunologically related to mammalian parathyroid hormone. In: Copp D, Talmage R (eds) Endocrinology of calcium metabolism pp 111–114

Penido MG, Alon US (2012) Phosphate homeostasis and its role in bone health. Pediatr Nephrol 27:2039–2048. https://doi.org/10.1007/s00467-012-2175-z

Persson P, Bjornsson BT, Takagi Y (1999) Characterization of morphology and physiological actions of scale osteoclasts in the rainbow trout. J Fish Biol 54:669–684. https://doi.org/10.1111/j.1095-8649.1999.tb00645.x

Piedecausa MA, Aguado-Giménez F, Cerezo Valverde J et al (2012) Influence of fish food and faecal pellets on short-term oxygen uptake, ammonium flux and acid volatile sulphide accumulation in sediments impacted by fish farming and non-impacted sediments. Aquac Res 43:66–74. https://doi.org/10.1111/j.1365-2109.2011.02801.x

Pinheiro PLC, Cardoso JCR, Gomes AS et al (2010) Gene structure, transcripts and calciotropic effects of the PTH family of peptides in *Xenopus* and chicken. BMC Evol Biol 10:373–387. https://doi.org/10.1186/1471-2148-10-373

Pinheiro PLC, Cardoso JCR, Power DM et al (2012) Functional characterization and evolution of PTH/PTHrP receptors: insights from the chicken. BMC Evol Biol 12:110. https://doi.org/10.1186/1471-2148-12-110

Potts JT (2005) Parathyroid hormone: past and present. J Endocrinol 187:311–325. https://doi.org/10.1677/joe.1.06057

Power DM, Ingleton PM, Flanagan JA et al (2000) Genomic structure and expression of parathyroid hormone-related protein gene (PTHrP) in a teleost, *Fugu rubripes*. Gene 250:67–76

Quarles LD, Drezner MK (2001) Pathophysiology of X-Linked Hypophosphatemia, Tumor-Induced Osteomalacia, and Autosomal Dominant Hypophosphatemia: A PerPHEXing Problem. J Clin Endocrinol Metab 86:494–496. https://doi.org/10.1210/jcem.86.2.7302

Quiros-Gonzalez I, Yadav VK (2014) Central genes, pathways and modules that regulate bone mass. Arch Biochem Biophys 561:130–136. https://doi.org/10.1016/j.abb.2014.06.005

Rotllant J, Redruello B, Guerreiro PM et al (2005) Calcium mobilization from fish scales is mediated by parathyroid hormone related protein via the parathyroid hormone type 1 receptor. Regul Pept 132:33–40. https://doi.org/10.1016/j.regpep.2005.08.004

Rowe P (2004) The wrickkened pathways of FGF23, MEPE and PHEX. Crit Rev Oral Biol Med 15:264–281. https://doi.org/10.1177/154411130401500503

Rowe P (2012) Regulation of bone - renal mineral and energy metabolism: the PHEX, FGF23, DMP1, MEPE ASARM pathway. Crit Rev Eukaryot Gene Expr 22:61–86

Roy PK, Lall SP (2003) Dietary phosphorus requirement of juvenile haddock (*Melanogrammus aeglefinus* L.). Aquaculture 221:451–468. https://doi.org/10.1016/S0044-8486(03)00065-6

Roy PK, Witten PE, Hall BK et al (2002) Effects of dietary phosphorus on bone growth and mineralisation of vertebrae in haddock (*Melanogrammus aeglefinus* L.). Fish Physiol Biochem 27:35–48. https://doi.org/10.1023/B:FISH.0000021778.99011.ce

Rubin DA, Hellman P, Zon LI et al (1999) A G Protein-coupled Receptor from Zebrafish Is Activated by Human Parathyroid Hormone and Not by Human or Teleost Parathyroid Hormone-related Peptide: Implication for the Evolutionary Conservation of Calcium-Regulating Peptides Hormones. J Biol Chem 274:23035–23042. https://doi.org/10.1074/jbc.274.33.23035
Ruchon AF, Marcinkiewicz M, Siegfried G et al (1998) Pex mRNA is localized in developing mouse osteoblasts and odontoblasts. J Histochem Cytochem 46:459–468
Sapir-Koren R, Livshits G (2011) Bone mineralization and regulation of phosphate homeostasis. IBMS Bonekey 8:286–300. https://doi.org/10.1138/20110516
Satoh S, Porn-Ngam N, Takeuchi T et al (1996) Influence of dietary phosphorus levels on growth and mineral availability in rainbow trout. Fish Sci 62:483–487. https://doi.org/10.2331/FISHSCI.62.483
Schein V, Cardoso JCR, Pinto PIS et al (2012) Four stanniocalcin genes in teleost fish: Structure, phylogenetic analysis, tissue distribution and expression during hypercalcemic challenge. Gen Comp Endocrinol 175:344–356. https://doi.org/10.1016/j.ygcen.2011.11.033
Schultz AG, Guffey SC, Clifford AM et al (2014) Phosphate absorption across multiple epithelia in the Pacific hagfish (*Eptatretus stoutii*). Am J Physiol Regul Integr Comp Physiol 307:R643–R652. https://doi.org/10.1152/ajpregu.00443.2013
Seeman E (2009) Bone modeling and remodeling. Crit Rev Eukaryot Gene Expr 19:219–233
Shahar R, Dean MN (2013) The enigmas of bone without osteocytes. Bonekey Rep 2:1–8. https://doi.org/10.1038/bonekey.2013.77
Shimada T, Mizutani S, Muto T et al (2001) Cloning and characterization of FGF23 as a causative factor of tumor-induced osteomalacia. Proc Natl Acad Sci U S A 98:6500–6505. https://doi.org/10.1073/pnas.101545198
Shimada T, Hasegawa H, Yamazaki Y et al (2003) FGF-23 Is a Potent Regulator of Vitamin D Metabolism and Phosphate Homeostasis. J Bone Miner Res 19:429–435. https://doi.org/10.1359/JBMR.0301264
Sims NA, Gooi JH (2008) Bone remodeling: Multiple cellular interactions required for coupling of bone formation and resorption. Semin Cell Dev Biol 19:444–451. https://doi.org/10.1016/j.semcdb.2008.07.016
Suarez-Bregua P, Torres-Nunez E, Saxena A et al (2017) Pth4, an ancient parathyroid hormone lost in eutherian mammals, reveals a new brain-to-bone signaling pathway. FASEB J 31:569–583. https://doi.org/10.1096/fj.201600815R
Sugiura SH, Hardy RW, Roberts RJ (2004) The pathology of phosphorus deficiency in fish-a review. J Fish Dis 27:255–265. https://doi.org/10.1111/j.1365-2761.2004.00527.x
Sullivan M, Hammond G, Roberts RJ et al (2007) Spinal deformation in commercially cultured Atlantic salmon, *Salmo salar* L.: a clinical and radiological study. J Fish Dis 30:745–752. https://doi.org/10.1111/j.1365-2761.2007.00889.x
Sundell K, Björnsson BT, Itoh H et al (1992) Chum salmon (*Oncorhynchus keta*) stanniocalcin inhibits in vitro intestinal calcium uptake in Atlantic cod (*Gadus morhua*). J Comp Physiol B 162:489–495
Suzuki N, Danks J, Maruyama Y et al (2011) Parathyroid hormone 1 (1-34) acts on the scales and involves calcium metabolism in goldfish. Bone 48:1186–1193. https://doi.org/10.1016/j.bone.2011.02.004
Takeda S (2008) Central control of bone remodelling. J Neuroendocrinol 20:802–807. https://doi.org/10.1111/j.1365-2826.2008.01732.x
Tenenhouse HS (1999) X-linked hypophosphataemia: a homologous disorder in humans and mice. Nephrol Dial Transplant 14:333–341
Terkeltaub RA (2001) Inorganic pyrophosphate generation and disposition in pathophysiology. Am J Physiol Cell Physiol 281:C1–11
To TT, Witten PE, Renn J et al (2012) Rankl-induced osteoclastogenesis leads to loss of mineralization in a medaka osteoporosis model. Development 139:141–150. https://doi.org/10.1242/dev.071035

Tseng D, Chou M, Tseng Y et al (2009) Effects of stanniocalcin 1 on calcium uptake in zebrafish (*Danio rerio*) embryo. Am J Physiol Regul Integr Comp Physiol 296:549–557

Urakawa I, Yamazaki Y, Shimada T et al (2006) Klotho converts canonical FGF receptor into a specific receptor for FGF23. Nature 444:770–774. https://doi.org/10.1038/nature05315

VanHouten J, Dann P, McGeoch G et al (2004) The calcium-sensing receptor regulates mammary gland parathyroid hormone-related protein production and calcium transport. J Clin Invest 113:598–608. https://doi.org/10.1172/JCI18776

Venkatakrishnan AJ, Deupi X, Lebon G et al (2013) Molecular signatures of G-protein-coupled receptors. Nature 494:185–194. https://doi.org/10.1038/nature11896

Wagner GF, Hampong M, Park CM et al (1986) Purification, characterization, and bioassay of teleocalcin, a glycoprotein from salmon corpuscles of Stannius. Gen Comp Endocrinol 63:481–491

Witten PE, Hall BK (2003) Seasonal changes in the lower jaw skeleton in male Atlantic salmon (*Salmo salar* L.): remodelling and regression of the kype after spawning. J Anat 203:435–450

Witten PE, Huysseune A (2009) A comparative view on mechanisms and functions of skeletal remodelling in teleost fish, with special emphasis on osteoclasts and their function. Biol Rev Camb Philos Soc 84:315–346. https://doi.org/10.1111/j.1469-185X.2009.00077.x

Witten PE, Hansen A, Hall BK (2001) Features of mono- and multinucleated bone resorbing cells of the zebrafish Danio rerio and their contribution to skeletal development, remodeling, and growth. J Morphol 250:197–207. https://doi.org/10.1002/jmor.1065

Witten PE, Owen MAG, Fontanillas R et al (2016) A primary phosphorus-deficient skeletal phenotype in juvenile Atlantic salmon *Salmo salar*: the uncoupling of bone formation and mineralization. J Fish Biol 88:690–708. https://doi.org/10.1111/jfb.12870

Yan Y-L, Bhattacharya P, He XJ et al (2012) Duplicated zebrafish co-orthologs of parathyroid hormone-related peptide (PTHrP, Pthlh) play different roles in craniofacial skeletogenesis. J Endocrinol 214:421–435. https://doi.org/10.1530/JOE-12-0110

Yeung BHY, Law AYS, Wong CKC (2012) Evolution and roles of stanniocalcin. Mol Cell Endocrinol 349:272–280. https://doi.org/10.1016/j.mce.2011.11.007

Yuan B, Takaiwa M, Clemens TL et al (2008) Aberrant Phex function in osteoblasts and osteocytes alone underlies murine X-linked hypophosphatemia. J Clin Invest 118:722–734. https://doi.org/10.1172/JCI32702

Chapter 10
Feeding and Development of Warm Water Marine Fish Larvae in Early Life

Jing Hu, Yibing Liu, Zhenhua Ma and Jian G. Qin

Abstract Mass mortality in the early stage of fish larvae frequently occurs in marine fish hatchery, and low survival has become a major bottleneck hindering fingerling production of marine fish. Inappropriate food supply is a key factor leading to fish mortality during the nursery period in a hatchery. The success of fish larvae to capture live prey in early life has become an important benchmark in hatchery production. Therefore, an understanding of feeding biology and development of marine fish larvae is essential to select appropriate feed and rearing environment to improve fish performance in the initial rearing period. In this chapter, we review both internal and external factors regulating the feeding and development of warm water marine fish larvae focusing on pompano *Trachinotus ovatus*, orange-spotted grouper *Epinephelus coioides*, and coral trout *Plectropomus leopardus* in attempts to understand the relationships of feeding, development, environmental requirements and performance of fish larvae.

Keywords Marine fish larvae · First feeding · Feeding behavior
Live food

10.1 Introduction

After hatching, most marine fish larvae still rely on yolk reserves for ontogenetic development before exogenous feeding. As yolk reserves gradually reduce, larval fish would experience the transition from endogenous feeding to a period of mixed feeding (Yúfera and Darias 2007). Before complete absorption of yolk reserves,

J. Hu · Z. Ma (✉)
Tropical Aquaculture Research and Development Centre, South China Sea Fisheries Research Institute, Chinese Academy of Fishery Sciences, Sanya 572018, China
e-mail: zhenhua.ma@hotmail.com

Y. Liu · Z. Ma · J. G. Qin
College of Science and Engineering, Flinders University, GPO Box 2100, Adelaide, SA 5001, Australia

M. Yúfera (ed.), *Emerging Issues in Fish Larvae Research*,
https://doi.org/10.1007/978-3-319-73244-2_10

larval fish start to capture and ingest live prey to obtain energy and nutrients to sustain growth and development (Ma et al. 2015; Yúfera and Darias 2007). However, the feeding apparatus and digestive function in most marine fish larvae are not fully developed at early life, and mass mortality in fish hatchery has become a hurdle of fingerling production (Qin 2013). Therefore, it is necessary to explore the factors affecting larval fish feeding and survival in early life history.

The success of larval fish feeding is the result of complex processes involving both internal and external factors (Yúfera and Darias 2007). Internally, development of feeding apparatus and vision is crucial because food intake requires coordination of food searching, detection, attract, capture, ingestion, digestion and evacuation (Rønnestad et al. 2013). Externally, larval fish growth and survival are related to food quality and quantity, feeding frequency (Ma et al. 2015) and feeding environment (Qin 2013). This chapter aims to use some species of warm water marine fish larvae that have aquaculture potential to illustrate the importance of internal and external factors in regulating fish growth and survival at early life stages. The internal factors affecting growth and survival of fish larvae include fish mouth size, vision, mechanoreceptors such as inter ear, taste buds and olfactory organ. The external factors include timing of initial feeding, prey density, tank wall color, prey color, water color and light intensity. This chapter will focus on the discussion of three marine fish species (pompano *Trachinotus ovatus*, orange-spotted grouper *Epinephelus coioides*, and coral trout *Plectropomus leopardus*) that have aquaculture potential in warm water regions, though we also include some other relevant species to support the discussion. It is our intention to update the recent research outcomes in larval fish feeding and to shed light on the improvement of technical management in larval rearing of warm water fish.

10.2 Internal Factors

Morphogenesis of sensory organs is closely related to the behavioral change of marine fish larvae. Rapid development of sensory organs related to the ability of feeding can improve fish survival, especially in the first a few days after hatch (Lim and Mukai 2014). In this section, we will review the functional development of organs that are associated with the success of first feeding.

10.2.1 Mouth

Mouth size could physically limit food particles that a fish can ingest and the allometric growth of mouth dimension to body size is commonly seen in most species of fish larvae. As mouth size sets the upper limit for a prey size (Dabrowski and Bardega 1984), the size of food provision must be synchronized with the change of fish size. Although fish larvae are able to ingest a prey with similar size to

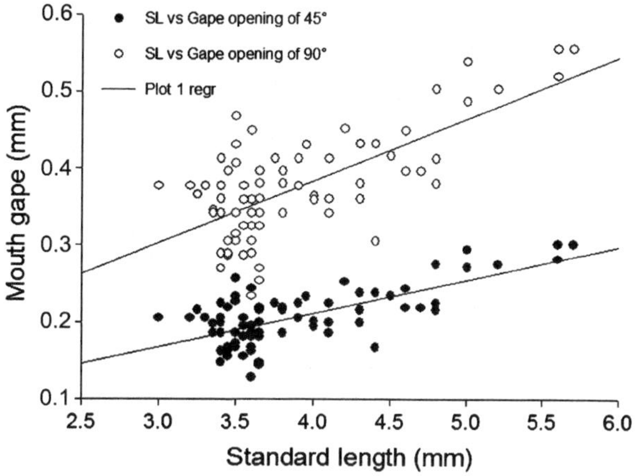

Fig. 10.1 Relationship between mouth gape and standard length (SL) of larval pompano *Trachinotus ovatus*. Reproduced from Ma et al. (2015)

mouth gape, they tend to ingest larger prey to acquire maximum energy gain per unit of feeding effort. The prey to gape size ratio varies among species and usually ranges from 25 to 60% (Cunha and Planas 1999; Østergaard et al. 2005). The size of prey that can be ingested is also determined by the diameter of esophagus (Busch 1996; Yúfera and Darias 2007), especially at the early stage when the presence of yolk sac prevents further expansion of the lumen size of esophagus (Busch 1996).

In pompano, Ma et al. (2015) measured the mouth gape of larval pompano based on larvae of 3.00–5.70 mm standard length (SL) (Fig. 10.1). The mouth gape at the openings of 45° and 90° increased with the increase in body length. The minimum mouth gape opened at 45° and 90° for a 3.00 mm SL fish was 205 and 377 µm, respectively. The maximum mouth gape openings at 45° and at 90° for a 5.70 mm SL fish was 302 and 557 µm, respectively.

10.2.2 *Vision Ontogeny*

Marine fish larvae are mainly visual feeders at first feeding. However, the vision function of most marine fish at hatch is underdeveloped and the eye can detect motion but is unable to functionally form an image (Yúfera and Darias 2007). Therefore, in order to reduce morality due to starvation, the visual system should be morphologically and functionally adapted to exogenous feeding soon after absorption of oil globules and able to search and detect live prey (Lim and Mukai 2014). Normally, the first feeding larvae possess small eyes with limited number and type of retinal tissues. The small eye usually contains a single type of photoreceptor such as cones, restricting the ability to forage under dim light because small eyes cannot accommodate parallel development of photopic and scotopic visual apparatus (Gisbert et al. 2013; Kotrschal et al. 1990). With the increase of eye size, there is more space available to accommodate the development of retinal tissues. Visual ontogeny commences with the enlargement of single cones to increase the ability of photon capture and visual acuity (Vandermeer 1994). The eye sensitivity increases with the addition of double cones that could improve resolution at dim light and develop rods associated with movement perception and scotopic vision (Pankhurst and Hilder 1998). Therefore, the ability to detect food at dim light would be improved by the formation of cone mosaics through rearrangement of cones, resulting in enhanced color resolution, contrast and visual acuity to perceive prey motion.

Lim and Mukai (2014) reported that at hatch the eyes of brown-marbled grouper *Epinephelus fuscoguttatus*, larvae comprised only a lens, but the retina was undifferentiated. At 1 day post hatching (dph), the retina differentiated into multiple layers, and the optic nerve connected to eyes with an optic tectum. The eyes were fully pigmented at 3 dph and the retinal layers containing an area of temporalis were clearly distinguishable (Fig. 10.2). At the start of 4 dph, the larvae possessed a pure-cone retina, and at 20 dph, the retinal rods first occurred.

Margulies (1997) compared the visual system in three scombrids species and found that the eye in early preflexion larvae of black skipjack tuna *Euthynnus lineatus*, bullet tuna *Auxis rochei*, frigate tuna *A. thazard* and Spanisch mackerel *Scombermorus sierra* consisted of a simplex retina with only cone cells in the visual cell layer. The external nuclear layer of the simplex retina contained a nuclei to cone cells ratio of 1:1. In all these species, the cone cell counts decreased as the growth rate of larvae was rapidly reduced when the larvae reached 2.5–10.0 mm SL. The retina in larvae (SL > 3.5 mm) had an area of high cone density, located in the temporal and ventro-temporal regions. At the preflexion stage, the cone cells of the high density region were normally undifferentiated and difficult to count. In contrast, at the flexion and post-flexion stages, the high density region of the retina was well differentiated and contained a large number of cone cells. In the above four fish species, the scattered double cones first appeared at the postflexion stage.

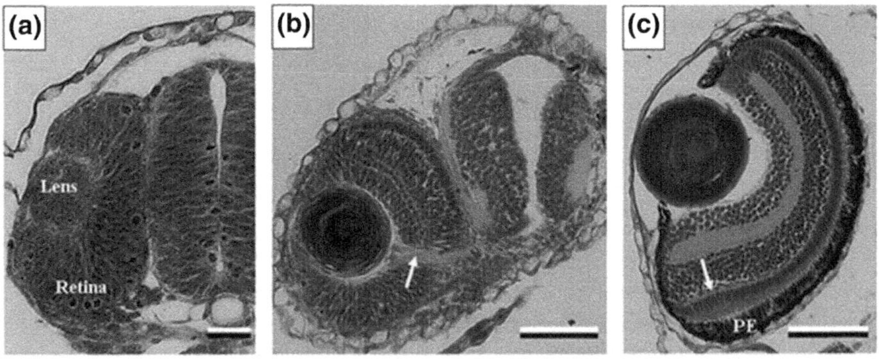

Fig. 10.2 Photomicrographs of the eyes of brown-marbled grouper *Epinephelus fuscoguttatus* larvae. **a** Newly hatched larvae; **b** 1-dph yolk-sac larvae with the arrow showing optic nerve; **c** 3-dph larvae. PE: pigmented epithelium with the arrow showing the area of temporalis. Scale bars, 20 μm in (**a**) and 50 μm in (**b**–**c**). Reproduced from Lim and Mukai (2014)

In all feeding larvae, the area of the optic tectum dominated the mesencephalon (midbrain). Visual accommodation in fish larvae is facilitated by the lens retractor muscle and the lens suspensory ligament. In pre-flexion larvae of the three tuna species and Spanish mackerel, these accommodative structures develop after larvae start feeding, indicating that first feeding larvae are probably restricted to near-sighted vision. It is possible that the larval lens has relatively great refractive power, thus negating the need for efficient adjustment during the initial feeding stage (Cobcroft and Pankhurst 2003; Margulies 1997).

10.2.3 Mechanoreception

Although photoreception in fish vision is the primary sensory modality in larval fish to detect and capture prey, mechanoreception is also an essential factor for larval fish feeding (Gisbert et al. 2013). This is because fish larvae would sense water movement through a mechanoreception organ within the skin, named the lateral line organ (Montgomery et al. 2002). The lateral line organ including the receptors with hair cells could produce nervous potential when hydrodynamic disturbance is detected and the receptors for signal detection are called neuromasts (Gisbert et al. 2013; Pankhurst 2008). The neuromasts would influence fish behavior such as schooling, prey localization, rheotaxis, predator evasion and obstacle detection (Gisbert et al. 2013). However, the development of mechanoreception organs at hatch is incomplete, and there is continuous addition of sensory cells and neurons along with larval fish ontogeny (Gisbert et al. 2013; Pankhurst 2008; Kasumyan 2003; Webb 1989).

Primary neuromasts develop directly from neurogenic placodes on the head region which give rise to both canal and free neuromasts (Fuiman et al. 2004). The primary neuromasts comprise a few support and receptor cells that erupt through the overlying ectoderm in a process involving both ectodermal retraction and upward projection of the neuromast (Gibbs 2004). A cupula is then secreted by the neuromast support cells. Neuromast eruption does not occur in an orderly sequence along the sensory ridge, but proceeds in successive waves, initially providing a dispersed neuromast array, and then filling in the gaps during development (Ledent 2002). At a later stage of hatch, several free neuromasts will be distributed over the head and body. During subsequent development, the free neuromasts can proliferate and there may be 100 or more neuromasts over the head in a lateral row on each myotome of the trunk and even on the caudal fin.

Canal formation commences first on the head at a late stage of larval ontogeny to form the supraorbital and infraorbital canals and then the trunk canals. The process starts with primary neuromasts to form dermal furrows. Ectodermal ridges are parallel to the long axis of the presumptive canals, which then eventually enclose the neuromasts. Myrberg and Fuiman (2002) suggest that the canal formation is delayed until the late larval period because of the constraint imposed by non-ideal flow properties of very small canals. While most primary neuromasts become canal neuromasts, some are destined to remain on the skin as “accessory” canal neuromasts. In the situation where canal formation is reduced, or canals are discontinuous, primary neuromasts have been referred as replacement neuromasts (Montgomery et al. 2002). On the basis of pharmacological and functional responses, all primary neuromasts are homologous with canal neuromasts and physiologically distinct from secondary neuromasts (Montgomery et al. 2002).

The developmental pattern of neuromasts in fish larvae is species dependent. For instance, Mukai and Lim (2016) reported that the newly hatched brown-marbled grouper had one pair of free neuromasts behind the eye. The free neuromasts increased with larval growth and were distributed on both sides of the trunk and around the eyes and nostrils. Until 10 dph, the number of neuromasts on the head was considerably greater in comparison with those in the trunk, whereas the neuromasts in the trunk began to increase after 20 dph. Kawamura and Munekiyo (1989) found that free neuromasts with cilia were found on the head of ribbonfish *Trichiurus leputus* larvae at 5.4–1.8 mm SL, and on the trunk at 16.5 mm SL. In 20.5 mm larvae, the free neuromasts on the head began to sink into the dermis. Canal organs were first found on the head of larvae at 51.0 mm and were completed on the head and trunk in 89 mm larvae.

10.2.4 The Inner Ear

Fish can use ears to detect acoustic stimuli. The inner ear in teleost fish is composed of three otolithic end organs, namely lagena, utricle and saccule to receive both auditory and vestibular stimuli. The otolithic end organs contain macular sensory hair cells that are coupled with an otolith, which is a biomineralized ear stone composed of calcium carbonate and protein and acts as an inertial mass Sound and head movement inside fish larvae can produce relative displacement between the otolith and hair cells due to the difference in their inertia. Although the three otolithic end organs are capable of detecting both inertial and acoustic stimuli, it is possible that the three end organs differ in contribution to motion detection and audition. The saccule and lagena are necessary for auditory perception and the utricle is essential for postural equilibrium (Popper and Schilt 2008).

In ribbonfish, the larvae at 5.4 mm have a well-developed and ossified inner ear with the basic elements of utriculus, sacculus and lagena, which have ciliated neuromasts and horizontal and vertical semicircular canals (Kawamura and Munekiyo 1989). In newly-hatched bluefin tuna larvae, the inner ear is on the ovate vesicle with two otoliths and an innervated ciliated epithelium (Fig. 10.3). Three maculae formed in 18 h after hatch, and the semicircular canals and crus commune formed after 2 dph. The canals are partly ossified at 8 dph and completely ossified at 12 dph when the notochord was fully flexed. The third otolith is not seen in larvae during the preflexion and flexion stages (Kawamura et al. 2003). Lim and Mukai (2014) found that at hatch, brown-marbled grouper larvae had otic vesicles with two otoliths. The otic epithelium was thickened in several regions at 1 dph, and the

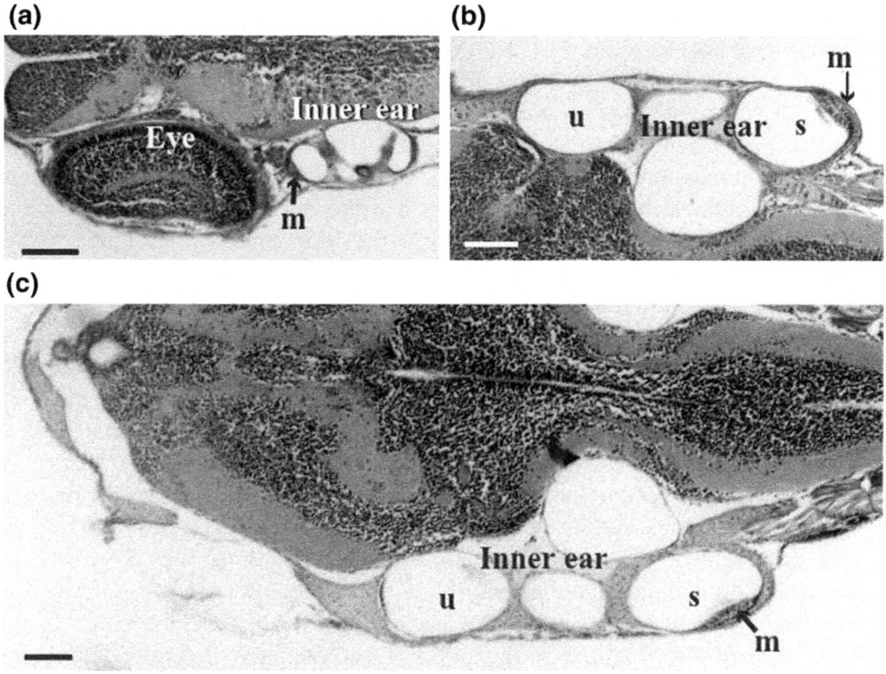

Fig. 10.3 Development of the inner ear in Pacific bluefin tuna larvae *Thunnus orientalis*: **a** larva 3.9 mm TL (2 dph) with three pockets in the inner ear; **b** larva 5.6 mm (8 dph) with partly ossified semicircular canals; **c** larva 8.2 mm (12 dph) with completely ossified semicircular canals. m, macula; s, saculus; u, utriculus. Scale bar, 50 μm. Reproduced from Kawamura et al. (2003)

crests and the three semicircular canals formed at 3 dph. The inner ear organs (utricule and saccule) of brown-marbled grouper formed as cavities with maculae, or patches of hair cells covered by the otolithic membrane at 10 and 20 dph, respectively. The lagena formed as a posterior pocket of the saccular cavity at 25 dph, and its macula formed at 30 dph (Lim and Mukai 2014).

10.2.5 Taste Buds

The gustatory sensory system is only involved in feeding behavior. Once a prey is captured, the action to swallow or reject the prey is based on sense from the gustatory system to taste food palatability. Therefore, gustation provides a penultimate sensory evaluation in the feeding process and sensation of taste at the consummatory phase of feeding. The taste is based on the detection of chemical stimuli (incitants, suppressants and stimulants) by means of taste buds. The gustatory system in fish is divided into two distinct subsystems, oral and extraoral, and both of them mediate the reaction to food items through contact with taste buds.

The taste buds are situated not only within the oral cavity, pharynx, esophagus and gills (oral gustatory system), but also on the lips, barbels, fins and over the entire body surface (i.e., the extraoral gustatory system) in many fish species. A typical taste bud has an ovoid or pear-shaped form and its long axis is oriented vertically to the surface of the epithelium. As a rule, the base of the taste bud is situated on top of a small ascending papilla of the dermis. Each taste bud is composed of two types of sensory cells, the electron lucent light cells and the electron denser dark cells that bear apical microvilli which together form the taste bud pore exposed to the environment. In addition, several basal cells are located below sensory cells with a network of nerves around them.

As the gustatory system plays an important role in feeding, and provides the ultimate sensory evaluation of a selected food item, several studies have compared the appearance of taste buds with the timing of onset exogenous feeding. Comparison of available data shows that in some species the first taste buds appear before the beginning of the exogenous feeding. The time when the estimation of taste quality of food becomes essential and necessary as in other species the onset of exogenous feeding takes place before taste bud formation.

The development of taste buds in fish larvae depends on species. For instance, there was no taste bud in 5.4–5.8-mm ribbonfish larvae, but taste buds appeared in the epithelium of the oral cavity of 6.2-mm fish, and became numerous in the pharyngeal region by 11.8 mm long and developed in pharyngeal teeth, and on the gill arches in a 16.5 mm fish (Kawamura and Munekiyo 1989). Kawamura et al. (2003) reported that the taste buds differentiated first in the upper pharynx at 8-dph larvae of Pacific bluefin tuna *Thunnus orientalis* and later in the epithelium of the oral cavity and gill arches at 10-dph larvae, when the canine-like teeth and pharyngeal teeth appeared. The taste buds progressively became denser in the upper pharynx and the epithelium of the oral cavity and gill arches but were not present on the lips until 33 dph. Lim and Mukai (2014) found that in the larvae of brown-marbled grouper, the taste buds developed rather late. They were first found inside the mouth cavity in larvae at 20 dph (Fig. 10.4). Between 30 and 50 dph, more taste buds appeared progressively towards the middle, then at the opening of the mouth cavity.

10.2.6 Olfactory Organ

Olfaction allows fish to detect a remote stimulus. During embryonic development, the olfactory organs gradually form. Olfactory placodes and pits are normally presented at the onset first feeding (Boglione et al. 2003; Kawamura et al. 2003), and further developed to form the deepening of the pits, nares, the folding of olfactory epithelium, and the olfactory lamellae by the late larval stage. When the olfactory organs of fish larvae are more functionally operated, olfaction of fish larvae is involved in detection of nursery areas (James et al. 2008; Lara 2008). Furthermore, evidence has clearly indicated that chemical signals have a relevant

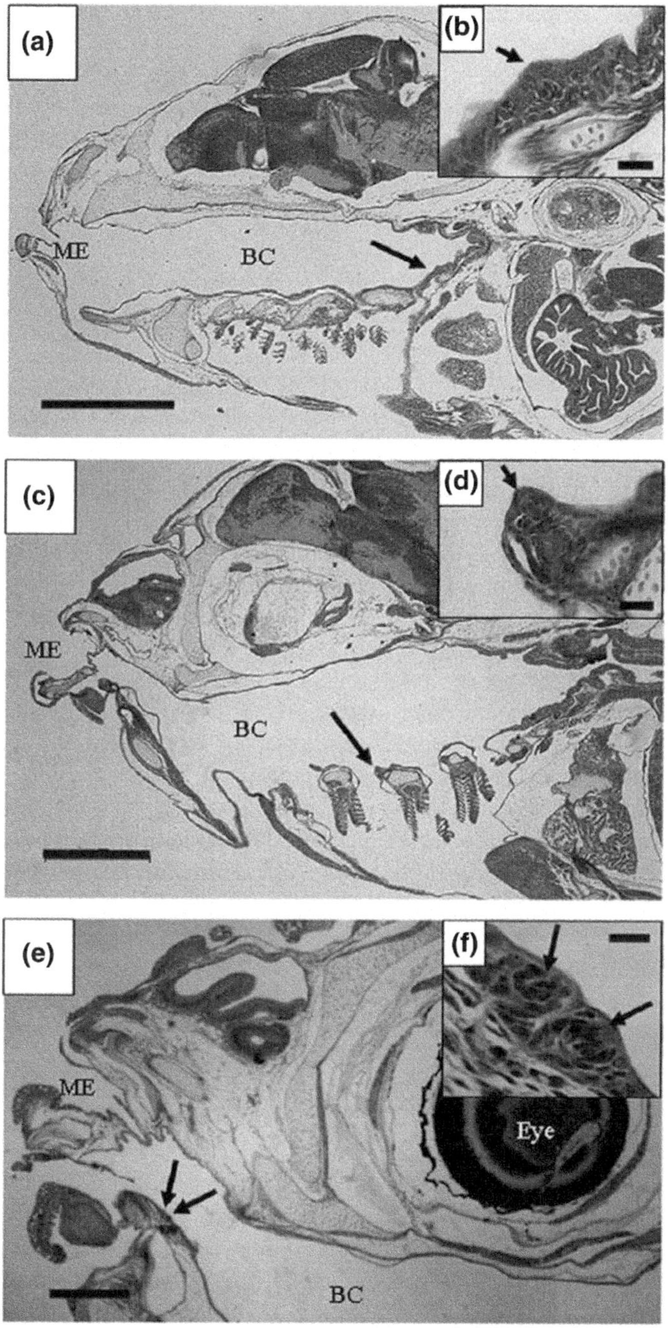

Fig. 10.4 Photomicrographs of longitudinal sections of the mouth cavity in brown-marbled grouper *Epinephelus fuscoguttatus* showing taste buds (at arrows) at two magnifications. **a**–**b** 20 dph larva; **c**–**d** 30 dph larva. ME: mouth entrance and BC: buccal cavity. Scale bars: **a** 200 μm, **b** 20 μm, **c** 500 μm, **d** 20 μm, **e** 1 mm and **f** 500 μm. Reproduced from Lim and Mukai (2014)

role in many species in searching for the appropriate habitat for settling after pelagic larval life (Lara 2008).

The organogenesis of the olfactory system is a complex process, and the formation of the olfactory epithelium varies greatly among teleost species. However, the most relevant ontogenetic events related to the development of the olfaction sense are as follows:

Stage 1 form the olfactory placodes from large fields of cells along the anterior-lateral neural plate during embryogenesis.

Stage 2 form the olfactory pits, and the development of olfactory receptor cells, microvillous receptor cells, supporting cells and non-sensory cells in the olfactory epithelium before hatching.

Stage 3 further development of the olfactory apparatus occurs after hatching during the larval phase when nares develop. The olfactory rosette differentiates and the olfactory bulb in the brain increases in volume.

The formation pattern of olfaction varies among fish species according to their life history, stage of development and onset of exogenous feeding. For instance, in Pacific bluefin tuna, although the olfactory pits were still closed at 13 h after hatch, the olfactory epithelium was already fused with the olfactory bulb of the brain, indicating early neural connection (Kawamura et al. 2003). The olfactory pits opened in 2-days old larvae, and the olfactory epithelium consisted of ciliated receptor cells, microvillous receptor cells, and ciliated nonsensory cells, whereas the olfactory tracts were elongated and the olfactory bulb and lobes were well developed in first feeding larvae (Kawamura et al. 2003). Anterior and posterior nares of Pacific bluefin tuna were not fully formed until the age of 16 days in concomitance with the folding of the olfactory epithelium (Kawamura et al. 2003).

In warm water fish species, Lim and Mukai (2014) found that at hatch the nares of larvae in brown-marbled grouper were open and ciliated, while the ciliated sensory receptor cells and non-sensory cilia were clearly evident at 1 dph. The olfactory epithelium was fully covered with non-sensory cilia at 10 dph. The first olfactory lamellae formed at 25 dph and the number increased in the later larval stage. Therefore, the discrepancy of olfaction formation between species should be carefully considered for feeding management in different fish species.

10.3 External Factors

In larval fish rearing, fish mortality in the early stage is generally associated with poor food supply under a favorable condition because fish development depends on adequate nutrition uptake (Ma et al. 2012). It is critically important for fish larvae to successfully capture live prey once mouth becomes functioning. In a warm water environment, high metabolism could significantly accelerate the time of fish larvae to reach the point of no return, and lead to mass mortality (Ma 2014). Therefore, the understanding of the impact of external factors on the feeding of warm water fish larvae in the early feeding phase would possibly improve the success of first feeding and ultimately improve fish survival.

Warm water marine fish larvae vary with their ability to capture different sizes and types of live prey at first feeding owing to differences in larval size at hatch, visual acuity, swimming patterns and abilities, and the size, swimming behavior, and other attributes of their prey. Environmental factors such as light intensity and temperature also affect capture rates. In the culture of warm water fish larvae, the color of tank and water is also known to affect capture rates of live prey. The consequence of this variability is that fish larvae are likely to vary significantly in energy requirements and efficiency of prey ingestion in their early life.

10.3.1 Timing of Initial Feeding

The time of initial feeding is critical for growth and survival of fish larvae (Mercier et al. 2004; Yufera and Darias 2007). Food deprivation or delayed food supply for first-feeding fish larvae can lead to poor growth, malformation, and even death (Blaxter and Hempel 1963; Chen et al. 2007; Yoseda et al. 2006). Therefore, it is important to identify appropriate time of exogenous feeding for newly-hatched fish larvae. Theoretically, this period usually starts within 24 h after mouth opening, and swim-up behaviour is usually used as a sign of commencing first feeding (Gisbert and Williot 1997; Ma et al. 2012; Twongo and MacCrimmon 1976). In warmer water fish species such as pompano (Ma et al. 2015), orange spotted grouper (Duray et al. 1996; Toledo et al. 2002; Zhang et al. 2015), and coral trout (Yoseda et al. 2008), the time of first feeding is normally observed on the second meal (afternoon feeding) or the third meal (night feeding) on the day of mouth opening.

10.3.2 Prey Density

Prey density is a fundamental factor affecting the success of first-feeding fish larvae in an artificial rearing environment. Previous studies have indicated that foraging success increases with the increase in prey density until an asymptote density is reached (Houde and Schekter 1980; Klumpp and Von Westernhagen 1986; Munk and Kiorboe 1985). However, evidence indicates that survival of fish larvae such as Atlantic cod *Gadus morhua* larvae increases to a maximum and then decreases with a further increase of prey density (Puvanendran and Brown 1999). The decrease of fish survival in a high prey density may be due to poor water quality by the release of metabolite from live prey (Houde 1975) or the reduction of prey capture ability in fish when the feeding "confusion effect" occurs due to fast prey movement (Laurel et al. 2001). In species such as yellowtail kingfish *Seriola lalandi* and fat snook *Centropomus parallelus*, survival is not affected by prey density when the prey reach to a certain threshold, but subsequent growth may be affected (Temple et al. 2004; Woolley and Partridge 2016).

Therefore, understanding of the relationship between prey density and feeding success of larval fish can identify the key factor that governs fish growth and survival in warm water. Our recent study indicates that prey density significantly affects the growth, survival and feeding performance of pompano larvae (Ma et al. 2015). The specific growth rates (SGRs) of fish fed 10 and 20 rotifers mL^{-1} were nearly two times higher than those of fish fed 1 and 40 rotifers mL^{-1} (Fig. 10.5). The survivals of fish larvae fed 10 and 20 rotifers mL^{-1} were significantly higher than those fed 1 and 10 rotifers mL^{-1} (Fig. 10.5). Low survival of fish larvae fed 40 rotifers mL^{-1} may be caused by poor water quality at high prey density or by low prey capture efficiency of fish larvae at an extremely high prey density (Laurel et al. 2001; Puvanendran and Brown 1999).

The relationship between prey density and food intake in fish larvae varies among species (Gotceitas et al. 1996; Ma et al. 2013a; Shaw et al. 2006). Fish feeding rates are normally limited by low prey density due to the increase of searching time (Houde and Schekter 1980). In 3-dph southern bluefin tuna *Thunnus maccoyii* larvae, Hilder et al. (2015) found that the increase of prey density significantly increased the proportion of feeding larvae with less feeding response at the lowest prey density (0.5 rotifers mL^{-1}) and more feeding response than expected at the highest prey density (25 rotifers mL^{-1}). In warm water fish species such as pompano larvae, when subjected to the same feeding time, fish fed 1 rotifer mL^{-1} ingested fewer rotifers than those provided with 10 and 20 rotifers mL^{-1} (Ma et al. 2015) (Fig. 10.6). Furthermore, the growth and survival of fish larvae in 1 rotifer mL^{-1} were significantly lower than fish held in 10 and 20 rotifers mL^{-1}. The appearance of such development pattern can be explained as larvae at low prey density need to spend more time and energy to search prey and swim, leading to low growth rate and survival (Ruzicka and Gallager 2006; Slembrouck et al. 2009).

Previous studies indicate that fish larvae tend to change foraging behavior to meet energy demand (Boujard and Medale 1994; Bromley and Adkins 1984; Grove et al. 1978) when they are physically able to eat large particles. However, recent evidence has indicated that prey densities can sway prey selectivity in fish larvae. According to the optimal foraging theory, selection for prey only happens when surplus food is offered in the ambient environment (O'Brien et al. 1976; Werner and Hall 1974). When fish are exposed to low prey density, food preference of fish larvae such as yellowtail kingfish can be affected in the subsequent co-feeding period (Ma et al. 2013b). In pompano larvae, food selection was affected by the increase of prey density from 10, 20 to 40 rotifers mL^{-1}. After co-feeding for 5 days, all fish larvae fed 10, 20 and 40 rotifers mL^{-1} adapted to feed on copepod nauplii. But larvae in 1 rotifer mL^{-1} feeding group still selected against copepod nauplii. This may suggest that the size of prey selected by pompano larvae increases as fish grow, and low prey density could delay food switch from small to large prey in fish ontogeny.

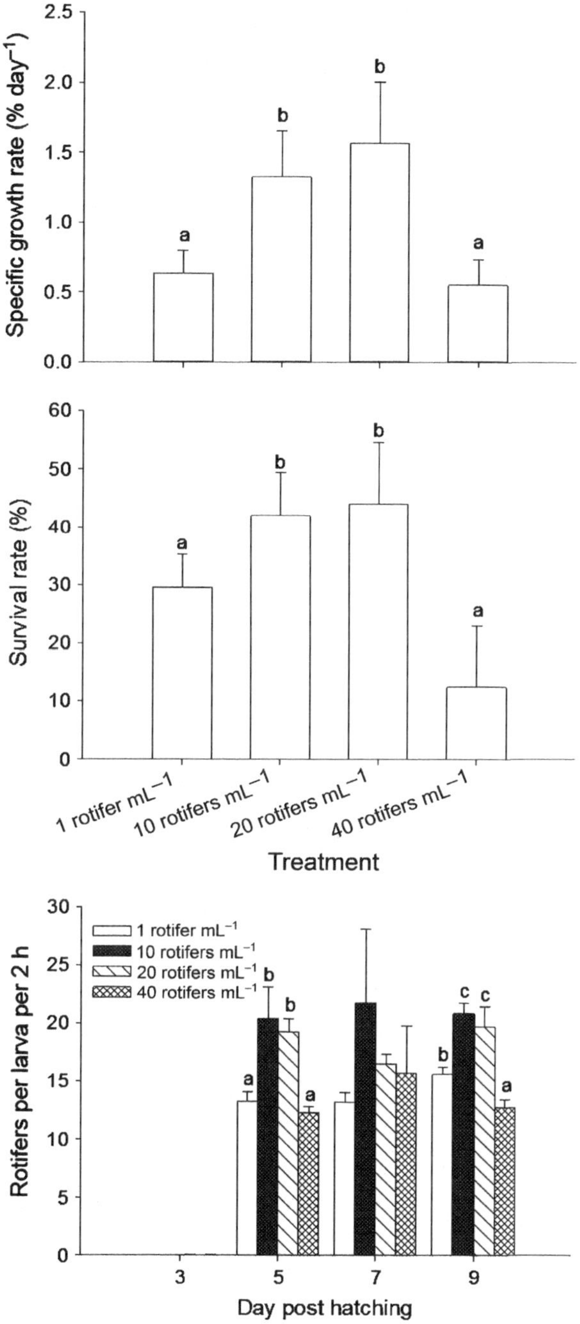

Fig. 10.5 Growth, survival and food ingestion of pompano *Trachinotus ovatus* larvae fed different rotifer densities. Different letters on the top of each bar represent significant difference. Reproduced from Ma et al. (2015)

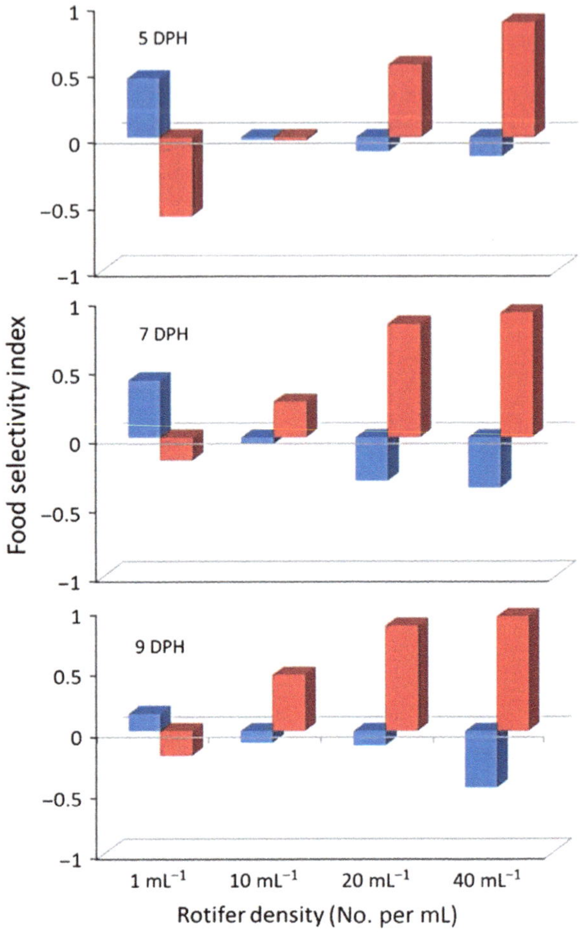

Fig. 10.6 Food selectivity indexes of pompano *Trachinotus ovatus* larvae. ▬: copepod nauplii. ▬: Rotifers. Reproduced from Ma et al. (2015)

10.3.3 Tank Wall Color

As marine fish larvae are visual feeders, the development of a pure cone retina during the first feeding requires light for feeding (Blaxter 1986). Under an artificial rearing condition, the color of tank walls can interact with light which can affect the success of prey capture by fish larvae (Monk et al. 2008). The impact of tank wall color on fish feeding success is species-specific (Ma et al. 2013a). For instance, species such as milkfish *Chanos chanos*, white bass *Morone chrysops* and striped bass *Morone saxatilis* prefer black color walls (Denson and Smith 1996, 1997; Martin-Tichaud and Peterson 1997), while the feeding of haddock *Melanogrammus aeglefinus* larvae does not perform well in black tanks (Downing and Litvak 2000). Tambaqui *Colossoma macropomum* larvae prefer light green tank walls (Pedreira and Sipauba-Taveres 2001). In pompano larvae, fish in the black and dark red wall

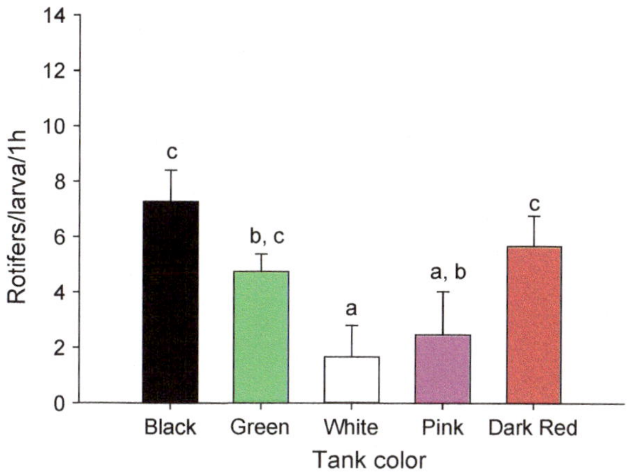

Fig. 10.7 Food ingestion of larval pompano *Trachinotus ovatus* in tanks with different wall color (on 5 dph). Different letters on the top of each bar represent significant difference. Reproduced from Ma et al. (2015)

tanks ingested more rotifers than in light-colored tanks, suggesting that this warm water fish at first feeding prefer dark walls (Fig. 10.7). Unlike pompano larvae, another warm water species orange-spotted grouper larvae in light blue wall bucket ingested significantly more rotifers than fish in black wall buckets (Zhang et al. 2015). This may suggest that the first feeding orange-spotted grouper prefer light color walls.

10.3.4 Prey Color

Apart from tank wall color, prey color can also affect the prey capture of marine fish larvae and the preference of prey colors in fish larvae is also species-specific (Zhang et al. 2015). For example, when yellowtail kingfish larvae were fed with brown-, pink-, pale white- and bight white-colored rotifers, a significantly higher feeding incidence was observed in fish fed brown-colored rotifers (Ma and Qin 2014). In pompano larvae, the number of rotifers ingested by fish larvae fed with green colored rotifers was significantly higher than those fed blue-, red- or purple-colored rotifers (Ma et al. 2015). In orange-spotted grouper larvae, the food ingestion of larvae fed green-colored rotifers was significantly higher than those fed red- or purple-colored rotifers (Fig. 10.8). This suggests that green food particles are more attractive to the first feeding orange-spotted grouper larvae than other colored food particles (Zhang et al. 2015). When orange-spotted grouper larvae were fed with artificial green colored rotifers and green microalgae-enriched rotifers, significantly higher food ingestion was observed in fish larvae fed rotifers enriched with green microalgae. Such circumstance is supported by the concept that green microalgae can stimulate the appetite of fish larvae by releasing some components acting as attractants (Stottrup et al. 1995).

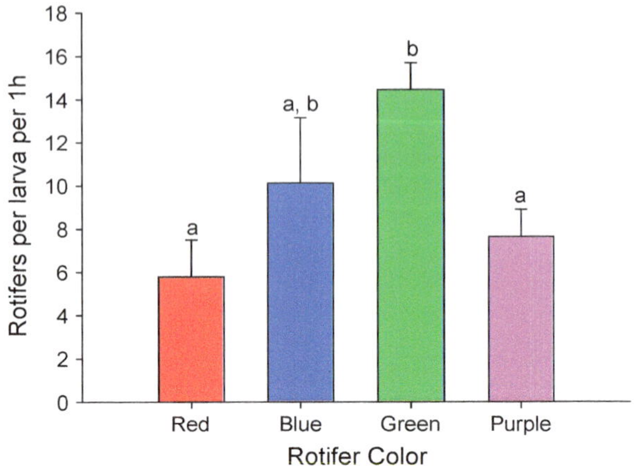

Fig. 10.8 Food ingestion of orange-spotted grouper *Epinephelus coioides* larvae fed with rotifers colored with red, blue, green, and purple. Different letters on the top of each bar represent significant difference. Reproduced from Zhang et al. (2015)

10.3.5 Water Color

As most marine fish larvae are visual feeders, water color can directly interact with light affecting prey capture by fish larvae. In developing hatchery rearing techniques for marine fish larvae, marine microalgae are widely used in the first feeding fish larvae (Reitan et al. 1993, 1997a). The presence of microalgae in artificial rearing environment can improve the growth and survival of fish larvae (Reitan et al. 1993; Stottrup et al. 1995). Microalgae are suggested to alter the microflora of the intestine of larval fish (Reitan et al. 1997b). Furthermore, microalgae in the rearing environment could affect the light milieu (Palmer et al. 2007). In order to understand the impact of water color on the feeding performance of fish larvae, we explored the impact of artificial water colors and microalgae water color on the feeding performance of orange-spotted grouper larvae. Water color significantly affected food ingestion of orange-spotted grouper larvae (Fig. 10.9), and the highest food ingestion was observed when fish larvae were held in a green water environment. Interestingly, when fish larvae were held in either artificially stained green water or water-stained green with green microalgae, fish larvae ingested significantly more rotifers in the algal-laden water than in artificial stained green water without the presence of green microalgae. These results may suggest that orange-spotted grouper larvae prefer a green color water environment, and green microalgae is a suitable additive to create green water and also promote food ingestion.

10.3.6 Light Intensity

Light is one of the abiotic factors that most likely affect feeding behavior of fish larvae (Kestemont and Baras 2001; Ma et al. 2013a) since most of the first feeding

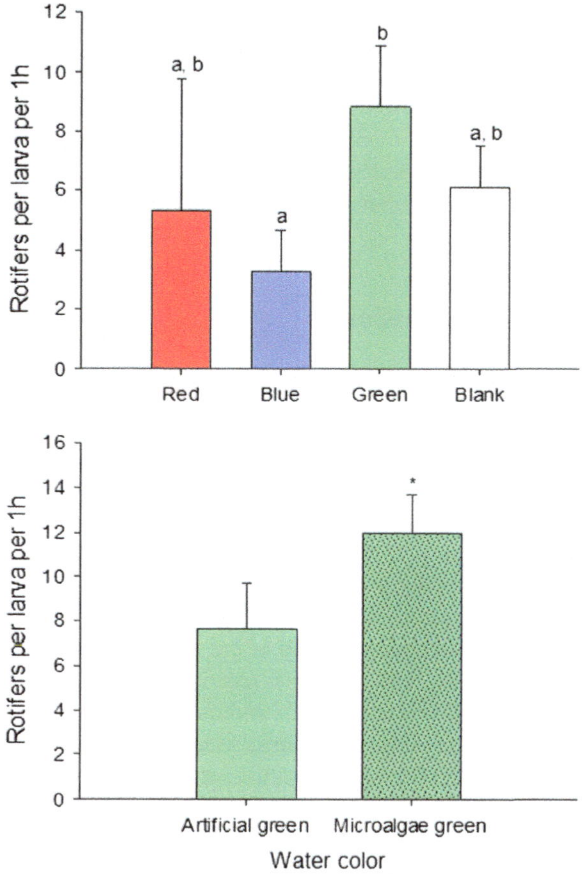

Fig. 10.9 Food ingestion of larval orange-spotted grouper *Epinephelus coioides* fed in different water colors. Different letters on the top of each bar represent significant difference. Reproduced from Zhang et al. (2015)

fish depend on vision for prey detection (Blaxter 1986). As we know that light is a complex factor that includes intensity, color spectrum and photoperiod. Light characteristics are very specific in an aquatic environment and light is extremely variable in nature (Boeuf and Le Bail 1999). It has been reported that both intensity and spectrum can affect feeding behavior (Fermin and Seronay 1997; Fermin et al. 1996). Light can indirectly influence fish feeding as light intensity and location may affect prey distribution and swimming pattern. Photoperiod is the most pervasive stimuli that could entrain daily rhythms (Kestemont and Baras 2001). In generally, a long photoperiod stimulates feeding activity, whereas a short photoperiod leads to a reduction in feeding.

Foraging ability is a key component for the survival and growth of fish larvae. Marine fish larvae are highly selective visual feeders, relying on visual stimuli to hunt and catch prey. The larval fish eyes contain rods and cones that respond to the wavelength range of the species in a particular environment niche. Therefore, the optimum light intensity for feeding is species-specific. In warm water marine fish larvae, the preferred light intensities are normally above 2000 lx. For instance, in commercial hatchery, the optimum light intensities of orange-spotted grouper and pompano are between 2000 and 3000 lx. In the first feeding leopard coral grouper *Plectropomus leopardus*, the average feeding success rate increased with the increasing light intensities from 0 to 3000 lx (Yoseda et al. 2008). Food intake increased with increasing light intensities, and the highest food intake was observed when the light intensity reaches 3000 lx (Yoseda et al. 2008).

10.4 Conclusion

Food supply is a key factor causing fish mortality during the nursery period in hatchery. The success of fish larvae to catch live prey during their early life has become an important benchmark in hatchery production. The timing to supply live feed to first-feeding larvae is critical to the survival and growth of warm water marine fish larvae. Live food should be supplied to fish larvae within 24 h after mouth opening. During fish ontogeny, important structural and functional changes occur in the body tissue, organs, and system. The quality and quantity of feed supply should change according to the ontogenetic development of fish larvae. Therefore, an understanding of the feeding of marine fish larvae in their early life is essential to select appropriate feed and rearing environment to improve fish survival in the initial rearing period. Although hatchery rearing techniques have been progressively developed in warm water fish, several key components governing fish growth and survival are still not well understood. Future research should be towards: (1) the understanding of timing of live food delivery to first-feeding warm water fish larvae, (2) quantify the light intensities required for first-feeding fish larvae, (3) the environmental requirements, especially temperature, that are suitable for warm water fish larval growth and survival.

References

Blaxter JHS (1986) Development of sense organs and behaviour of teleost larvae with special reference to feeding and predator avoidance. Trans Am Fish Soc 115:98–114

Blaxter JHS, Hempel G (1963) The influence of egg size on herring larvae (*Clupea harengus L.*). ICES J Mar Sci 28:211–240

Boeuf G, Le Bail PY (1999) Does light have an influence on fish growth? Aquaculture 177:129–152

Boglione C, Giganti M, Selmo C et al (2003) Morphoecology in larval fin-fish: a new candidate species for aquaculture, *Diplodus puntazzo* (Sparidae). Aquac Int 11:17–41

Boujard T, Medale F (1994) Regulation of voluntary feed intake in juvenile rainbow trout fed by hand or by self-feeders with diets containing two different protein/energy ratios. Aquat Living Resour 7:211–215

Bromley PJ, Adkins TC (1984) The influence of cellulose filler on feeding, growth and utilization of protein and energy in rainbow trout (*Salmo gairdneri*). J Fish Biol 24:235–241

Busch A (1996) Transition from endogenous to exogenous nutrition: larval size parameters determining the start of external feeding and size of prey ingested by Ruegen spring herring *Clupea harengus*. Mar Ecol Prog Ser 130:39–46

Chen BN, Qin JG, Carragher JF et al (2007) Deleterious effects of food restrictions in yellowtail kingfish *Seriola lalandi* during early development. Aquaculture 271:326–335

Cobcroft JM, Pankhurst PM (2003) Sensory organ development in cultured striped trumpeter larvae *Latris lineata*: implications for feeding behaviour. Mar Freshw Res 54:669–682

Cunha I, Planas M (1999) Optimal prey size for early turbot larvae (Scophthalmus maximus L.) based on mouth and ingested prey size. Aquaculture 175:103–110

Dabrowski K, Bardega R (1984) Mouth size and predicted food size preferences of larvae of three cyprinid fish species. Aquaculture 40:41–46

Denson MR, Smith TIJ (1996) Larval rearing and weaning techniques for white bass, *Morone chrysops*. J World Aquac Soc 27:194–201

Denson MR, Smith TIJ (1997) Diet and light intensity effects on survival, growth and pigmentation of southern flounder *Paralichthys lethostigma*. J World Aquac Soc 28:366–373

Downing G, Litvak MK (2000) The effect of photoperiod, tank color and light intensity on growth of larval haddock. Aquac Int 7:369–382

Duray MN, Estudillo CB, Alpasan LG (1996) The effect of background color and rotifer density on rotifer intake, growth and survival of the grouper (*Epinephalus suillus*) larvae. Aquaculture 146:217–224

Fermin AC, Seronay GA (1997) Effects of different illumination levels on zooplankton abundance, feeding periodicity, growth and survival of the Asian sea bass, Lates calcarifer (Bloch), fry in illuminated floating nursery cages. Aquaculture 157:227–237

Fermin AC, Bolivar ME, Gaitan A (1996) Nursery rearing of the Asian sea bass, Lates calcarifer, fry in illuminated floating net cages with different feeding regimes and stock densities. Aquat Living Resour 9:43–49

Fuiman LA, Higgs DM, Poling KR (2004) Changing structure and function of the ear and lateral line system of fishes during development. Symp Am Fish Soc 40:117–144

Gibbs MA (2004) Lateral line receptors: where do they come from developmentally and where is our research going? Brain Beha Evolut 64:163–181

Gisbert E, Williot P (1997) Larval behaviour and effect of the timing of initial feeding on growth and survival of Siberian sturgeon (*Acipenser baeri*) larvae under small scale hatchery production. Aquaculture 156:63–76

Gisbert E, Morais S, Moyano FJ (2013) Feeding and digestion. In: Qin JG (ed) Larval Fish Aquaculture. Nova Science Publishers, New York, pp 73–123

Gotceitas V, Puvanendran V, Leader L et al (1996) An experimental investigation of the "match/mismatch" hypothesis using larval Atlantic cod. Mar Ecol Prog Ser 130:29–37

Grove DJ, Loizides LJ, Nott J (1978) Satiation amount, frequency of feeding and gastric emptying rate in *Salmo gairdneri*. J Fish Biol 12:507–516

Hilder PI, Cobcroft JM, Battaglene SC (2015) The first-feeding response of larval southern Bluefin tuna, *Thunnus maccoyii* (Castelnau, 1872), and yellowtail kingfish, *Seriola lalandi* (Valenciennes, 1833), to prey density, prey size and larval density. Aquac Res 46:2736–2751

Houde ED (1975) Effects of stocking density and food density on survival, growth and yield of laboratory reared larvae of sea bream *Archosargus rhomboidalis* (L.) (Sparidae). J Fish Biol 7:115–127

Houde DD, Schekter RC (1980) Feeding by marine fish larvae: developmental and functional responses. Environ Biol Fish 5:315–334

James NC, Cowley PD, Whitfield AK et al (2008) Choice chamber experiments to test the attraction of postflexion Rhabdosargus holubi larvae to water of estuarine and riverine orgin. Estuar Coast Shelf Sci 77:143–149

Kasumyan AO (2003) The lateral line in fish: structure, function, and role in behavior. J Ichthyol 43(Suppl. 2):S175–S213

Kawamura G, Munekiyo M (1989) Development of the sense organs of ribbonfish Trichiurus lepturus larvae and juveniles. Nippon Suisan Gakk 55:2075–2078

Kawamura G, Masuma S, Tezuka N et al (2003) Morphogenesis of sense organs in the Bluefin tuna *Thunnus orientalis*. In: Browman HI, Skiftesvik AB (eds) The big fish bang proceedings of the 26th annual larval fish conference. Institute of Marine Research, Bergen, pp 125–135

Kestemont P, Baras E (2001) Environmental factors and feed intake: mechanisms and interactions. In: Houlihan D, Boujard T, Joblin M (eds) Food intake in Fish. Blackwell Science, Cornwall, pp 131–156

Klumpp DW, Von Westernhagen H (1986) Nitrogen balance in marine fish larvae: Influence of developmental stage and prey density. Mar Biol 93:189–200

Kotrschal K, Adam H, Brandstatter R et al (1990) Larval size constraints determine directional ontogenic shifts in the visual system of teleosts-a mini review. Z Zool Syst Evol 28:166–182

Lara MR (2008) Development of the nasal olfactory organs in the larvae, settlement-stages and some adults of 14 species of Caribbean reef fishes (*Labridae Scaridae*, Pomacentridae). Mar Biol 154:51–64

Laurel BJ, Brown JA, Anderson R (2001) Behaviour, growth and survival of redfish larvae in relation to prey availability. J Fish Biol 59:884–901

Ledent V (2002) Postembryonic development of the posterior lateral line in zebrafish. Development 129:597–604

Lim L, Mukai Y (2014) Morphogenesis of sense organs and behavioural changes in larvae of the brown-marbled grouper *Epinephelus fuscoguttatus* (Forsskål). Mar Freshw Behav Phy 47:313–327

Ma Z (2014) Food ingestion, prey selectivity, feeding incidence, and performance of yellowtail kingfish *Seriola lalandi* larvae under constant and varying temperatures. Aquac Int 22:1317–1330

Ma Z, Qin JG (2014) Replacement of fresh algae with commercial formulas to enrich rotifers in larval rearing of yellowtail kingfish *Seriola lalandi* (Valenciennes, 1833). Aquac Res 45:949–960

Ma Z, Qin JG, Nie Z (2012) Morphological Changes of Marine Fish Larvae and Their Nutrition Need. In: Pourali K, Raad VN (eds) Larvae Morphology, Biology and Life Cycle. Nova Science Publishers Inc, New York, USA, pp 1–20

Ma Z, Guo H, Zhang N, Bai Z (2013a) State of art for larval rearing of grouper. Int J Aquacul 3:63–72

Ma Z, Qin JG, Hutchinson W et al (2013b) Optimal live food densities for growth, survival, food selection and consumption of yellowtail kingfish *Seriola lalandi* larvae. Aquac Nutr 19:523–534

Ma Z, Guo H, Zhang D et al (2015) Food ingestion, consumption, and selectivity of pompano, *Trachinotus ovatus* (Linnaeus 1758) under different rotifer densities. Aquac Res 46:2593–2603

Margulies D (1997) Development of the visual system and inferred performance capabilities of larval and early juvenile scombrids. Mar Freshw Behav Phy 30:75–98

Martin-Tichaud DJ, Peterson RH (1997) Factors affecting swim bladder inflation success in larval striped bass (*Morone saxatilis*). Bul Aquac Assoc Can, Special Publication 2:111

Mercier L, Audet C, de la Noue J et al (2004) First feeding of winter flounder (*Pseudopleuronectes americanus*) larvae: use of Brachionus plicatilis acclimated at low temperature as live prey. Aquaculture 229:361–376

Monk J, Puvanendran V, Brown JA (2008) Does different tank bottom color affect the growth, survival and foraging behaviour of Atlantic cod (*Gadus morhua*) larvae. Aquaculture 277: 197–202

Montgomery JC, Macdonald F, Baker CF et al (2002) Hydrodynamic contributions to multimodal guidance of prey capture behaviour in fish. Brain Behav Evolut 59:190–198

Mukai Y, Lim LS (2016) Morphogenesis of free neuromasts in the larvae of brown-marbled grouper *Epinephelus fuscoguttatus*. Mar Freshw Behav Phy 49:159–171

Munk P, Kiorboe T (1985) Feeding behavior and swimming activity of larval herring (*Clupea Harengus*) in relation to density of copepod nauplii. Mar Ecol-Prog Ser 24:15–21

Myrberg AA, Fuiman LA (2002) The sensory world of coral reef fishes In: PF S (ed) Coral reef fishes: dynamics and diversity in a complex ecosystem. Academic Press, Amsterdam, pp 123–148

O'Brien JW, Slade NA, Winyard GL (1976) Apparent size as the determinant of prey selection by blugill sunfish (*Lepomis macrochirus*). Ecology 57:1304–1310

Østergaard P, Munk P, Janekarn V (2005) Contrasting feeding patterns among species of fish larvae from the tropical Andaman Sea. Mar Biol 146:595–606

Palmer PJ, Burke MJ, Palmer CJ et al (2007) Developments in controlled green-water larval culture technologies for estuarine fishes in Queensland, Australia and elsewhere. Aquaculture 272:1–21

Pankhurst PM (2008) Mechanoreception. In: Finn RN, Kapoor BG (eds) Fish larval physiology. Science Publishers, Enfield, pp 305–329

Pankhurst PM, Hilder PE (1998) Effect of light intensity on feeding of striped trumpeter *Latris lineata* larvae. Mar Freshw Res 49:363–368

Pedreira MN, Sipauba-Taveres LH (2001) Effect of light green and dark brown colored tanks on survival and development of Tambiqui larvae, *Colossoma macropomum* (Osteichthyes, Serrasalmidae). Acta Sci 23:521–525

Popper AN, Schilt CR (2008) Hearing and acoustic behaviour (basic and applied). Webb JF, Fay RR, Popper AN(ed) Fish bioacoustics. Springer Science Business Media, New York, pp 17–48

Puvanendran V, Brown JA (1999) Foraging, growth and survival of Atlantic cod larvae reared in different prey concentrations. Aquaculture 175:77–92

Qin JG (2013) Larval Fish Aquaculture. Nova Science Publishers, New York, p 230

Reitan KI, Rainuzzo JR, Øie G et al (1993) Nutritional effects of algal addition in first-feeding of turbot (*Scophthalmus maximus* L.) larvae. Aquaculture 118:257–275

Reitan KI, Rainuzzo JR, Øie G et al (1997a) A review of the nutritional effects of algae in marine fish larvae. Aquaculture 155:211–225

Reitan KI, Rainuzzo JR, Øie G et al (1997b) A review of the nutritional effects of algae in marine fish larvae. Aquaculture 155:207–221

Rønnestad I, Yúfera M, Ueberschär B et al (2013) Feeding behaviour and digestive physiology in larval fish: current knowledge, and gaps and bottlenecks in research. Rev Aquac 5:S59–S98

Ruzicka JJ, Gallager SM (2006) The importance of the cost of swimming to the foraging behavior and ecology of larval cod (Gadus morhua) on Georges Bank. Deep-Sea Res Part II 53:2708–2734

Shaw GW, Pankhurst PM, Battaglene SC (2006) Effect of turbidity, prey density and culture history on prey consumption by greenback flounder Rhombosolea tapirina larvae. Aquaculture 253:447–460

Slembrouck J, Baras E, Subagja J et al (2009) Survival, growth and food conversion of cultured larvae of *Pangasianodon hypophthalmus*, depending on feeding level, prey density and fish density. Aquaculture 294:52–59

Stottrup JG, Gravningen K, Norsker NH (1995) The role of different algae in the growth and survival of turbot larvae (*Scophthalmus maximus* L.) in intensive rearing system. ICES. Marine Science Symposia 201:173–186

Temple S, Cerqueira VR, Brown JA (2004) The effects of lowering prey density on the growth, survival and foraging behaviour of larval fat snook (*Centropomus parallelus poey* 1860). Aquaculture 233:205–217

Toledo JD, Caberoy NB, Quinitio GF et al (2002) Effects of salinity, aeration and light intensity on oil globule absorption, feeding incidence, growth and survival of early-stage grouper *Epinephelus coioides* larvae. Fisheries Sci 68:478–483

Twongo TK, MacCrimmon HR (1976) Significance of the timing of initial feeding in hatchery rainbow trout (*Salmo gairdneri*). J Fish Res Board Can 33:1914–1921

Vandermeer HJ (1994) Ontogenic change of visual thresholds in the cichlid fish *Haplochromis sauvagei*. Brain Behav Evolut 44:40–49

Webb JF (1989) Developmental constraints and evolution of the lateral line system in teleost fishes. In: Coombs S, Görner P, Münz H (eds) The mechanosensory lateral line-neurobiology and evolution. Springer, New York, pp 79–97

Werner EE, Hall DJ (1974) Optimal foraging and selection of prey by the bluegill sunfish (*Lepomis macrochirus*). Ecology 55:1042–1052

Woolley LD, Partridge GJ (2016) The effect of different rotifer feeding regimes on the growth and survival of yellowtail kingfish *Seriola lalandi* (Valenciennes, 1833) larvae. Aquac Res 47:2723–2731

Yoseda K, Dan S, Sugaya T et al (2006) Effects of temperature and delayed initial feeding on the growth of Malabar grouper (*Epinephelus malabaricus*) larvae. Aquaculture 256:192–200

Yoseda K, Yamamoto K, Asami K et al (2008) Influence of light intensity on feeding, growth, and early survival of leopard coral grouper (*Plectropomus leopardus*) larvae under mass-scale rearing conditions. Aquaculture 279:55–62

Yufera M, Darias MJ (2007) The onset of exogenous feeding in marine fish larvae. Aquaculture 268:53–63

Zhang J, Guo H, Ma Z et al (2015) Effects of prey color, wall color and water color on food ingestion of larval orange-spotted grouper *Epinephelus coioides* (Hamilton, 1822). Aquac Int 23:1377–1386

Printed by Printforce, the Netherlands